Selected Topics in Algebra

Mathematics and Its Applications (East European Series)

(D. Reidel Publishing Company).

Ionel Bucur

Selected Topics in Algebra

and its Interrelations with Logic, Number Theory and Algebraic Geometry

Translated from the Romanian by

Mihnea Moroianu

Polytechnical Institute, Bucharest

EDITURA ACADEMIEI
Bucureşti, Romania

D. Reidel Publishing Company
A MEMBER OF THE KLUWER ACADEMIC PUBLISHERS GROUP

Dordrecht / Boston / Lancaster

Library of Congress Cataloging in Publication Data

Bucur, Ion.

Selected topics in algebra and its interrelations with logic, number theory, and algebraic geometry.

(Mathematics and its applications. East European series)
Translation of: Capitole speciale de algebră.
Includes index.
1. Algebra. 2. Geometry, Algebraic. 3. Logic, Symbolic and mathematical. I. Title. II. Title: Selected topics in algebra. III. Series: Mathematics and its applications (D. Reidel Publishing Company). East European series.
QA155.B8313 1984 512 83–24609
ISBN 90–277–1671–4 (Reidel)

Distributors for the U.S.A. and Canada
Kluwer Academic Publishers,
190 Old Derby Street, Hingham, MA 02043, U.S.A.

Distributors for Albania, Bulgaria, Chinese People's Republic, Cuba, Czechoslovakia, German Democratic Republic, Hungary, Korean People's Republic, Mongolia, Poland, Romania, the U.S.S.R., Vietnam, and Yugoslavia
Editura Academiei, Bucharest, Romania

Distributors for all remaining countries
Kluwer Academic Publishers Group,
P.O. Box 322, 3300 AH Dordrecht, Holland.

Original title: *Capitole speciale de algebră*
First edition published in 1980 by
Editura Academiei, 125, Calea Victoriei, Bucharest, Romania
First English Edition published in 1984 by
Editura Academiei and D. Reidel Publ. Co.

Printed in Romania

Table of Contents

Editor's Preface vii

CHAPTER I. *Quadratic Forms and Arithmetical Symbols* 1

I. Elements of Quadratic Forms Theory
Quadratic Forms over a Field k of Characteristic $\neq 2$ 1

II. Symbols. Elements of the Theory of the Group $K_2(A)$ 29

§1. Symbols 29
§2. The Group of the Units of p-adic Fields 40
§3. Quadratic Forms over Discrete Valuation Fields 48
§4. A Formula for the Hilbert "Symbol" in the Case of p-adic Fields 67
§5. The Product Formula (Hilbert) 79
§6. Applications. Comments 82
§7. The Theorem of Minkowski-Hasse 85

CHAPTER II. *Methods of Logic and Algorithmic Methods in Algebra*

§1. Countable Sets 103
§2. The Notion of Lattice 108
§3. The Algebra of Propositions 119
§4. Formal Systems 123
§5. Primitive Recursive Functions 156
§6. Recursive Functions 161
§7. Turing Machines 168
§8. The Concept of Recursive (Algorithmic) Decidability 177
§9. On the Formalist and the Constructive Point of View in Mathematics 213
§10. Derived Terms in the Theory of Formal Systems 218
§11. The Constructive Approach to Set Theory 222

CHAPTER III. *Introduction to Modern Algebraic Geometry* 229

§1. Generalities 229
§2. The Spectrum of a Commutative Ring 235
§3. Schemes. Chevalley Schemes 251
§4. Elements of Projective Geometry 263
§5. The Chevalley Scheme Associated to an Integral and Irreducible Scheme 269

CHAPTER IV. *Theory of Topoi* 271

I. General Theory of Topoi 271
§1. Introduction 271
§2. Some Definitions 278
§3. Elements of Descent Theory 291
§4. Grothendieck Topologies 304

II. Theory of Lawvere-Tierney Topoi 316

CHAPTER V. *Elements of the Theory of Elliptic Curves* 341

I. Elliptic Curves Defined over the Complex Field 341

II. Divisors 350

CHAPTER VI. *Algebraic Varieties over a Finite Field* 365

§1. The Zeta Function 365
§2. Cohomology Theories 370
§3. Weil's Diophantine Conjectures 372

CHAPTER VII. *Elementary Theory of Non-Standard Real Numbers* 383

APPENDIX. *On Local Rings with the Approximation Property* 403

INDEX 405

Editor's Preface

Approach your problems from the right end and begin with the answers. Then one day, perhaps you will find the final question.

'The Hermit Clad in Crane Feathers' in R. van Gulik's *The Chinese Maze Murders*.

It isn't that they can't see the solution. It is that they can't see the problem.

G. K. Chesterton. The Scandal of Father Brown 'The Point of a Pin'.

Growing specialization and diversification have brought a host of monographs and textbooks on increasingly specialized topics. However, the "tree" of knowledge of mathematics and related fields does not grow only by putting forth new branches. It also happens, quite often in fact, that branches which were thought to be completely disparate are suddenly seen to be related.

Further, the kind and level of sophistication of mathematics applied in various sciences has changed drastically in recent years: measure theory is used (non-trivially) in regional and theoretical economics; algebraic geometry interacts with physics; the Minkowsky lemma, coding theory and the structure of water meet one another in packing and covering theory; quantum fields, crystal defects and mathematical programming profit from homotopy theory; Lie algebras are relevant to filtering; and prediction and electrical engineering can use Stein spaces. And in addition to this there are such new emerging subdisciplines as "completely integrable systems", "chaos, synergetics and large-scale order", which are almost impossible to fit into the existing classification schemes. They draw upon widely different sections of mathematics.

This program, *Mathematics and Its Applications*, is devoted to such (new) interrelations as *exempli gratia:*

— a central concept which plays an important role in several different mathematical and/or scientific specialized areas;

— new applications of the results and ideas from one area of scientific endeavor into another;
— influences which the results, problems and concepts of one field of enquiry have and have had on the development of another.

The *Mathematics and Its Applications* programme tries to make available a careful selection of books which fit the philosophy outlined above. With such books, which are stimulating rather than definitive, intriguing rather than encyclopaedic, we hope to contribute something towards better communication among the practitioners in diversified fields.

Because of the wealth of scholarly research being undertaken in the Soviet Union, Eastern Europe, and Japan, it was decided to devote special attention to the work emanating from these particular regions. Thus it was decided to start three regional series under the umbrella of the main MIA programme.

Algebra, understood to include algebraic geometry, is one of the areas of mathematics which is growing fastest in terms of applications (both actual and potential) and its interrelations with other fields in mathematics. This book offers a likely sample of such interrelations. It discusses quadratic forms and the part of algebraic K-theory having to do with "symbols", relations of algebra with logic and algorithmic questions, the theory of topoi and elementary non-standard theory (more relations with logic), elliptic functions and curves, parts of arithmetic algebraic geometry and rings of continuous functions. Hardly a collection of topics one finds in a standard textbook on algebra, but a tribute to the power of algebraic modes of thinking about problems in mathematics.

The unreasonable effectiveness of mathematics in science...

Eugene Wigner

Well, if you knows of a better 'ole, go to it.

Bruce Bairnsfather

What is now proved was once only imagined.

William Blake

As long as algebra and geometry proceeded along separate paths, their advance was slow and their applications limited.
But when these sciences joined company they drew from each other fresh vitality and thenceforward marched on at a rapid pace towards perfection.

Joseph Louis Lagrange

Amsterdam, August 1983

MICHIEL HAZEWINKEL

CHAPTER I

Quadratic forms and arithmetical symbols

I. ELEMENTS OF QUADRATIC FORMS THEORY QUADRATIC FORMS OVER A FIELD k OF CHARACTERISTIC $\neq 2$

DEFINITION 1.1. (a) A quadratic *module* (over the field k) is a pair $V, Q)$, where:

V is a finite dimensional vector space over the field k;

Q is a quadratic form on V, i.e. a mapping

$$Q: V \to k,$$

for which a symmetric bilinear form

$$q: V \times V \to k$$

exists, such that

$$Q(x) = q(x, x), \qquad \forall x \in V.$$

(b) If (V, Q), (V', Q') are two quadratic modules, a morphism of (V, Q) in (V', Q') is a linear mapping

$$f: V \to V'$$

such that $Q' \cdot f = Q$.

An isomorphism of quadratic forms will also be called an *isometry*.

1.2. *The main elements, associated with a quadratic module*, are the following:

1.2.1. *The matrix of the quadratic form* Q with respect to a basis $e_1, e_2, \ldots, e_n$ of the vector space V is the symmetric matrix $A = (a_{ij})_{1 \leqslant i,j \leqslant n}$ whose elements, belonging to the field k, are defined by

$$a_{ij} = q(e_i, e_j)$$

If $x = \sum_{i=1}^{n} \xi_i e_i$, $(\xi_i \in k)$, is an element of V, we have

$$Q(x) = \sum_{i,j} a_{ij} \xi_i \xi_j.$$

If instead of (e_i) we consider the basis (e_i'), and if X is the transition matrix from the basis (e_i) to the basis (e_i'), then the matrix A' of the quadratic form Q with respect to the basis (e_i') is written

$$A' = X \cdot A \cdot t_X \tag{1.2.1.1}$$

where t_X stands for the transposed matrix of X.

1.2.2. *The discriminant* of a quadratic form Q with respect to the basis (e_i) is, by definition, $\det(A)$. From (1.2.1.1) it follows

$$\det(A') = \det(A) \cdot \det(X)^2$$

which shows that the discriminant of a quadratic form is determined up to multiplication by an element from k^{*2} (k^* = the multiplicative group, subjacent to k).

1.2.3. *The radical (or the kernel)* of the quadratic module (V, Q), denoted by $\operatorname{rad}(V)$, is, by definition, the subspace of V, defined by

$$\operatorname{rad}(V) = \{x \in V \mid q(x, y) = 0, \ \forall\, y \in V\}.$$

1.2.4. *The rank* of the quadratic form Q is, by definition, the codimension of the subspace $\operatorname{rad}(V)$.

1.2.5. *The canonical morphism* $q_U : V \to U^* = \operatorname{Hom}_k(U, k)$, U being a subspace of V, is defined by

$$q_U(x)(y) = q(x, y), \ \forall\, x \in V, \ \forall\, y \in U.$$

Obviously, $\operatorname{rad}(V) = \operatorname{Ker} q_V$.

The matrix A, associated to the quadratic form Q with respect to the basis (e_i) coincides with the matrix of the linear mapping q_V with respect to the bases (e_i), (e_i^*), where (e_i^*) is the basis of V^*, dual to the basis (e_i) of V.

1.3. *The main definitions concerning a quadratic module* are the following:

1.3.1. A quadratic module (V, Q) is *non-degenerate* if it satisfies one of the following, obviously equivalent, conditions:

(α) $\mathrm{rad}(V) = 0$;

(β) q_V is an isomorphism;

(γ) the discriminant of the quadratic form Q with respect to every basis of the space V is non-vanishing.

If the quadratic module (V, Q) is non-degenerate, the quadratic form Q will be also called non-degenerate.

If (V, Q) is a non-degenerate quadratic module, the discriminants associated with Q with respect to different bases of V define a unique element of the group k^*/k^{*2}; this element will be denoted by disc (Q) and will be referred to as the discriminant of the non-degenerate quadratic form Q.

1.3.2. Consider a quadratic module (V, Q) and two elements x, y of V. We say that these elements are *orthogonal* if $q(x, y) = 0$.

For any subset H of V, we denote by H^0 the set of all elements of V which are orthogonal to every element of H:

$$H^0 = \{y \in V \mid q(x, y) = 0, \ \forall\, x \in H\}.$$

Clearly, $\mathrm{rad}(V) = V^0$.

If V_1, V_2 are two subspaces of V, we say that they are orthogonal, if they satisfy one of the following, obviously equivalent, conditions:

(α) $\qquad V_1 \subset V_2^0.$

(β) $\qquad V_2 \subset V_1^0.$

(γ) $\qquad q(x, y) = 0, \ \forall\, x \in V_1, \ \forall\, y \in V_2.$

1.3.3. *Orthogonal direct sum.* Let $U_1, U_2, \ldots, U_m$ be m subspaces of the vector space V. We say that V is the orthogonal direct sum of $U_1, U_2, \ldots, U_m$ and write

$$V = U_1 \perp U_2 \perp \ldots \perp U_m,$$

if:

(I) V is the direct sum of the subspaces $U_1, U_2, \ldots, U_m$:

$$V = U_1 \oplus U_2 \oplus \ldots \oplus U_m.$$

(II) $U_1, U_2, \ldots, U_m$ are mutually orthogonal:

$$q(x_i, x_j) = 0, \ \forall\, x_i \in U_i,\ x_j \in U_j,\ i \neq j.$$

We say that a basis $(e_1, e_2, \ldots, e_n)$ of a quadratic module (V, Q) is orthogonal if $V = (ke_1) \perp (ke_2) \perp \ldots \perp (ke_n)$.

Example 1.3.3.1. If U is a supplement of $\mathrm{rad}(V)$ in V, then

$$V = \mathrm{rad}(V) \perp U.$$

1.3.4. *Isotropic vectors and spaces.* An element x of a quadratic module (V, Q) is called *isotropic* if $Q(x) = 0$.

A subspace U of V is called *isotropic* if all its elements are isotropic.

1.3.5. *Hyperbolic plane.* We say that a quadratic module (V, Q) is a *hyperbolic plane* if it has a basis consisting of two isotropic elements x, y such that $q(x, y) \neq 0$.

A quadratic module V is called *hyperbolic* if it is an orthogonal direct sum of hyperbolic planes $H_1, H_2, \ldots, H_r$,

$$V = H_1 \perp H_2 \perp \ldots \perp H_r.$$

1.3.6. We say that the quadratic form Q, or the quadratic module (V, Q), are *defined*, if $Q(x) \neq 0$ for any $x \in V$, $x \neq 0$.

1.4. *Elementary properties of quadratic modules*

PROPOSITION 1.4.1. *Any quadratic module (V, Q) has an orthogonal basis.*

Proof. We argue by induction on the dimension of V, the statement being clear for $\dim_k V = 1$. Since any basis of an isotropic space is orthogonal, we may suppose that V is not isotropic. Let then $x \in V$ be such that $Q(x) \neq 0$ and let $x, x_2, \ldots, x_n$ be a basis of V. Under these assumptions, the elements

$$x, x_2 - \frac{q(x, x_2)}{Q(x)} x, x_3 - \frac{q(x, x_3)}{Q(x)} x, \ldots, x_n - \frac{q(x, x_n)}{Q(x)} x,$$

form a basis of V such that

$$q\left(x, x_i - \frac{q(x, x_i)}{Q(x)} x\right) = 0, \quad i = 2, 3, \ldots, n,$$

i.e. x is orthogonal to the subspace W generated by the vectors

$$x_2 - \frac{q(x, x_2)}{Q(x)} x, x_3 - \frac{q(x, x_3)}{Q(x)} x, \ldots, x_n - \frac{q(x, x_n)}{Q(x)} x.$$

Now, by the induction hypothesis, it suffices to take a basis of the quadratic module $(W, Q|_W)$.

PROPOSITION 1.4.2. *If $(U, Q|_U)$ is a non-degenerate quadratic submodule of the quadratic module (V, Q), then*

$$V = U \perp U^0.$$

Proof. Let $x_1, x_2, \ldots, x_p$ be an orthogonal basis of $(U, Q|_U)$:

$$U = (kx_1) \perp (kx_2) \perp \ldots \perp (kx_p).$$

Since $(U, Q|_U)$ is non-degenerate, we have $Q(x_i) \neq 0$ for $i = 1, 2, \ldots, p$. First we show that $V = U + U^0$. To this end, let z be any element of V and let $y \in U$ be defined by

$$y = \frac{q(z, x_1)}{Q(x_1)} x_1 + \frac{q(z, x_2)}{Q(x_2)} x_2 + \ldots + \frac{q(z, x_p)}{Q(x_p)} x_p.$$

It suffices to show that $z - y \in U^0$, i.e.

$$q(x_i, z - y) = 0, \quad i = 1, 2, \ldots, p.$$

But

$$q(x_i, z - y) = q(x_i, z) - \sum_{j=1}^{p} \frac{q(z, x_j)}{Q(x_j)} q(x_j, x_i) = 0.$$

On the other hand, we have the general formula

$$U \cap U^0 = \mathrm{rad}(U)$$

and $(U, Q|_U)$ being non-degenerate, we get $U \cap U^0 = 0$, which completes the proof.

PROPOSITION 1.4.3. *If U is any subspace of a quadratic non-degenerate module (V, Q), the following relations hold:*

(1.4.3.1) $\quad \dim U + \dim U^0 = \dim V$

(1.4.3.2) $\quad U^{00} = U.$

Proof. Using the notation from 1.2.5 and the hypothesis, we get

$$\dim U^0 = \dim (q_V(U^0)).$$

But $q_V(U^0)$ is the subspace of V^* consisting of all linear forms of V, vanishing on U, i.e.

$$q_V(U^0) = \mathrm{Hom}_k(V/U, k) = (V/U)^*.$$

Hence

$$\dim U^0 = \dim(V/U)^* = \dim V/U = \dim V - \dim U.$$

In order to prove (1.4.3.2), we remark that

$$U \subset U^{00}$$

and on the other hand

$$\dim U = \dim U^{00}$$

from which (1.4.3.2) follows.

COROLLARY 1.4.3′. *Under the assumptions of Proposition 1.4.3, if* $U_1 \subsetneqq U_2$, *then*

$$U_1^0 \supsetneqq U_2^0.$$

PROPOSITION 1.4.4. *Any two hyperbolic planes are isomorphic.*

Proof. Remark first, that if (H, Q) is a hyperbolic plane, then we can find elements $u, v \in H$ such that

$$Q(u) = Q(v) = 0,$$

$$q(u, v) = 1.$$

Indeed, by hypothesis, there exist $x, y \in H$ such that

$$Q(x) = Q(y) = 0,$$

$$q(x, y) \neq 0.$$

(cf. 1.3.5).

We put

$$u = x,$$

$$v = \frac{1}{q(x, y)}\, y.$$

If (H', Q') is another hyperbolic plane and $u', v' \in H'$ are such that

$$Q'(u') = Q'(v') = 0$$

$$q'(u', v') = 1$$

it will suffice to consider the linear mapping

$$\sigma : H \to H'$$

defined by

$$\sigma(u) = u', \ \sigma(v) = v'.$$

PROPOSITION 1.4.5. *If x is a non zero isotropic element of a non-degenerate quadratic module (V, Q), there exists a subspace U of V, containing x, which is a hyperbolic plane.*

Proof. Since V is non-degenerate, it contains an element z, such that $q(x, z) = 1$. Then $y = 2z - q(z, z)x$ is isotropic and $q(x, y) = 2$, hence the subspace $kx + ky$ satisfies the required properties.

PROPOSITION 1.4.6. *Let (V, Q) be a non-degenerate quadratic module; and F a subspace of V. Let*

$$F = \mathrm{rad}(F) \perp U.$$

(cf. example 1.3.3.1).
Suppose that $u_1, u_2, \ldots, u_s$ is a basis of rad(F). *Then we can find elements $v_1, v_2, \ldots, v_s$ in V, satisfying the following properties:*

(I) v_i *is orthogonal to the subspace U* for $i = 1, 2, \ldots, s$.

(II) *The subspace* H_i *generated by the system* $\mathrm{u}_\mathrm{i}, \mathrm{v}_\mathrm{i}$ *is a hyperbolic plane.*

(III) *We have the direct orthogonal sum* $U \perp H_1 \perp H_2 \perp \ldots \perp H_s$.

Proof. We argue by induction on s. Assume $s = 1$. In this case, we have

$$F = (ku_1) \perp U.$$

From the strict inclusion $U \subsetneqq F$ we get (Corollary 1.4.3'): $U^0 \supsetneqq F^0$.

Hence, there exists an element $v_1 \in U^0$ which does not belong to F^0. In particular, we shall have: $q(u_1, v_1) \neq 0$. Since $Q(u_1) = 0$, it follows that the subspace H_1 generated by the elements u_1, v_1 is a hyperbolic plane and in addition we have the orthogonal direct sum

$$U \perp H_1.$$

Consider now the case of an arbitrary s. Put $U_1 = ((ku_2) \oplus (ku_3) \oplus \oplus \ldots \oplus (ku_s)) \perp U$. From the relation $U_1 \subsetneqq F$ we get as above (Corollary 1.4.3′) $U_i^0 \supsetneqq F^0$, therefore we can find an element $u_1 \in V$ such that

$$q(u_1, v_1) \neq 0,$$

$$v_1 \in U_1^0.$$

Consequently, if H_1 is the hyperbolic plane determined by u_1, v_1, we get the following orthogonal direct sum:

$$U' = ((ku_2) \oplus (ku_3) \oplus \ldots \oplus (ku_s)) \perp U \perp H_1.$$

Since $\operatorname{rad}(U') = (ku_2) \oplus (ku_3) \oplus \ldots \oplus (ku_s)$, by the induction hypothesis, the required orthogonal direct sum

$$U \perp H_1 \perp H_2 \perp \ldots \perp H_s, \ H_i \ni u_i.$$

follows.

THEOREM 1.4.7. *(Witt). If (V, Q) and (V', Q') are isomorphic and non-degenerate, then every injective morphism*

$$s: F \to V'$$

of a vector subspace F of V can be extended to an isomorphism of V on V'.

Proof. Obviously we can suppose $V = V'$. Now assume that $(F, Q|_F)$ is a degenerate quadratic module. We shall show that in this case s can always be extended to an injective morphism $s': F \to V$. In fact, $(F, Q|_F)$ being degenerate, Proposition 1.4.6 yields the following orthogonal direct sums

$$F = \mathrm{rad}(F) \perp U, \quad s(F) = \mathrm{rad}(s(F)) \perp s(U)$$

$$F \subset U \perp H_1 \perp H_2 \perp \ldots \perp H_s,$$

$$S(F) \subset S(U) \perp H_1' \perp H_2' \perp \ldots \perp H_s'$$

which show that s can be extended to $U \perp H_1 \perp \ldots \perp H_s$.

Hence we can assume that $(F, Q|_F)$ is non-degenerate. By induction on the dimension of F, we show that in this case s can be extended to an automorphism of V.

If $\dim F = 1$, F is generated by a non-isotropic element x, and putting $y = s(x)$, we have $Q(y) = Q(x) \neq 0$. Hence

$$Q(x + y) + Q(x - y) = 2Q(x) + 2Q(y) = 4Q(x),$$

consequently either $Q(x + y)$ or $Q(x - y)$ are non zero. Choose then $\varepsilon = \pm 1$ such that $z = x + \varepsilon y$ be non-isotropic: $Q(x + \varepsilon y) \neq 0$. We have then

$$V = (kz) \perp H,$$

where H is a hyperplane (Proposition 1.4.2). Let $\sigma: V \to V$ be the symmetry with respect to H, in other words:

$$\sigma(u) = \begin{cases} u & \text{if } u \in H \\ -u & \text{if } u = z. \end{cases}$$

But $x - \varepsilon y \in H$:

$$q(x + \varepsilon y, x - \varepsilon y) = Q(x) - Q(y) + q(x, y) -$$

$$- \varepsilon q(y, x) = 0.$$

Hence

$$\sigma(x - \varepsilon y) = x - \varepsilon y \text{ and } \sigma(x + \varepsilon y) = - x - \varepsilon y,$$

consequently

$$2\sigma(x) = - 2\varepsilon y,$$

i.e.

$$\sigma(x) = -\varepsilon y.$$

and the automorphism $-\varepsilon\sigma(x)$ is an extension of s.

Now, assume $\dim F > 1$. One can find for F a splitting of the form

$$F = F_1 \perp F_2, \; F_1, F_2 \neq 0.$$

Let s_1 be the restriction of s to F_1. By the induction hypothesis, s_1 can be extended to an (isometric) automorphism

$$\sigma_1 \colon V \to V$$

Let $t \colon F \to V$ be defined by $t = \sigma_1^{-1} \circ s$.

The restriction of t to F_1 is the identity map of F_1. Consequently, t maps F_2 into the orthogonal complement V_1 of F_1. Hence, by the induction hypothesis, $t|_{F_2}$ can be extended to an automorphism $\tau_1 \colon V_1 \to$ $\to V_1$. Let $\tau \colon V \to V$ be defined as follows:

$$\tau|_{F_1} = \mathrm{id}_{F_1},$$

$$\tau|_{V_1} = \tau_1.$$

We have $\tau|_F = t$. Writing $\sigma = \sigma_1\tau$, σ is an extension of s.

COROLLARY 1.4.8. *If U_1, U_2 are two isomorphic subspaces of a non-degenerate quadratic module (V, Q), their orthogonal complements U_1^0, U_2^0 are isomorphic too.*

Proof. Let $f \colon U_1 \to U_2$ be an isomorphism; this isomorphism can be extended to an automorphism $F \colon V \to V$, which maps U_1^0 isomorphically into U_2^0.

DEFINITION 1.4.9. A subspace U of the quadratic module (V, Q) is a *null subspace* if $Q|_U = 0$. If U is a null subspace of the quadratic module Q, then

$$q(x, y) = 0, \; \forall \, x, y \in U.$$

In fact,

$$4q(x, y) = Q(x + y) - Q(x - y)$$

and $\mathrm{char}(k) \neq 2$.

COROLLARY 1.4.10. *If W_1 and W_2 are two maximal zero subspaces of the non-degenerate quadratic module (V, Q), then W_1 and W_2 are isomorphic.*

Proof. Assuming that $\dim W_1 \leqslant \dim W_2$, choose an injective morphism

$$W_1 \to W_2$$

which can be extended to an automorphism of the quadratic module (V, Q). Consequently, by the maximality hypothesis,

$$\dim W_1 = \dim W_2.$$

1.4.11. *The Witt splitting of a quadratic module.* Let (V, Q) be a quadratic module. A *Witt splitting* of this module is an orthogonal direct sum of the form

$$V = \operatorname{rad}(V) \perp H \perp D,$$

where: H is a hyperbolic subspace of (V, Q); (D, Q) is a definite quadratic module, i.e. D possesses no isotropic vector other than the zero vector: $Q(x) \neq 0$ for every $x \in D$, $x \neq 0$.

PROPOSITION 1.4.12. *Every quadratic module has a Witt splitting. If*

$$V = \operatorname{rad}(V) \perp H \perp D$$

$$V = \operatorname{rad}(V) \perp H' \perp D'$$

are two Witt splittings of the same quadratic module, then H and H' (respectively D and D') are isomorphic.

Proof. In view of example 1.3.3.1 we have the splitting

$$V = \operatorname{rad}(V) \perp U$$

where $(U, Q|_U)$ is obviously a non-degenerate quadratic module. Let W be a maximal null subspace of U and let $w_1, w_2, \ldots, w_r$ be a basis of W. Using Proposition 1.4.6 we get the splitting

$$U = H_1 \perp H_2 \perp \ldots \perp H_r \perp D,$$

where H_i are hyperbolic planes, $H_i \ni w_i$ and $(D, Q|_D)$ is obviously a definite quadratic module. This yields a Witt splitting of the quadratic module (V, Q).

Now, consider another Witt splitting

$$V = \mathrm{rad}(V) \perp H' \perp D',$$

$$H' = H_1' \perp H_2' \perp \ldots \perp H_s', \; H_i' = \text{hyperbolic plane.}$$

Let $U' = H' \perp D'$. Obviously, $(U', Q|_{U'})$ is a non-degenerate quadratic module, and this module is isomorphic to $(U, Q|_U)$, since both are isomorphic to the quotient quadratic module $(V/\mathrm{rad}\,(Q), Q)$. In these conditions, applying Corollary 1.4.8, we see that D and D' are isomorphic.

COROLLARY 1.4.13. *Let (V, Q) be a non-degenerate quadratic module possesing at least one non-zero isotropic vector. For any element $a \in k$, there is at least an element $z \in V$ such that $Q(z) = a$.*

Proof. Using Corollary 1.4.12, we can assume that (V, Q) is a hyperbolic plane. Choose elements $x, y \in V$ such that $Q(x) = Q(y) = 0$, $q(x, y) = 1$. It will suffice to put $z = x + \dfrac{a}{2}\, y$.

1.5. *The Witt group.* Let (V_1, Q_1), (V_2, Q_2) be two quadratic modules over the field k. On the vector space $V = V_1 \perp V_2$ one can define a quadratic form Q by putting

$$Q(x_1 + x_2) = Q_1(x_1) + Q_2(x_2),$$

respectively

$$q(x_1 + x_2,\, y_1 + y_2) = q_1(x_1, y_1) + q_2(x_2, y_2).$$

The quadratic module (V, Q) defined in this way will be called the *direct sum of the quadratic modules* (V_1, Q_1), (V_2, Q_2) and will be denoted by $(V_1 \perp V_2,\; Q_1 + Q_2)$.

Now, consider the class $\mathscr{V}(k)$ of all non-degenerate quadratic modules over the field k. An equivalence relation on $\mathscr{V}(k)$ will be introduced as follows:

1.5.1. $(V, Q) \sim (V', Q')$ if and only if there are two hyperbolic spaces (H, h), (H', h') such that we have an isomorphism

$$V \perp H \simeq V' \perp H'.$$

Let Witt(k) be the corresponding *quotient set.*

PROPOSITION 1.5.2. *The direct sum of quadratic modules induces on Witt (k) an Abelian group structure.*

Proof. If H is a hyperbolic space, its class obviously represents a unit element for the composition law defined on Witt(k).

In order to prove the existence of an inverse for the class of the quadratic module (V, Q), consider the direct sum

$$(V \perp V, Q + (-Q)).$$

We show that this quadratic module is a hyperbolic one. In fact, its diagonal space

$$W = \{(x, x) | x \in V\}$$

is a null space of the same dimension as V. But this implies the existence of a hyperbolic subspace H of $V \perp V$, of dimension equal to 2 dim V, hence $H = V \perp V$.

Examples:

1.5.3. If the field k is algebraically closed, then Witt$(k) = \mathbb{Z}/2\mathbb{Z}$.

1.5.4. Witt$(\mathbb{R}) \simeq \mathbb{Z}$.

In fact, let (V, Q) be a non-degenerate quadratic module over $\mathbb{R}$. One can find a basis of V with respect to which the quadratic form Q writes

$$Q(x) = \xi_1^2 + \xi_2^2 + \ldots + \xi_r^2 - \xi_{r+1}^2 - \ldots - \xi_{r+s}^2$$

$$r, s \geqslant 0,\ r + s = \dim_{\mathbb{R}} V$$

where the integers r, s are uniquely determined by Q (the inertia law). Let us associate to the quadratic module (V, Q) its signature $\sigma(V, Q)$, i.e. the difference $r - s$.

If $(V, Q) \sim (V', Q')$, then clearly $\sigma(V, Q) = \sigma(V', Q')$ and, in addition, $\sigma((V_1, Q_1) \oplus (V_2, Q)) = \sigma(V_1, Q_1) + \sigma(V_2, Q_2)$. This yields a group homomorphism

$$\sigma\colon \mathrm{Witt}(\mathbb{R}) \to \mathbb{Z}$$

which is clearly an isomorphism.

1.6. *The Grothendieck group associated to a commutative monoid.*

PROPOSITION 1.6.1. *Let M be a commutative monoid. A commutative group $K(M)$, called the Grothendieck group associated to the monoid M, and a monoid morphism*

$$\gamma\colon M \to K(M)$$

exist, such that the following conditions be fulfilled:

(i) *If* L *is an Abelian group and $u\colon M \to L$ is a monoid morphism, a unique group homomorphism $v\colon K(M) \to L$ exists, such that the diagram*

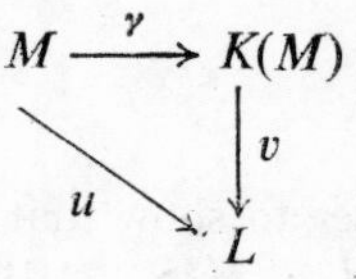

$$\begin{array}{ccc} M & \xrightarrow{\gamma} & K(M) \\ & \searrow^{u} & \downarrow v \\ & & L \end{array}$$

commutes.

(ii) *If M has the reduction property*

$$(x, y, z \in M \text{ and } x + z = y + z) \Rightarrow x = y,$$

then γ is injective.

Proof. Let $F(M)$ be the free Abelian group having the set M as a basis, and let

$$i\colon M \to F(M)$$

be the canonical inclusion. Denote by A the subgroup of $F(M)$ generated by the elements of the form

$$i(x + y) - i(x) - i(y), \quad x, y \in M.$$

It will suffice to put

$$K(M) = F(M)/A.$$

We shall take for γ the composition of i with the canonical morphism

$$F(M) \to F(M)/A.$$

In order to prove *ii*, it will suffice to show the existence of an Abelian group $Z(M)$ and of an injection

$$j: M \to Z(M)$$

which is a monoid morphism.

The construction of $Z(M)$ is completely similar to that employed for defining the group $\mathbb{Z}$ starting with $\mathbb{N}$. In fact, we consider first the product $M \times M$, on which we introduce the following equivalence relation:

$$(x, y) \sim (x', y') \Leftrightarrow x + y' = y + x'.$$

The reduction property is used in order to show that this is actually an equivalence relation. Let $Z(M) = M \times M/\sim$ be the quotient set. An algebraic operation is introduced on $Z(M)$ by putting

$$\text{class}\,(x, y) + \text{class}(x_1, y_1) = \text{class}(x + x_1, y + y_1).$$

We obtain in this way an Abelian group.

Put

$$j(x) = \text{class}(x, 0).$$

The morphism j is injective:

$$j(x) = j(y) \Rightarrow (x, 0) \sim (y, 0),\ x = y.$$

Remark. If the reduction property does not hold in M, one can introduce on M the equivalence relation

$$(x, y) \sim (x', y') \Leftrightarrow \exists\, z \in M,\ z + x + y' = z + y + x'.$$

One obtains as above a group, which is actually isomorphic to $K(M)$. The morphism γ will be defined by

$$\gamma(x) = \text{class}\,(x, 0)$$

but, in general, it will be no longer injective.

1.7. *The Witt-Grothendieck group associated to a commutative field.*

DEFINITION 1.7.1. *Let $\mathfrak{M}(k)$ be the set of isomorphism classes of quadratic modules over the field k.*

A monoid structure can be introduced on $\mathfrak{M}(k)$ by putting.

$\text{class}(V_1, Q_1) + \text{class}(V_2, Q_2) = \text{class}(V_1 \perp V_2, Q_1 + Q_2)$.

The monoid obtained in this way is commutative and, in view of Corollary 1.4.8, it satisfies the reduction condition 1.6.1. *ii.*

The Grothendieck group associated to the monoid $\mathfrak{M}(k)$ (cf. 1.6) is called *the Witt-Grothendieck group of the field k* and is denoted by $WG(k)$. In particular, there are canonical group morphisms

$$WG(k) \to \text{Witt}(k) \tag{1.7.1}$$

$$WG(k) \xrightarrow{d} k^*/k^{*2}$$

where d is induced by the discriminant of a quadratic form.

We also have the following exact sequence:

$$0 \to \mathbb{Z} \to WG(k) \to \text{Witt}(k) \to 0. \tag{1.7.2}$$

1.8. *Elements of $\mathcal{K}$-theory.* We intend to show that the exact sequence 1.7.2 is a fragment of a general exact sequence which can be described in terms of $\mathcal{K}$-theory.

DEFINITION 1.8.1. We say that a category $\mathscr{C}$ is a groupoid if every morphism of $\mathscr{C}$ is an isomorphism.

Example. To every category $\mathscr{C}$ one can obviously associate a groupoid having the objects of $\mathscr{C}$ as objects, and the isomorphisms of $\mathscr{C}$ as morphisms.

Definition 1.8.2. We say that a product is defined on a category $\mathscr{C}$ if a functor

$$\perp : \mathscr{C} \times \mathscr{C} \to \mathscr{C}$$

is defined satisfying the conditions of associativity and commutativity up to an isomorphism (cf. Bucur: *Lecture Notes*, Springer, 108).

Proposition 1.8.3. *(Existence of the Grothendieck group associated to a groupoid with products). Let* $(\mathscr{C}, \perp)$ *be a groupoid with products. There is an Abelian group* $K_0(\mathscr{C})$ *and a mapping* $\gamma: \mathrm{Ob}(\mathscr{C}) \to K_0(\mathscr{C})$ *such that the following conditions be fulfilled*

(a) $X \sim Y \Rightarrow \gamma(X) = \gamma(Y)$ *for every* $X, Y \in \mathrm{Ob}(\mathscr{C})$.

(b) $\gamma(X \perp Y) = \gamma(X) + \gamma(Y)$ *for every* $X, Y \in \mathrm{Ob}(\mathscr{C})$.

(c) *If* G *is an Abelian group and* $u: \mathrm{Ob}(\mathscr{C}) \to G$ *is a mapping satisfying conditions a and b, then there exists a unique group homomorphism* $v: K_0(\mathscr{C}) \to G$ *such that the diagram*

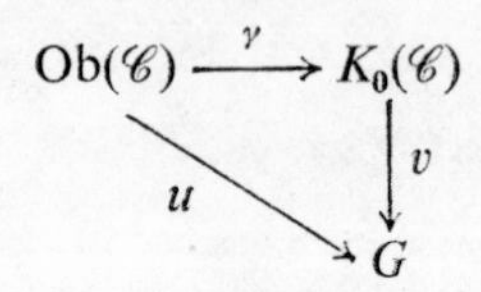

commutes. The group $K_0(\mathscr{C})$ *is called the Grothendieck group associated with the groupoid with products* $(\mathscr{C}, \perp)$.

The proof is straightforward and we leave it to the reader.

Definition 1.8.4. Let $(\mathscr{C}, \perp)$ be a groupoid with products. We define a new groupoid with products $(\underline{\mathrm{Aut}}(\mathscr{C}), \perp)$ in the following way:

$\mathrm{Ob}(\underline{\mathrm{Aut}}(\mathscr{C})) = \{(A, \alpha) \mid A \in \mathrm{Ob}(\mathscr{C});\ \alpha$ an automorphism of $A\}$;

$\mathrm{Hom}_{\underline{\mathrm{Aut}}(\mathscr{C})}((A, \alpha), (B, \beta)) = \{f \in \mathrm{Hom}_{\mathscr{C}}(A, B)$ such that

$$\begin{array}{ccc} A & \xrightarrow{f} & B \\ {\scriptstyle\alpha}\downarrow & & \downarrow{\scriptstyle\beta} \\ A & \xrightarrow{f} & B \end{array}$$

commutes$\}$.

Obviously we put $(A, \alpha) \perp (B, \beta) = (A \perp B, \alpha \perp \beta)$.

PROPOSITION 1.8.5. *(Existence of the Whitehead group associated to a groupoid with products). Let* $(\mathscr{C}, \perp)$ *be a groupoid with products. There exists an Abelian group* $K_1(\mathscr{C})$ *and a mapping* $\delta: \mathrm{Ob}(\underline{\mathrm{Aut}}(\mathscr{C})) \rightarrow$ $\rightarrow K_1(\mathscr{C})$ *such that the following conditions be fulfilled:*

(a) $(A, \alpha) \sim (A', \alpha') \Rightarrow \delta(A, \alpha) = \delta(A', \alpha')$ *for every* $(A, \alpha), (A', \alpha') \in$ $\in \mathrm{Ob}(\underline{\mathrm{Aut}}(\mathscr{C}))$;

(b) $\delta((A, \alpha) \perp (B, \beta)) = \delta((A, \alpha)) + \delta((B, \beta))$ *for every* $(A, \alpha), (B, \beta) \in$ $\in \mathrm{Ob}(\underline{\mathrm{Aut}}(\mathscr{C})))$;

(c) $((A, \alpha_1, \alpha_2)) = \delta((A, \alpha_1)) + \delta((A, \alpha_2))$ *for every pair* α_1, α_2 *of automorphisms of* A.

(d) *if* G *is an Abelian group and* $u: \mathrm{Ob}(\underline{\mathrm{Aut}}(\mathscr{C})) \rightarrow K_1(\mathscr{C})$ *is a mapping satisfying conditions a), b), c), then there exists a unique group homomorphism* v *such that the following diagram commutes:*

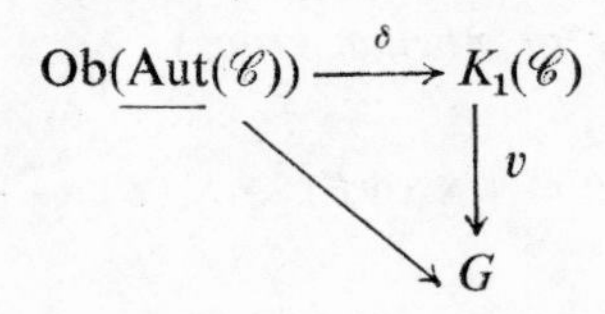

The group $K_1(\mathscr{C})$ is called the *Whitehead group* associated to the groupoid $(\mathscr{C}, \perp)$.

Proof: left to the reader.

Example: Let $(\mathscr{V}, \oplus)$ be the monoid of finite dimensional vector spaces over the field k. We have

$$K_0(\mathscr{V}) = \mathbb{Z}, \ \gamma(V) = \dim_k(V),$$

$$K_1(\mathscr{V}) = k^*, \ \delta(V, \mu) = \det(\mu).$$

DEFINITION 1.8.6. Let $\mathscr{C}, \mathscr{C}'$ be two groupoids with products and let $F: \mathscr{C} \rightarrow \mathscr{C}'$ be a product preserving functor. We define a new groupoid with products ΦF in the following way:

the objects of ΦF are triplets of the form (A, α, B), where $A, B \in \mathrm{Ob}(\mathscr{C})$ and $\alpha: F(A) \rightarrow F(B)$;

a morphism $(A, \alpha, B) \to (A', \alpha', B')$ in ΦF is a pair (f, g) consisting of morphisms $f: A \to A'$, $g: B \to B'$ in $\mathscr{C}$, such that the diagram

$$\begin{array}{ccc} F(A) & \xrightarrow{F(f)} & F(A') \\ \alpha \downarrow & & \downarrow \alpha' \\ F(B) & \xrightarrow{F(g)} & F(B') \end{array}$$

commutes;

$$(A, \alpha, B) \perp (A', \alpha', B') = (A \perp A', \alpha \perp \alpha', B \perp B').$$

DEFINITION 1.8.7. A functor $F: \mathscr{C} \to \mathscr{C}'$ between two categories with products is called *cofinal* if for every object $A' \in \mathrm{Ob}(\mathscr{C}')$ there exists an object $A \in \mathrm{Ob}(\mathscr{C})$ and an object $B' \in \mathrm{Ob}(\mathscr{C}')$ such that $A' \perp B' \simeq F(A)$.

PROPOSITION 1.8.8. *If $F: \mathscr{C} \to \mathscr{C}'$ is a cofinal functor of groupoids with products, then the sequence of Abelian groups*

$$K_1(\mathscr{C}) \xrightarrow{K_1(F)} K_1(\mathscr{C}') \xrightarrow{d'} K_0(\Phi F) \xrightarrow{d} K_0(\mathscr{C}) \xrightarrow{K_0(F)} K_0(\mathscr{C}') \tag{1.8.8.1}$$

is exact.

Proof. (Hyman Bass, *Lectures on Topics in Algebraic K-Theory*, Chap. I, Theorem 4.6, p. 30). We shall limit ourselves at indicating the definition of the morphisms d', d.

In order to define d', we have only to consider the mapping $\mathrm{Ob}(\underline{\mathrm{Aut}}(\mathscr{C}')) \to K_0(\Phi F)$

$$(A', \alpha') \to \gamma(A, \alpha' \perp 1_{B'}, A),$$

where $A \in \mathscr{C}$ is such that $A' \perp B' = F(A)$, while d is obtained from the mapping

$$\mathrm{Ob}(\Phi F) \to K_0(\mathscr{C}),$$

$$(A, \alpha, B) \mapsto \gamma(A) - \gamma(B).$$

Remark. In order to deduce the exact sequence (1.7.2) from the exact sequence (1.8.8.1), consider the groupoids with products

$$(\mathscr{C}, \perp) = (\mathrm{Vec}(k), \oplus),$$

$$(\mathscr{C}', \perp) = (\mathrm{Quad}(k), \perp),$$

together with the "hyperbolic" functor

$$H: \mathscr{C} \to \mathscr{C}'$$

which associates the hyperbolic module

$$H(V) = (V \oplus V^*, Q),$$

to every vector space V, where $q((x,f), (y,g)) = f(x) + g(y)$.

Remark that $\operatorname{Witt}(k) = \operatorname{Coker}(K_0(H))$ and $WG(k) = K_0(\operatorname{Quad}(k))$.

1.8.9. The tensor product of two quadratic modules induces on the groups $\operatorname{Witt}(k)$, $WG(k)$ structures of commutative rings with unit elements.

Denoting by $\langle \gamma \rangle$ the class (V, Q), where $\dim_k V = 1$, $Q(e) = \gamma$ $e \neq 0$, $e \in V$, the multiplication in $WG(k)$ is characterized by the identity

$$(1.8.9.1) \qquad \langle a \rangle \langle b \rangle = \langle ab \rangle.$$

The canonical map (1.7.1)

$$(1.8.9.2) \qquad WG(k) \to \operatorname{Witt}(k)$$

is obviously a ring homomorphism.

Definition 1.8.10. Let

$$\operatorname{rk}: \operatorname{Ob}(\operatorname{Quad}(k)) \to \mathbb{Z}$$

be the additive function which associates its rank

$$\operatorname{rk}(V, Q) = \dim_k(V).$$

o every non-degenerate quadratic module.

This function induces a ring homomorphism which will be still denoted by rk:

$$(1.8.10.1) \qquad \operatorname{rk}: WG(k) \to \mathbb{Z}.$$

Let $J(k)$ be the ideal Ker(rk). Since the canonical map (1.8.9.2) is surjective, the image of $J(k)$ through this map is an ideal of Witt (k), which will be denoted by $I(k)$.

PROPOSITION 1.8.11. *The underlying additive group of the ideal $J(k)$ is generated by the elements of $WG(k)$ of the form*

$$(\langle a\rangle - \langle 1\rangle),\ a \in k^*.$$

Proof. Every $x \in WG(k)$ can be written as

$$x = \sum_{i=1}^{n} (\mathrm{cl}(M_i) - \mathrm{cl}(N_i)), \tag{1.8.11.1}$$

where

$$\mathrm{cl}\colon \mathrm{Ob}(\mathrm{Quad}(k)) \to WG(k)$$

is the canonical map 1.8.3. The assumption $\mathrm{rk}(x) = 0$ implies

$$\sum_{i=1}^{r} \mathrm{rk}(N_i) = \sum_{i=1}^{r} \mathrm{rk}(N_i). \tag{1.8.11.2}$$

Suppose that

$$M_i = \langle \mu_{i1}\rangle + \langle \mu_{i2}\rangle + \ldots + \langle \mu_{im_i}\rangle,$$

$$N_i = \langle \nu_{i1}\rangle + \langle \nu_{i2}\rangle + \ldots + \langle \nu_{in_i}\rangle.$$

Relations (1.8.11.1), (1.8.11.2) yield

$$\chi = \sum_{\substack{1 \leqslant i \leqslant r\\ 1 \leqslant j \leqslant m_i}} (\langle \mu_{ij}\rangle - \langle 1\rangle) - \sum_{\substack{1 \leqslant i \leqslant r\\ 1 \leqslant j \leqslant n_i}} (\langle \nu_{ij}\rangle - \langle 1\rangle).$$

COROLLARY 1.8.12. *The underlying additive group of the ideal $I(k)$ is generated by the elements of Witt (k) of the form*

$$(\langle a\rangle - \langle 1\rangle),\ a \in k^*.$$

PROPOSITION 1.8.13. If $c \in k^*$, *we continue to denote by* $\langle c \rangle$ *the element of* $Witt(k)$ *which corresponds to the element* $\langle c \rangle$ *of* $WG(k)$ *(cf. section 1.8.9) by the canonical homomorphism*

$$WG(k) \to \mathrm{Witt}(k).$$

For every two elements $a, b \in k^*$, *the following relation holds in the ring* $Witt(k)$:

$$\langle a + b \rangle = \langle a \rangle + \langle b \rangle - \langle ab(a + b) \rangle. \tag{1.8.13.1}$$

Proof. Consider the quadratic form

$$Q = a\xi^2 + b\eta^2 - (a + b)\,\zeta^2.$$

The vectors $(1, 1, 1)$, $(1, -1, 1)$ generate in k^3 a hyperbolic plane. Since the vector $\left(\frac{a+b}{a}, 0, 1\right)$ is orthogonal to this plane, in the ring $\mathrm{Witt}(k)$ we have

$$\mathrm{cl}(k^3, Q) = \left\langle Q\left(\frac{a+b}{a}, 0, 1\right)\right\rangle =$$

$$= \left\langle \frac{(a+b)^2}{a} - a + b \right\rangle =$$

$$= \left\langle \frac{(a+b)}{a}(a + b - a)\right\rangle = \left\langle \frac{a+b}{a} \cdot b \right\rangle =$$

$$= \left\langle \frac{(a+b)\,ab}{a^2}\right\rangle = \langle (a+b)\,ab \rangle.$$

1.9. *Quadratic modules over finite fields* (char $(\mathbb{F}_q) \neq 2$)

PROPOSITION 1.9.1. *For every finite field* $\mathbb{F}_q$, *the subgroup* $\mathbb{F}_q^{*2}$ *of* $\mathbb{F}_q^*$ *has the index 2.*

Proof. Let Ω be an algebraic closure of $\mathbb{F}_q$ and let

$$\varphi: \Omega^* \to \Omega^*$$

be the group homomorphism defined by

$$\varphi(x) = x^{\frac{q-1}{2}}.$$

We have

$$\text{(1.9.1.1)} \qquad \operatorname{Ker} \varphi = \mathbb{F}_q^{*2}$$

In fact, put $\xi \in \mathbb{F}_q^{*2}$. We shall have $\xi = a^2$, $a \in \mathbb{F}_q^*$. Consequently, $a^{q-1} = 1$, hence $\xi^{\frac{q-1}{2}} = 1$ i.e. $\xi \in \operatorname{Ker} \varphi$.

Now, consider $\xi \in \operatorname{Ker} \varphi$. We have to show that $a \in \mathbb{F}_q^{*2}$. Let $a \in \Omega$ be such that $\xi = a^2$. It is sufficient to prove that $a \in \mathbb{F}_q^*$, i.e. $a^{q-1} = 1$. But $a^{q-1} = \xi^{\frac{q-1}{2}} = 1$.

On the other hand, since the roots of the polynomial $X^{\frac{q-1}{2}} - 1$ are simple in Ω, we get

$$\operatorname{Card}(\operatorname{Ker} \varphi) = \frac{q-1}{2}.$$

Since

$$\operatorname{Card} \mathbb{F}_q^* = q - 1$$

the assertion follows.

PROPOSITION 1.9.2. *Let (V, Q) be a non-degenerate quadratic module of rank 2 over $\mathbb{F}_q$ and $c \in \mathbb{F}_q^*$. Then there exists at least one element $x \in V$ such that $Q(x) = c$.*

Proof. It is sufficient to prove that if $a, b \in \mathbb{F}_q^*$, then the equation

$$\text{(1.9.2.1)} \qquad a\xi^2 + b\eta^2 = c$$

has at least one solution in $\mathbb{F}_q^*$.

But according to Proposition 1.9.1, we have

$$\text{Card}\,\{a\xi^2 | \xi \in \mathbb{F}_q\} = \text{Card}\,\{c - b\eta^2 | \eta \in \mathbb{F}_q\} = \frac{q+1}{2}.$$

Consequently,

$$\{a\xi^2 | \xi \in \mathbb{F}_q\} \cap \{c - b\eta^2 | \eta \in \mathbb{F}_q\} \neq \emptyset$$

hence (1.9.2.1) has at least one solution.

Remark 1.9.2.2. The assumption $c \neq 0$ plays actually no rôle in the proof of Proposition 1.9.2. On the other hand, in the case $c = 0$, equation 1.9.2.1 may have no other solution besides the trivial one.

Example. Take $q=2$, i.e. $\mathbb{F}_3 = \mathbb{Z}/3\mathbb{Z}$ and consider the quadratic form $x^2 + y^2$; obviously, in $\mathbb{F}_3$, the only solution of equation $x^2 + y^2 = 0$ is the trivial one.

Remark 1.9.2.3. The problem of representing an element c of a field K as $Q(x) = c$, where Q is a quadratic form, is an old problem of arithmetics. If $K = Q$, we have the following result: Every rational uumber is the sum of four squares.

PROPOSITION 1.9.3. *If (V, Q) is a non-degenerate quadratic module of rank $\geqslant 3$ over $\mathbb{F}_q$, then, for every $c \in \mathbb{F}_q$, there exists at least one element $x \in V$ such that $Q(x) = c$.*

Proof. Straightforward consequence of Corollary 1.4.13 and of the fact that every finite field has the property C_1.

Remark. For $c \in \mathbb{F}_q^*$, the proof of Proposition 1.9.3 uses only Proposition 1.9.2.

PROPOSITION 1.9.4. *Let (V, Q) be an n-dimensional non-degenerate quadratic module over $\mathbb{F}_q$. Then, there exists an orthogonal basis $(e_1, e_2, \ldots, e_n)$ of V such that*

$$Q(x) = \begin{cases} \xi_1^2 + \xi_2^2 + \ldots + \xi_n^2 & \text{if disc } (Q) \text{ is a square} \\ \xi_1^2 + \xi_2^2 + \ldots + a\xi_n^2 & \text{if disc}(Q) \text{ is not a square.} \end{cases}$$

$$(x = \xi_1 e_1 + \xi_2 e_2 + \ldots + \xi_n e_n),\ (a \neq 1).$$

The proof of this proposition uses the following:

LEMMA 1.9.5. *If (V, Q) is a non-degenerate quadratic module over the field k and if $a \in k^*$, the following properties are equivalent:*

(i) There exists $x \in V$ such that $Q(x) = a$;

(ii) There exists an orthogonal basis $(e_1, e_2, \ldots, e_n)$ of V such that $Q(e_1) = a$.

Proof of Lemma 1.9.5. The implication $ii \Rightarrow i$ is obvious. In order to prove the implication $i \Rightarrow ii$, let $x \in V$ be such that $Q(x) = a$. The restriction of Q to the subspace kx induces a non-degenerate quadratic module, whence (Proposition 1.4.2) the decomposition:

$$V = V' \perp (kz).$$

In order to get the required basis $(e_1, e_2, \ldots, e_n)$ of V, it is sufficient to take $e_1 = x$ and subsequently to consider an orthogonal basis of V' (Proposition 1.4.1).

Proof of Proposition 1.9.4. We argue by induction on the dimension n of V. The proof is obvious for $n = 1$. If $n \geqslant 2$, Proposition 1.9.3 together with Lemma 1.9.5 for $a = 1$ show the existence of a decomposition of V in the form

$$V = V_1 \perp (e_1),\ Q(e_1) = 1,$$

whence, by the induction assumption, the existence of the required basis follows.

COROLLARY 1.9.6. *In order for two quadratic modules (V, Q), (V', Q') over $\mathbb{F}_q$ to be isomorphic, it is necessary and sufficient that the following conditions be fulfilled:*

(α) $\dim V = \dim V'$,

(β) $\operatorname{disc}(Q) = \operatorname{disc}(Q')$

COROLLARY 1.9.7. *We have* $WG(\mathbb{F}_q) \simeq \mathbb{Z} \times (\mathbb{F}_q^*/\mathrm{F}_q^{*2})$.

Proof. The map

$$(V, Q) \to (\dim_{\mathbb{F}_q} V,\ \operatorname{disc}(Q))$$

induces a homomorphism

$$WG(\mathbb{F}_q) \to \mathbb{Z} \times \mathbb{F}_q^*/\mathbb{F}_q^{*2}$$

which, by Corollary 1.9.6, is an isomorphism.

COROLLARY 1.9.8. *If*

$$\mathbb{Z} \xrightarrow{\varphi} \mathbb{Z} \times (\mathbb{F}_q^*/\mathbb{F}_q^{*2})$$

is the homomorphism defined by the condition $\varphi(1) = (2, -1)$, *then* $\mathrm{Witt}(k) \simeq \mathrm{Coker}\,\varphi$.

1.10. *The Clifford algebra of a quadratic module*

DEFINITION 1.10.1. Let (V, Q) be a quadratic space over the field k. A Clifford algebra associated to the quadratic space (V, Q) is a pair $(C(Q), \rho)$ consisting of an algebra $C(Q)$ over the field k and a linear map $\rho: V \to C(Q)$ such that the following conditions be fulfilled:

(1) For every $x \in V$, we have $(\rho(x))^2 = Q(x).1$.

(2) If L is an algebra over k and $f: f \to L$ is a linear map such that

$$(f(x))^2 = Q(x) \cdot 1)$$

for every $x \in V$, then there exists a *unique* algebra homomorphism

$$F: C(Q) \to L$$

such that the following diagram be commutative:

$$\begin{array}{ccc} V & \xrightarrow{\rho} & C(Q) \\ & {}_{f}\searrow \quad \swarrow_{F} & \\ & L & \end{array}$$

Obviously, two Clifford algebras associated to the same quadratic module are canonically isomorphic.

PROPOSITION 1.10.2. *For every quadratic module* (V, Q), *there exists at least one Clifford algebra associated to it. Moreover,* ρ *is an injection and* $\dim_k C(Q) = 2^n$, *where* $n = \dim_k V$.

Proof. Let $T(V)$ be the tensor algebra associated with the vector space V and I its bilateral ideal spanned by the elements.

$$(x \oplus x - Q(x) \cdot 1)_{x \in V}.$$

One can take $C(Q) = T(V)/I$. As for the map ρ, it is obtained by composing the canonical maps:

$$V \to T(V) \to T(V)/I.$$

In order to get a different description of the Clifford algebra associated to a quadratic module, let, as above, $n = \dim_k V$ and $e_1, e_2, \ldots, e_n$ be an orthogonal basis of V. We denote by M the set $\{1, 2, \ldots, n\}$ and by $\mathscr{P}(M)$ the collection of its subsets (Card $(\mathscr{P}(M) = 2^n)$.

Let $C = k^{\mathscr{P}(M)}$ and $(e_S)_{S \in \mathscr{P}(M)}$ be its canonical basis. A structure of k-algebra can be introduced on C as follows:

$$e_S \cdot e_T = \alpha(S, T)\, e_{S+T},$$

where — $S + T$ is the subset of M consisting of the elements of the union $S \cup T$ belonging to one and only one of the subsets S, T;

$$-\ \alpha(S, T) = \prod_{\substack{s \in S \\ t \in T}} (s, t) \prod_{i \in S \cap T} (Qe_i) \text{ and}$$

$$(s, t) = \begin{cases} 1 \text{ if } s \leqslant t, \\ -1 \text{ if } s > t. \end{cases}$$

The injection ρ is defined as follows:

$$\rho(e_i) = e_{\{i\}}, \quad i = 1, 2, \ldots, n.$$

Example 1.10.3. Let (V, Q) have the dimension 2, let e_1, e_2 be an orthogonal basis of V and $\alpha = Q(e_1)$, $\beta = Q(e_2)$: $Q = \alpha\xi_1^2 + \beta\xi_2^2$. The Clifford algebra constructed above has a basis $e_\varnothing, e_{\{1\}}, e_{\{2\}}, e_{\{1,2\}}$ and the multiplication table is given by 1.10.3.1. This algebra is denoted by $\left(\dfrac{\alpha, \beta}{k}\right)$.

Table 1.10.3.1

	$e_{\emptyset}$	$e_{\{1\}}$	$e_{\{2\}}$	$e_{\{1,2\}}$
$e_{\emptyset}$	$e_{\emptyset}$	$e_{\{1\}}$	$e_{\{2\}}$	$e_{\{1,2\}}$
$e_{\{1\}}$	$e_{\{1\}}$	$\alpha e_{\emptyset}$	$e_{\{1,2\}}$	$\alpha e_{\{2\}}$
$e_{\{2\}}$	$e_{\{2\}}$	$-e_{\{1,2\}}$	$\beta e_{\emptyset}$	$-\beta e_{\{1\}}$
$e_{\{1,2\}}$	$e_{\{1,2\}}$	$-e_{\{2\}}$	$-\alpha e_{\{1\}}$	$-\alpha\beta e_{\emptyset}$

The algebra $\left(\frac{1, -1}{k}\right)$ is isomorphic to $\mathcal{M}(2, k)$, the algebra of 2×2 matrices with elements in k.

The algebra $\left(\frac{-1, -1}{\mathbb{R}}\right)$ is the classical quaternions algebra.

II. SYMBOLS. ELEMENTS OF THE THEORY OF THE GROUP $K_2(A)$

§ 1. *Symbols*

DEFINITION 2.1.1. Let A be a commutative ring, denote by A^* the group of its units and let G be an Abelian group. A G-valued symbol defined in A is a map

$$(,): A^* \times A^* \to G$$

satisfying the following conditions:

(*i*) $(aa', b) = (a, b) + (a', b)$,

(*ii*) $(a, bb') = (a, b) + (a, b')$,

(*iii*) if $a \in A^*$, $1 - a \in A^*$, then $(a, 1 - a) = 0$,

(*iv*) $(a, -a) = 0$.

Remark 2.1.1.1. If A is a field, then *iv* follows from *i*, *ii* and *iii*.

Indeed, we have

$$(a, -a) = (a, -a) - (a, 1-a) = (a, -a) +$$

$$+\left(a, \frac{1}{1-a}\right) = \left(\frac{1}{a}, \frac{1-a}{-a}\right) =$$

$$= \left(\frac{1}{a}, 1, -\frac{1}{a}\right) = 0$$

PROPOSITION 2.1.2. *If*

$$(,): A^* \times A^* \to G$$

is a symbol, then, for every $a \in A^*$, $b \in A^*$, *we have:*

$$(2.1.2.1) \qquad (a, b) + (b, a) = 0.$$

Proof. Indeed, using conditions *iv*, *i* and *ii* from Definition 2.1.1, we get

$$0 = (ab, -ab) = (a, -ab) + (b, -ab) = (a, (-a)b) +$$

$$+ (b, a(-b)) = (a, -a) + (a, b) + (b, a) +$$

$$+ (b, -b) = (a, b) + (b, a)$$

which is the desired equality.

PROPOSITION 2.1.3. *For every commutative ring* A *there exists a symbol:*

$$\{,\},: A^* \times A^* \to \mathrm{Symb}\,(A) = K_2(A)$$

such that for every symbol $(,): A^* \times A^* \to G$ *there is a unique group homomorphism*

$$\mathrm{Symb}(A) \to G$$

making commutative the diagram

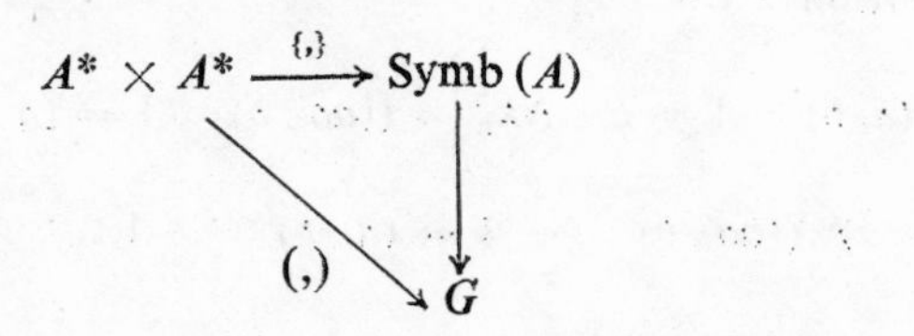

Proof. It is sufficient to take for Symb (A) the quotient of the free Abelian group whose basis is the set $A^* \times A^*$, with respect to the subgroup generated by relations *i—iv* from Definition 1.1.

DEFINITION 2.1.4. *(Hilbert "symbol").* Let k be a commutative field and let G be the group of invertible elements of the ring $\mathbb{Z}$. If $a, b \in k^*$, we put

$$(a, b) = \begin{cases} 1 \text{ if } z^2 - ax^2 - by^2 = 0 \text{ has a root } \neq 0 \text{ in } k^3; \\ -1 \text{ otherwise.} \end{cases}$$

PROPOSITION 2.1.5. *For every commutative field k, the Hilbert "symbol" satisfies the conditions:*

(2.1.5.1) $(a, -a) = 1, \ \forall a \in k^*;$

(2.1.5.2) $(a, 1 - a) = 1, \ \forall a \in k^*, \ 1 - a \in k^*;$

(2.1.5.3) $(a, b) = (b, a), \ \forall a, b \in k^*;$

(2.1.5.4) $(a, c^2) = 1, \ \forall a, c \in k^*;$

(2.1.5.5) $(a, b) = 1 \Rightarrow (aa', b) = (a', b), \ \forall a, a', b \in k^*;$

(2.1.5.6) $(a, b) = (a, -ab) = (a, (1 - a)b), \ \forall a, b \in k^*,$

$1 - a \in k^*.$

Proof. In order to prove (2.1.5.1) we remark that (1, 1, 0) is a solution of the equation $z^2 - ax^2 + ay^2 = 0$.

In order to get (2.1.5.2), we remark that (1, 1, 1) is a solution of the equation $z^2 - ax^2 - (1 - a)y^2 = 0$.

Relations (2.1.5.3)—(2.1.5.4) are obvious. In order to prove (2.1.5.5) we use Proposition 2.1.6:

$$(a, b) = 1 \Rightarrow a \in Nk_b^* \Rightarrow ((aa', b) = 1 \Leftrightarrow (a', b) = 1) \wedge$$

$$\wedge ((aa', b) = -1 \Leftrightarrow (a', b) = -1.$$

In order to get (2.1.5.6), we use (2.1.5.1), (2.1.5.2), (2.1.5.3), (2.1.5.5):

$$(a, -a) = 1 \Rightarrow (-a, a) = 1 \Rightarrow (-ab, a) = (a, -ab) =$$

$$= (a, b),$$

$$(a, 1-a) = 1 \Rightarrow (1-a, a) = 1 \Rightarrow ((1-a)b, a) =$$

$$= (b, a) \Rightarrow (a, b) = (a, (1-a)b)).$$

PROPOSITION 2.1.6. *Let $a, b \in k^*$ and $k_b = k(\sqrt{b})$ be the decomposition field of the polynomial $X^2 - b$. In order for (a, b) to be equal to 1, it is necessary and sufficient for a to belong to the group Nk_b^* of the norms of the elements from k_b^*.*

Proof. First, assume that $k_b = k$. In this case $Nk_b^* = k^*$ and b is the square of an element c in k:

$$b = c^2.$$

We have to prove that $(a, b) = 1$ for every $a \in k^*$. But $(0, 1, c)$ is obviously a root of the polynomial $z^2 - ax^2 - by^2$.

Now, assume that $k_b \supsetneq k$. There exists an element $\beta \in k_b$ such that

$$k(\beta) = k_b,$$

$$\beta^2 = b.$$

$\forall \xi \in k_b$ there exists $\zeta, \eta \in k$ such that $\zeta + \beta\eta = \xi$. Obviously, we shall have:

$$N(\xi) = \zeta^2 - b\eta^2.$$

Consequently, if $a \in Nk_b$ there are $\zeta, \eta \in k$ such that

$$a = \zeta^2 - b\eta^2$$

hence $(1, \zeta, \eta)$ is a non-trivial root of the quadratic form $z^2 - ax^2 - by^2$.

Conversely, if $(a, b) = 1$, then this quadratic form has a non-trivial root (ζ, ξ, η). But ξ cannot be zero, since otherwise b would be the square of some element in k. Consequently, a is the norm of the element $\frac{\zeta}{\xi} + \beta \frac{\eta}{\xi}$.

In the following, we shall prove that if k is the field $\mathbb{Q}_p$ of p-adic numbers, Hilbert's symbol satisfies also conditions *i, ii* from Definition 1.1 (for *iii, iv*, see Proposition 2.1.5), in other words, it defines a symbol

$$(,): \mathbb{Q}_p^* \times \mathbb{Q}_p^* \to \{\pm 1\}$$

For a deeper understanding of Hilbert "symbol", we introduce a new symbol, namely the Legendre "symbol".

DEFINITION 2.1.7. *(Legendre "symbol").* For every prime integer $p > 2$, Proposition 1.9.1 yields the following exact sequence of group homomorphisms:

$$(2.1.7.1) \qquad 0 \to \mathbb{F}_p^{*2} \to \mathbb{F}_p^* \xrightarrow{\lambda} \{\pm 1\} \to 0.$$

We recall that the homomorphism λ is defined by $\lambda(x) = x^{\frac{p-1}{2}}$.

Let $\qquad \pi_p: \mathbb{Z} \to \mathbb{F}_p = \mathbb{Z}/_p\mathbb{Z}$

be the canonical homomorphism associating with every integer x its equivalence class modulo p.

Now we define a map

$$(\bar{p}): \mathbb{Z} \to \{\pm 1, 0\}$$

given by

$$(2.1.7.2) \qquad \left(\frac{x}{p}\right) = \begin{cases} \lambda(\pi_p(x)), & \text{if } \pi_p(x) \text{ is non-zero} \\ 0, & \text{otherwise.} \end{cases}$$

The element $\left(\frac{x}{p}\right)$ is called the Legendre mod p symbol associated with the integer x.

If x is an integer such that $\pi_p(x) \neq 0$, then $\left(\frac{x}{p}\right) = 1$ iff $\pi_p(x) \in \mathbb{F}_p^{*2}$.

It should be understood that the Abelian multiplicative group $\{\pm 1\}$ from the exact sequence (2.1.7.1) is a subgroup of $\mathbb{F}_p^*$. It follows that for every $x \in \mathbb{Z}$ such that $\pi_p(x) \neq 0$, the following relation

$$\pi_p(x^{\frac{p-1}{2}}) = \left(\frac{x}{p}\right).$$

holds in $\mathbb{F}_p^*$.

2.1.8. *Examples.* (a) We have

$$\left(\frac{1}{11}\right) = \left(\frac{3}{11}\right) = \left(\frac{4}{11}\right) = \left(\frac{5}{11}\right) = \left(\frac{9}{11}\right) = 1,$$

$$\left(\frac{2}{11}\right) = \left(\frac{6}{11}\right) = \left(\frac{7}{11}\right) = \left(\frac{8}{11}\right) = \left(\frac{10}{11}\right) = -1.$$

Indeed, $F_{11}^{*2} = \{\dot{1}, \dot{3}, \dot{4}, \dot{5}, \dot{9}\}$.

(b) We have $\left(\frac{3}{13}\right) = 3^6 \bmod 13 = 1$, $\left(\frac{10}{17}\right) = 10^8 \bmod 17 = (-2)^4 \bmod 17 = 16 \bmod 17 = -1$.

PROPOSITION 2.1.9. *We have the formula*

$$\left(\frac{-1}{p}\right) = \begin{cases} +1, & \text{if } p \equiv 1 \bmod 4 \\ -1, & \text{if } p \equiv 3 \bmod 4. \end{cases}$$

in other words,

$$\left(\frac{-1}{p}\right) = (-1)^{\varepsilon(p)}$$

where, for every odd integer n, we have put

$$\varepsilon(n) \equiv \frac{n-1}{2} \pmod 2 \doteq \begin{cases} 0 \ \textit{if } n \equiv 1 \bmod 4 \\ 1 \ \textit{if } n \equiv 3 \bmod 4. \end{cases}$$

Proof. Let $\left(\frac{-1}{p}\right) = 1.$

In view of formula (2.1.7.2), we have

$$(-1)^{\frac{p-1}{2}} \equiv 1 \bmod p \text{ i.e. } (-1)^{\frac{p-1}{2}} = 1.$$

Consequently, $\frac{p-1}{2} = 2r$, hence $p \equiv 1 \bmod 4$. Conversely, if $p \equiv 1$ mod 4, then clearly $\left(\frac{-1}{p}\right) = 1.$

Now suppose that $\left(\frac{-1}{p}\right) = -1$. Then, as above, we get $\frac{p-1}{2} = 2r + 1$, i..e. $p \equiv 3 \bmod 4$ and conversely, if $p \equiv 3 \bmod 4$, then $\left(\frac{-1}{p}\right) = -1.$

Application 2.1.10. If $p \equiv 1 \bmod 4$, then $\mathrm{Witt}(\mathbb{F}_p) \simeq \mathbb{Z}/_2\mathbb{Z} \times \mathbb{F}_p^*/\mathbb{F}_p^{*2})$.

Proof. Using the notation from Corollary 1.9.7, we deduce from Proposition 2.1.9 that

$$\mathrm{Im}\, \varphi = \{2n \times \dot{1} \mid n \in \mathbb{Z}\}$$

consequently $\mathrm{Witt}\,(\mathbb{F}_p) = \mathrm{Coker}(\varphi) = \mathbb{Z}/_2\mathbb{Z} \times (\mathbb{F}_p^*/\mathbb{F}_p^{*2})$.

PROPOSITION 2.1.11. *(Quadratic reciprocity law). If* 1 *and p are two distinct prime integers different from* 2, *then*

$$\left(\frac{l}{p}\right) = (-1)^{\varepsilon(l)\varepsilon(p)} \left(\frac{p}{l}\right).$$

The proof of this proposition is an easy consequence of the following:

LEMMA 2.1.12. *Let Ω be an algebraic closure of $\mathbb{F}_p$ and $w \in \Omega$ a primitive root of order 1 of the unity and consider "Gauss' sum":*

$$(2.1.12.1) \qquad S = \sum_{x \in \mathbb{F}_l^*} \left(\frac{z}{l}\right) w^x.$$

Then the following relations are true

$$(2.1.12.2) \qquad S^2 = \left(\frac{-1}{l}\right) l, \qquad (2.1.12.3) \qquad S^{p-1} = \left(\frac{p}{l}\right).$$

(By definition, S is an element of $\mathbb{F}_p$ and in (2.1.12.2) we have denoted by l the canonical image of l in $\mathbb{F}_p$ by the canonical homomorphism

$$\mathbb{Z} \to \mathbb{F}_p = \mathbb{Z}/p\mathbb{Z}.$$

Proof. From the definition of S, it follows

$$S^2 = \sum_{t, z \in \mathbb{F}_l^*} \left(\frac{tz}{l}\right) w^{t+z} = \sum_{t, z \in \mathbb{F}_l^*} \left(\frac{t(tz)}{l}\right) w^{t+tz} =$$

$$= \sum_{t, z \in \mathbb{F}_l^*} \left(\frac{z}{l}\right) w^{t(z+1)} = \sum_{t \in \mathbb{F}_l^*} \left(\frac{-1}{l}\right) w^0 +$$

$$+ \sum_{\substack{z \in \mathbb{F}_l^* \\ z \neq -1}} \left[\left(\frac{z}{l}\right)\left(\sum_{t \in \mathbb{F}_l^*} w^{t(z+1)}\right)\right].$$

But for every $z \in \mathbb{F}_l$, $z \neq -1$:

$$\sum_{t \in \mathbb{F}_l^*} w^{t(z+1)} = w^{z+1} + (w^{z+1})^2 + \ldots + (w^{z+1})^{l-1} = -1$$

since every l-th order root ζ of the unity different from 1 satisfies the condition

$$1 + \zeta + \zeta^2 + \ldots + \zeta^{l-1} = 0.$$

Hence we get

$$S^2 = \left(\frac{-1}{l}\right)(l-1) + (-1)\sum_{\substack{z\in\mathbb{F}_l^*\\ z\neq -1}} \left(\frac{z}{l}\right).$$

But since clearly

$$\sum_{z\in\mathbb{F}_l^*} \left(\frac{z}{l}\right) = 0$$

this implies relation (2.1.12.2).

In order to prove (2.1.12.3) we use the fact that the field has the characteristic p, hence

$$S^p = \sum_{x\in\mathbb{F}_l^*} \left(\frac{x}{l}\right) w^{xp} = \sum_{x\in\mathbb{F}_l^*} \left(\frac{xp^{-1}}{l}\right) w^x =$$

$$= \left(\frac{p^{-1}}{l}\right) S = \left(\frac{p}{l}\right) S$$

which implies $S^{p-1} = \left(\frac{p}{l}\right)$.

Proof of Proposition 2.1.11. Using Proposition 2.1.9 and Lemma 2.1.12, we have

$$\left(\frac{(-1)^{\varepsilon(l)}\, l}{p}\right) = \left(\frac{S^2}{p}\right) = (S^2)^{\frac{p-1}{2}} = S^{p-1} = \left(\frac{p}{l}\right).$$

On the other hand, from Proposition 2.1.9 we get

$$\left(\frac{(-1)^{\varepsilon(l)}}{p}\right) = (-1)^{\varepsilon(l)\varepsilon(p)}$$

from which the quadratic reciprocity law follows.

2.1.13. *Historical note.* The statement of the quadratic reciprocity law has been independently discovered by the Swiss mathematician Leonhard

Euler (1707—1783) and by the French mathematician Adrien-Marie Legendre (1752—1833). The first complete proof belongs to the German mathematician Karl-Friedrich Gauss (1777—1855) who exposed it in his monumental work *Disquisitiones arithmeticae* (1801).

PROPOSITION 2.1.14. *We have the formula:*

$$\left(\frac{2}{p}\right)=\begin{cases} 1 \text{ if } p \equiv \pm 1 \bmod 8 \\ -1 \text{ if } p \equiv \pm 5 \bmod 8 \end{cases}$$

Proof. Let α be an 8-th order primitive root of the unity in an algebraic closure Ω of $\mathbb{F}_p$. The element $y = \alpha + \frac{1}{\alpha}$ satisfies the relation $y^2 = 2$ since $\alpha^4 + 1 = 0$ and hence $\alpha^2 + \frac{1}{\alpha^2} = 0$. In this way, the statement of the proposition is equivalent to the following relations:

$$2^{\frac{p-1}{2}} = y^{p-1} = \begin{cases} 1 \text{ if } p \equiv \pm 1 \bmod 8 \\ -1 \text{ if } p \equiv \pm 5 \bmod 8 \end{cases}$$

At any rate, we have $y^p = \alpha^p + \frac{1}{\alpha^p}$, hence $p \equiv \pm 1 \bmod 8$, $y^p = y$, consequently $y^{p-1} = 1$,

$$p \equiv \pm 5 (\bmod 8),\ \ y^p = \alpha^5 + \frac{1}{\alpha^5} = -\left(\alpha + \frac{1}{\alpha}\right) = -y.$$

In fact,

$$\alpha\left(\alpha^5 + \frac{1}{\alpha^5} + \alpha + \frac{1}{\alpha}\right) = \alpha(\alpha^{10} + 1 + \alpha^6 + \alpha^4) =$$

$$= \alpha + \alpha^3 + \alpha^5 + \alpha^7 = \alpha + \alpha^3 - \alpha - \alpha^3 = 0.$$

PROPOSITION 2.1.15. *The notation being as in* (1.8.10) *and* (2.1.3), *there exists a surjective homomorphism*

$$\varphi: K_2(k)/2K_2(k) \to I^2(k)/I^3(k).$$

Proof. Consider the group homomorphism

$$\mu: k^* \oplus_{\mathbb{Z}} k^* \to I^2(k)/I^3(k)$$

defined by

$$(a \oplus b) \mapsto (\langle a\rangle - \langle 1\rangle)(\langle b\rangle \langle -1\rangle) \bmod I^3(k).$$

Such a homomorphism exists since

$$\langle a\rangle - \langle 1\rangle + \langle b\rangle - \langle 1\rangle \equiv \langle ab\rangle - \langle 1\rangle) \bmod I^2(k).$$

If a, $1 - a \in k^*$, we have

$$\mu(a, 1 - a) = 0. \tag{2.1.15.1}$$

Indeed, by (1.8.9.1), (1.8.13.1), we have:

$$\mu(a, 1 - a) = (\langle a\rangle - \langle 1\rangle)(\langle 1 - a\rangle - \langle 1\rangle) =$$

$$= \langle a(1 - a)\rangle - \langle a\rangle - \langle 1 - a\rangle + \langle 1\rangle$$

$$\langle 1\rangle = \langle a + (1 - a)\rangle = \langle a\rangle + \langle 1 - a\rangle - \langle a(1 - a)\rangle.$$

Applying Proposition 2.1.3 and the fact that when A is a field, Axiom *iv* from the definition of the symbols is a consequence of the first three axioms (Remark 2.1.1.1) it follows that μ generates a homomorphism

$$K_2(k) \to I^2(k)/I^3(k).$$

This homomorphism vanishes on the elements of $2K_2(k)$. In order to establish this fact, it is sufficient to show that $\mu(2(a \otimes b)) = 0$. But, since $\langle a^2\rangle = \langle 1\rangle$, we have

$$\mu(2(a \otimes b)) = \mu(2a \otimes b) = (\langle a^2\rangle - \langle 1\rangle)(\langle b\rangle -$$

$$- \langle 1\rangle) = 0$$

The homomorphism φ is surjective since $I(k)$ is additively generated by the elements of the form $\langle a\rangle - \langle 1\rangle$ (cf. Corollary 1.8.12).

§ 2. *The group of the units of p-adic fields*

2.2. *The filtration of the group of the units of a complete field of discrete valuation.* Let K be a field endowed with a discrete valuation with respect to which K is complete. We denote by A the ring of the valuation, by $\underline{m}$ its maximal ideal, by k its residual field, and by U the group of the units of A, i.e. the multiplicative group of the invertible elements of A:

$$A = \{x \in K |\ v(x) \geqslant 0\},\ \underline{m} = \{x \in K | v(x) \geqslant 1\},$$

$$k = A/\underline{m},\ U = A - \underline{m}.$$

We define a descending filtration

$$U^{(0)} \supset U^{(1)} \supset \ldots \supset U^{(i)} \supset U^{(i+1)} \supset \ldots$$

by putting: $U^{(0)} = U$; $U^{(i)} = 1 + \underline{m}^{(i)}$ for every $i \geqslant 1$.

Every $U^{(i)}$ is actually a subgroup of U. Let us prove, for instance, that if $\xi \in U^{(i)}$, then its inverse η also belongs to $U^{(i)}$. We have $\xi = 1 + x$, $x \in \underline{m}^i$ and $(1 + x)\eta = 1$, i.e. $\eta = 1 - x\eta \in 1 + \underline{m}^i$.

In the $\underline{m}$-adic topology, each subgroup $U^{(i)}$ is closed. First of all, it is clear that U itself is closed, since $U = A - \underline{m}$ and $\underline{m}$ is open. For an arbitrary i, let

$$\rho: A \to A/\underline{m}^i$$

be the canonical continuous map. The set ($1 \bmod \underline{m}^i$) is closed in the $\rho(\underline{m})$-adic topology of $A/\underline{m}^i$. Consequently, its inverse image by ρ is closed; but this inverse image is identical with $U^{(i)}$.

The subgroups $U^{(i)}$ are also open in the $\underline{m}$-adic topology. This is obvious for $i \geqslant 1$, and, for $i = 0$, it follows from the fact that if $\xi \in U$, then $\xi + \underline{m} \subset U$.

We also have

$$\bigcap_{i \geqslant 0} U^{(i)} = \{1\}. \tag{2.2.0}$$

PROPOSITION 2.2.1. *The canonical homomorphism*

$$\pi: U^{(n)} \to \varprojlim_i U^{(n)}/U^{(i)}$$

is an isomorphism.

Proof. The injectivity of π is a consequence of (2.2.0).

In order to prove its surjectivity, let $\mu = (\mu_i)_{i \geqslant n} \in \varprojlim_i U^{(n)}/U^{(i)}$ and let $(u_i)_{i \geqslant n}$ be a sequence of elements of $U^{(n)}$ such that for every i, the image of u_i by the canonical morphism

$$U^{(n)} \to U^{(n)}/U^{(i)}$$

be μ_i. This sequence is convergent to a limit $u \in U^{(n)}$. Obviously, $\pi(u) = \mu$.

PROPOSITION 2.2.2. (a) *There is a canonical isomorphism*

$$U^{(0)}/U^{(1)} \rightleftarrows k^*$$

(b) *For every $i \geqslant 1$, there exists a canonical isomorphism*

$$U^{(i)}/U^{(i+1)} \rightleftarrows \underline{m}^i/\underline{m}^{i+1}$$

The group m^i/m^{i+1} is (in a, generally speaking, non-canonical way) isomorphic with the additive group k^* subjacent to the residual field k.

Proof. The isomorphism asserted in a sends every element u mod $U^{(1)}$ of $U^{(0)}/U^{(1)}$, $(u \in U)$, into the element u mod $\underline{m}$ of k.

The isomorphism asserted in b sends every element x mod $\underline{m}^{i+1}$, $x \in \underline{m}^i$, into the element $(1 + x)$ mod $U^{(i+1)}$.

Since $\underline{m}^i/\underline{m}^{i+1}$ is the Abelian group subjacent to a 1-dimensional vector space over k, for every choice of a basis in $\underline{m}^i/\underline{m}^{i+1}$ (in particular for every uniformizing element π in A) we get an isomorphism between $\underline{m}^i/\underline{m}^{i+1}$ and k.

In the particular case when $K = \mathbb{Q}_p$, there exists a canonical uniformizing element: $\pi = p$.

COROLLARY 2.2.3. *If $k = \mathbb{F}_q$, $q = p^r$, then for every $i, j \geqslant 1$, the group $U^{(i)}/U^{(i+j)}$ is a p-group. More precisely, it is a group of order $q^j = p^{rj}$.*

Proof. Making an induction on $\underline{j}$, it is sufficient to consider the exact sequence

$$1 \to U^{(i+j)}/U^{(i+j+1)} \to U^{(i)}/U^{(i+j+1)} \to U^{(i)}/U^{(i+j)} \to 0.$$

Proposition 2.2.4. *Assume that the residual field k is finite:* $k = \mathbb{F}_q$, char $(k) = p$.

(a) *There exists a unique subgroup* V *of* U, $V = \{x \in U \mid x^{p-1} = 1\}$ * *such that* U *is isomorphic with* $\mathbb{F}_p^*$.

(b) *We have the relation*

$$U = V \oplus U^{(1)}$$

and the canonical projection $\mathrm{pr}_V : U \to V$ *can be identified with the morphism obtained by restricting the morphism* $A \to K$.

Proof. We essentially use the following group-theoretic lemma, which is a straightforward consequence of the primary decomposition theorem:

Lemma 2.2.4.1. *Let*

$$0 \to A \to E \to B \to 0$$

be an exact sequence of Abelian groups, with A, B *finite of relatively prime orders* a, b. *If* B' *is the subgroup of* E *consisting of all elements* x' *for which* $bx' = 0$, *then* E *is the direct sum of* A *and* B', *and* B' *is the only subgroup of* E *which is isomorphic to* B.

In order to prove Proposition 2.2.4, consider the exact sequence

$$1 \to U^{(1)}/U^{(n)} \to U/U^{(n)} \to F_q^* \to 1.$$

The assumptions of Lemma 2.2.4.1 are fulfilled. Indeed, the order of $\mathbb{F}_q^*$ is $q - 1$, while the order of $U^{(1)}/U^{(n)}$ is a power of p (Corollary 2.2.3). It follows that $U/U^{(n)}$ has a unique subgroup isomorphic with $\mathbb{F}_q^*$. Moreover, the projection

$$U/U^{(n)} \to U/U^{(n-1)}$$

maps V_n isomorphically into V_{n-1}**.

* If $q = 2$, then $\mathbb{F}_2^* = 1$ and in this case it is obvious that $U = U^{(1)}$. Indeed, if $x \in U$, then $x \bmod \underline{m} = 1$, hence $x \in U^{(1)}$.

** If $f : G_1 \to G_2$ is a homomorphism between two torsion groups, and if

$$G_1 = \bigoplus_p G_1(p), \quad G_2 = \bigoplus_p G_2(p)$$

are the primary decompositions of these groups, then $f = \bigoplus_p f(p)$, where $f(p) : G_1(p) \to G_2(p)$ is the morphism induced by f on the primary components.

By passing to the limit, we get a subgroup V of U, such that

$$U = V \oplus U^{(1)}.$$

V is the only subgroup of U which is isomorphic with $\mathbb{F}_q^*$. In fact, assuming the converse, U would contain an element whose order would be a divisor of $q - 1$ and whose image by at least one of the canonical projections

$$U \to U/U^{(n)}$$

would not belong to V_n, thus contradicting the uniqueness of V_n.

COROLLARY 2.2.5. *The field K contains the roots of order $q - 1$ of the unity.*

A proposition analogous to Proposition 2.2.4 can be proved in the case of discrete valuation ring with a perfect residual field. In the particular case of a formal series ring in one variable with coefficients in an arbitrary field k, we have:

PROPOSITION 2.2.6. *If $A = k[[T]]$ is the formal series ring in one variable with coefficients in a field k and $G_m(k)$ is the algebraic multiplicative group associated with the field k, then for every $n \geqslant 1$ there exists an isomorphism*

$$U/U^{(n)} \xrightarrow{\rho_n} G_m(k) \times U^{(1)}/U^{(n)}$$

such that the following diagram

$$\begin{array}{ccc} U/U^{(n)} & \xrightarrow{\rho_n} & G_m(k) \times U^{(1)}/U^{(n)} \\ \uparrow{\scriptstyle \sigma_n} & & \uparrow{\scriptstyle \mathrm{id}_{G_m(k)} \times \sigma_n^1} \\ U/U^{(n+1)} & \xrightarrow{\rho_{n+1}} & G_m(k) \times U^{(1)}/U^{(n+1)} \end{array}$$

is commutative. In this diagram, the two morphisms σ_n, σ_n^1 are induced by the canonical injections

$$U^{(n)} \hookrightarrow U^{(1)} \hookrightarrow U.$$

Proof. The isomorphism ρ_n is obtained by passing to the quotient in the morphism

$$U \to G_m(k) \times U^{(1)}/U^{(n)}$$

$$\sum_{i=0}^{\infty} a_i T^i \mapsto \left(a_0, \sum_{i \geq 0} \frac{a_i}{a_0} T^i \right) \bmod U^{(n)}$$

The following results of this paragraph concern the p-adic field $\mathbb{Q}_p$.

LEMMA 2.2.7. *Let* $K = \mathbb{Q}_p$ *be the field of p-adic numbers and* $\alpha \in U^{(i)} - U^{(i+1)}$ *with* $i \geq 1$ *if* $p \neq 2$, *and* $i \geq 2$ *if* $p = 2$. *Under these assumptions,* $\alpha^p \in U^{(i+1)} - U^{(i+2)}$.

Proof. From the hypothesis it follows that

$$\alpha = 1 + kp^i, \ k \in p\mathbb{Z}_p,$$

hence, applying the formula of the binomial, we get

$$\alpha^p = (1 + kp^i)^p = 1 + kp^{i+1} + \ldots + k^p p^{ip},$$

where the omitted powers of p are greater than $2i + 1$, hence than $i + 2$. On the other hand, we have in any case $ip \geq i + 2$, hence

$$\alpha^p \equiv 1 + kp^{i+1} \bmod p^{i+2}\mathbb{Z}_p$$

i.e. $\alpha^p \in U^{(i+1)} - U^{(i+2)}$.

PROPOSITION 2.2.8. (a) *Let* $K = \mathbb{Q}_p$, $p \neq 2$, $\alpha \in U^{(i)} - U^{(i+1)}$, $i \geq 1$. *For every* $n \geq 0$, *there exists an isomorphism*

$$\theta_{\alpha,n}: \mathbb{Z}/p^n\mathbb{Z} \to U^{(i)}/U^{(i+n)}$$

satisfying the following conditions:

(i) *The image by* $\theta_{\alpha,n}$ *of the canonical generator of* $\mathbb{Z}/p^n\mathbb{Z}$ *coincides with* $\alpha_n = \alpha \bmod (U^{(i+n)})$:

$$\theta_{\alpha,n}(z) = \alpha_n^z, \ z \in \mathbb{Z}/p^n\mathbb{Z}$$

(ii) $\theta_{\alpha,n}(p^j\mathbb{Z}/p^n\mathbb{Z}) = U^{(i+j)}/U^{(i+n)}$

(iii) The diagram

$$\begin{array}{ccc} \mathbb{Z}/p^{n+1}\mathbb{Z} & \xrightarrow{\theta_{\alpha,n+1}} & U^{(i)}/U^{(i+n+1)} \\ \downarrow & & \downarrow \\ \mathbb{Z}/p^{n}\mathbb{Z} & \xrightarrow{\theta_{\alpha,n}} & U^{(i)}/U^{(i+n)} \end{array}$$

is commutative.

(b) *The isomorphisms* $(\theta_{\alpha_n})_{n\geqslant 0}$ *induce an isomorphism*

$$\theta_\alpha\colon \mathbb{Z}_p \to U^{(i)}.$$

(c) *Let* $K = \mathbb{Q}_2$, $\alpha \in U^{(i)} - U^{(i+1)}$, $i \geqslant 2$. *For every* $n \geqslant 0$, *there exists an isomorphism*

$$\theta_{\alpha,n}\colon \mathbb{Z}/2^n\mathbb{Z} \to U^{(i)}/U^{(i+n)}$$

satisfying conditions analogous to i, ii, iii.

(d) *The isomorphisms* $(\theta_{\alpha,n})_{n\geqslant 0}$ *induce an isomorphism*

$$\theta\colon \mathbb{Z}_2 \to U^{(i)}.$$

(e) *The extension*

$$0 \to U^{(2)} \to U^{(1)} \to U^{(1)}/U^{(2)} \to 0$$

is splitted by a homomorphism

$$U^{(1)}/U^{(2)} \to U^{(1)}$$

which maps the group $U^{(1)}/U^{(2)}$ *isomorphically onto the subgroup* $\{\pm 1\}$ *of* $U^{(1)}$, *hence*

$$U^{(1)} = (\pm 1) \times U^{(2)}.$$

Proof. It is sufficient to show that $\alpha_n^{p^{n-1}} \neq 1$ and $\alpha_n^{p^n} = 1$, which follows from Lemma 2.2.7.

As for point e, the only one deserving a special comment, assume, by absurd, that $1 = -1 \bmod U^{(2)}$. This would imply $-1 \in 1 + 2^2\mathbb{Z}_2$, which is impossible. (Obviously, $-1 \in U^{(1)}$ since $-1 = 1 - 2$.)

PROPOSITION 2.2.9. *We have the following isomorphisms:*

$$\mathbb{Q}_p^* \simeq \begin{cases} \mathbb{Z} \times \mathbb{Z}_p \times \mathbb{Z}/(p-1)\mathbb{Z} & \text{if } p \neq 2 \\ \mathbb{Z} \times \mathbb{Z}_2 \times \mathbb{Z}/2\mathbb{Z} & \text{if } p = 2 \end{cases}$$

Proof. Since every element $x \in \mathbb{Q}_p^*$ can be uniquely expressed as

$$x = p^n u, \ n \in \mathbb{Z}, \ u \in U$$

we get the decomposition

$$\mathbb{Q}_p^* \simeq \mathbb{Z} \times U.$$

In order to obtain the required isomorphisms, it is sufficient to use Propositions 2.2.4 and 2.2.8.

DEFINITION 2.2.10. Let $\mathbb{Z}_p$, $p \neq 2$, be the ring of p-adic integers and let

$$\pi_p\colon \mathbb{Z}_p \to \mathbb{F}_p$$

be the canonical homomorphism. For every $x \in \mathbb{Z}_p$, by analogy and according to 2.1.7, we define the "Legendre symbol" $\left(\dfrac{x}{p}\right)$ by the formula

$$\left(\frac{x}{p}\right) = \begin{cases} \lambda(\pi_p(x)) & \text{if } \pi_p(x) \neq 0, \\ 0 & \text{if } \pi_p(x) = 0. \end{cases} \tag{2.2.10.1}$$

PROPOSITION 2.2.11. Let $p \neq 2$, $x = p^n u \in \mathbb{Q}_p^*$, $n \in \mathbb{Z}$, $u \in U$. *In order for x to be a square, it is necessary and sufficient that n be even and* $\left(\dfrac{u}{p}\right) = 1$.

Proof. The condition is clearly necessary. In order to prove its sufficiency, we use the isomorphism from Proposition 2.2.9

$$\mathbb{Q}_p^* \simeq \mathbb{Z} \times \mathbb{Z}_p \times \mathbb{Z}/(p-1)\mathbb{Z}.$$

It follows that, in order for $x = (n, \mu_x, \nu_x)$ to be a square, it is sufficient that

$$n = 2n_0, \ \mu_x = 2\mu_0, \ \nu_x = \nu_0^2.$$

But in $\mathbb{Z}_p$ every element is of the form 2μ, 2 being invertible in $\mathbb{Z}_p$ for $p \neq 2$. On the other hand, from Proposition 2.2.4 it follows that $\nu_x = \pi_p(u)$, which shows that the previous condition is equivalent to $\left(\frac{u}{p}\right) = 1$.

PROPOSITION 2.2.12. *In order for an element* $x = 2^n u$ *of* $\mathbb{Q}_2^*$, $u \in U$, *to be a square it is necessary and sufficient that* n *be even and* $u \equiv 1 \bmod 8$.

Proof. Using the decomposition

$$U = U^{(1)} = \{\pm 1\} \times U^{(2)}$$

which follows from Propositions 2.2.8e and 2.2.9, we see that x is a square if and only if n is even and $u \in U^{(2)}$ is a square. But the morphism θ_α from Proposition 2.2.8 maps the subgroup $2\mathbb{Z}_2$ onto $U^{(3)}$. Consequently, $u \in U^{(2)}$ is a square if and only if $u \in U^{(3)}$, i.e. $u \equiv 1 \bmod 8$.

PROPOSITION 2.2.13. *If* $p \neq 2$, *we have the isomorphism*

$$\mathbb{Q}_p^*/\mathbb{Q}_p^{*2} \to \mathbb{Z}/2\mathbb{Z} + \mathbb{Z}/2\mathbb{Z}$$

and we can find a system of representatives of the form

$$(1, p, \mu, \mu p) \left(\frac{\mu}{p}\right) = -1.$$

Proof: straightforward.

PROPOSITION 2.2.14. *We have the isomorphism*

$$\mathbb{Q}_2^*/\mathbb{Q}_2^{*2} \simeq \mathbb{Z}/2\mathbb{Z} \times \mathbb{Z}/2\mathbb{Z} + \mathbb{Z}/2\mathbb{Z}$$

and the representatives $\{\pm 1, \pm 5, \pm 2, \pm 10\}$.

Proof: straightforward.

§ 3. *Quadratic forms over discrete valuation fields*

2.3.1. Let K be a discrete valuation field, denote by v its valuation, by A the valuation ring, by $\underline{m}$ the maximal ideal of A, by π an uniformizing element ($v(\pi) = 1$) and let $k = A/\underline{m}$ be the residual field.

PROPOSITION 2.3.2. *Let (V, Q) be a non-degenerate quadratic module over the discrete valuation field K. Then there exists a basis $e_1, e_2, \ldots, e_n$ of V such that*

$$Q(x) = \varepsilon_1\xi_1^2 + \varepsilon_2\xi_2^2 + \ldots + \varepsilon_r\xi_r^2 + \pi(\varepsilon_{r+1}\xi_{r+1}^2 + + \varepsilon_{r+2}\xi_{r+2}^2 + \ldots + \varepsilon_n\xi_n^2) \tag{2.3.2.0}$$

where (a) $\varepsilon_i \in U$, $i = 1, 2, \ldots, n$; (b) $x = \xi_1 e_1 + \xi_2 e_2 + \ldots + \xi_n e_n$.

Proof. Let $f_1, f_2, \ldots, f_n$ be an orthogonal basis of (V, Q). With respect o this basis, we have

$$Q(x) = \alpha_1\eta_1^2 + \alpha_2\eta_2^2 + \ldots + \alpha_n\eta_n^2,\ \alpha_i \in K,\ \alpha_i \neq 0, \tag{2.3.2.1}$$

$$x = \eta_1 e_1 + \eta_2 e_2 + \ldots + \eta_n e_n.$$

Renumbering, if necessary, the vectors $f_1, f_2, \ldots, f_n$ of the basis, we can suppose that

$$\alpha_i = \pi^{2k_i}\varepsilon_i,\ k_i \in \mathbb{Z},\ \varepsilon_i \in U,\ i = 1, 2, \ldots, r$$

$$\alpha_i = \pi^{2k_i+1}\varepsilon_i,\ k_i \in \mathbb{Z},\ \varepsilon_i \in U,\ i = r+1, r+2, \ldots, n.$$

Now consider the orthogonal basis given by the formulae

$$f_i = \pi^{k_i} e_i, \quad i = 1, 2, \ldots, n.$$

If $(\xi_1, \xi_2, \ldots, \xi_n)$ are the components of x with respect to the basis $(e_1, e_2, \ldots, e_n)$, we have

$$\xi_i = \pi^{k_i}\eta_i. \tag{2.3.2.2}$$

In view of (2.3.2.1) and (2.3.2.2), we get

$$Q(x) = \sum_{i=1}^{r} \alpha_i \eta_i^2 + \sum_{i=r+1}^{n} \alpha_i \eta_i^2 = \sum_{i=1}^{r} \varepsilon_i (\pi^{k_i} \eta_i)^2 +$$

$$+ \pi \sum_{i=r+1}^{n} \varepsilon_i (\pi^{k_i} \eta_i)^2 = \sum_{i=1}^{r} \varepsilon_i \xi_i^2 + \pi \sum_{i=r+1}^{n} \varepsilon_i \xi_i^2.$$

Remark 2.3.3. If (V, Q) is a non-degenerate quadratic module over the discrete valuation field K and $e_1, e_2, \ldots, e_n$ is a basis of V such that (2.3.2.0) holds, then the quadratic module $(V, \pi Q)$ has a basis $e_1', e_2', \ldots, e_n'$ such that

$$\pi Q(x) = \pi \sum_{i=1}^{r} \varepsilon_i (\xi_i')^2 + \sum_{i=r+1}^{n} \varepsilon_i (\xi_i')^2.$$

2.3.4. HENSEL LEMMA. *Let A be a complete valuation ring and let $P \in A[X_1, X_2, \ldots, X_n]$ be a polynomial in n variables. Assume that there exists an element $a = (a_1, a_2, \ldots, a_n)$ A^n and an integer $d \geqslant 0$ such that the following conditions be fulfilled:*

(I) $P(a) \equiv 0 \pmod{\underline{m}^{2d+1}}$;

(II) *there exists at least an $i \in \{1, 2, \ldots, n\}$ such that*

$$\frac{\partial P}{\partial X_i}(a) \not\equiv 0 \bmod \underline{m}^{d+1}.$$

Then there exists an element $b \in A^n$ such that

(i) $P(b) = 0$;

(ii) $b \equiv a \pmod{\underline{m}^{d+1}}$.

Proof. Consider the polynomial in one variable $R \in A[X]$, defined by the condition

$$R = P(a_1, a_2, \ldots, a_{i-1}, X, a_{i+1}, \ldots, a_n).$$

It is sufficient to prove the Hensel Lemma in the case $n = 1$, $a = (a_i)$ and $P = R$. In this particular case, the lemma can be restated by adding a uniqueness assertion, as follows.

LEMMA. *Let A be a complete discrete valuation ring and suppose $P \in A[X]$. Assume that there exists an element $a \in A$ and a positive integer $d \geqslant 0$ such that the following two conditions be fulfilled*

(I) $P(a) \equiv 0 \pmod{\underline{m}^{2d+1}}$,

(II) $P'(a) \not\equiv 0 \pmod{\underline{m}^{d+1}}$.

Then there exists a unique element $b \in A$ such that

(i) $P(b) = 0$,

(ii) $b \equiv a (\text{mod } \underline{m}^{d+1})$.

Proof. Existence. We construct by induction a sequence $b_1, b_2, \ldots, b_n$ of elements of the ring A such as to have:

$$(1) \qquad b_1 = a;$$

$$(2) \qquad b_{n+1} \equiv b_n (\text{mod } \underline{m}^{d+n}) \qquad \text{for every } n \geqslant 1;$$

$$(3) \qquad P(b_n) \equiv 0 \text{ mod } \underline{m}^{2d+n} \qquad \text{for every } n \geqslant 1.$$

Assume that we have already constructed the first n ($n \geqslant 1$) terms $b_1, b_2, \ldots, b_n$ with the desired properties. Then

$$P'(b_n) \not\equiv 0 (\text{mod } \underline{m}^{d+1}).$$

Indeed, for every $i \in \{1, 2, \ldots, n-1\}$, we have

$$b_{i+1} \equiv b_i (\text{mod } \underline{m}^{d+i}),$$

$$P'(b_{i+1}) \equiv P'(b_i) (\text{mod } \underline{m}^{d+i})$$

and since $\underline{m}^{d+i} \subset \underline{m}^{d+1}$, this implies

$$b_{i+1} \equiv b_i (\text{mod } \underline{m}^{d+i})$$

$$(4) \qquad P'(b_{i+1}) \equiv P'(b_i) (\text{mod } \underline{m}^{d+1})$$

from which the required relation follows recalling that $b_1 = a$ and $P'(a) = 0$ mod $\underline{m}^{d+1}$. In particular, we get $P'(b_n) \neq 0$.

Let π be a uniformizing element of the discrete valuation ring A and consider in the field of quotients of A the element x defined by

$$x = -\frac{P(b_n)}{\pi^{d+n}P'(b_n)}.$$

Actually, the element x belongs to the ring A. In fact,

$$v(x) = v\left(-\frac{P(b_n)}{\pi^{d+n}P'(b_n)}\right) = v(P(b_n)) - (d+n) -$$

$$- v(P'(b_n)) \geqslant (2d+n) - (d+n) - v(P'(b_n)) =$$

$$= d - v(P'(b_n)) \geqslant 0,$$

the relation $P'(b_n) \not\equiv 0 \bmod \underline{m}^{d+1}$ being equivalent to

(5) $$v(P'(b_n)) < d+1, \ (v(P'(b_n)) \leqslant d).$$

Now, for every $n \geqslant 1$, we define

(6) $$b_{n+1} = b_n + \pi^{d+n}x.$$

Condition (2) is clearly satisfied. In order to verify condition (3), we evaluate the element $P(b_{n+1})$ using the Taylor formula:

$$P(b_{n+1}) = P(b_n + \pi^{d+n}x) = P(b_n) + \pi^{d+n}xP'(b_n) +$$

$$+ \pi^{2d+2n}x^2y,$$

for some $y \in A$. Since our assumption on n ($n \geqslant 1$) implies that $2d + 2n \geqslant 2d + n + 1$, we get

$$P(b_{n+1}) \equiv P(b_n) + \pi^{d+n}xP'(b_n) \bmod \underline{m}^{2d+n+1}.$$

But from the definition of the element x, we deduce

$$P(b_n) + \pi^{d+n}xP'(b_n) = 0$$

from which relation (3) follows.

Condition (2) together with the completeness of A show that the sequence $b_1, b_2, \ldots, b_n, \ldots$ has a limit, say b.

From (3) we get $P(b) = 0$, and from (4), $b \equiv a \bmod \underline{m}^{d+1}$.

Uniqueness. Let $b' \in A$ be such that $P(b') = 0$ and $b' \equiv a \bmod \underline{m}^{d+1}$. We show by induction that for every $n \geqslant 1$ we have

$$(7) \qquad b' \equiv b_n \bmod \underline{m}^{d+n}$$

This relation is clearly satisfied for $n = 1$ since $b_1 = a$.
Now assume that (7) is proved; we show that this implies

$$(8) \qquad b' \equiv b_{n+1} \bmod \underline{m}^{d+n+1}.$$

To this aim, consider the following relation obtained by applying the Taylor formula:

$$0 = P(b') = P(b_n + (b'-b_n)) = P(b_n) + (b' - b_n)P'(b_n) + $$

$$+ (b' - b_n)^2 d_n, \ (d_n \in A).$$

This leads to the equality

$$b' = b_n - \frac{P(b_n)}{P'(b_n)} - \frac{(b' - b_n)^2}{P'(b_n)} d_n$$

i.e. using the recursive definition of b_{n+1},

$$b' - b_{n+1} = - \frac{(b' - b_n)^2}{P'(b_n)} d_n .$$

But we have

$$v(b' - b_{n+1}) = v((b' - b_n)^2) + v(d_n) - v(P'(b_n))$$

i.e. using (7) and (5),

$$v(b' - b_{n+1}) \geqslant 2(d + n) - d = d + 2n \geqslant d + n + 1.$$

Relation (7) being proved for every $n \geqslant 1$, this implies the uniqueness of b.

2.3.4.1. *Remark.* The recursive definition of the sequence $(b_n)_{n\geqslant 1}$, which actually can be written as

$$b_{n+1} = b_n - \frac{P(b_n)}{P'(b_{n+1})}$$

is inspired by the classical Newton method for approximizing the real roots of a polynomial with real coefficients (Fig. 2.3.4.1.1).

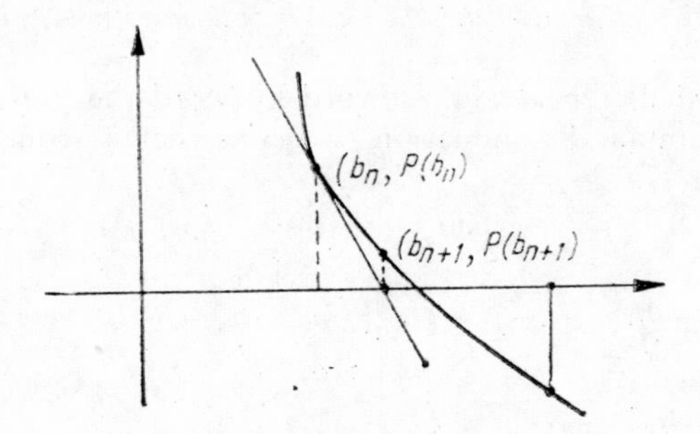

Fig. 2.3.4.1.1

The Hensel Lemma can be generalized to the case of r polynomials $P_1, P_2, \ldots, P_r \in A[X_1, \ldots, X_r]$ in r variables with coefficients in the complete discrete valuation ring A. In order to formulate this general form of the Hensel Lemma, we denote by

$$J_a(P_1, \ldots, P_r) = \left(\frac{\partial(P_1, P_2, \ldots, P_r)}{\partial(X_1, X_2, \ldots, X_r)}\right)_a$$

the Jacobian matrix of the polynomials $P_1, P_2, \ldots, P_r$ at the point $a = (a_1, a_2, \ldots, a_r)$.

PROPOSITION 2.3.4.2. *Let $P_1, P_2, \ldots, P_r$ be r polynomials in r variables with coefficients in the complete discrete valuation ring A and assume* *that there exist an element* $a = (a_1, \ldots, a_r) \in A^n$ *and a positive integer* $d \geqslant 0$, *such that the following two conditions*

$$\text{(I)} \qquad P_i(a_1, a_2, \ldots, a_r) \equiv 0 \ (\text{mod } \underline{m}^{2d+1}), \ i = 1, 2, \ldots, r;$$

$$\text{(II)} \qquad \det (J_a(P_1, \ldots, P_r)) \not\equiv 0 \ (\text{mod } \underline{m}^{d+1})$$

be fulfilled.

Then there exists a unique element $b = (b_1, b_2, \ldots, b_r) \in A^r$ *such that*

$$(i) \qquad P_i(b_1, b_2, \ldots, b_r) = 0, \ i = 1, 2, \ldots, r;$$

$$(ii) \qquad b \equiv a (\text{mod } \underline{m}^{d+1}) \text{ i.e. } (b_i \equiv a_i \text{ mod } \underline{m}^{d+1}, i=1, 2, \ldots, r).$$

Proof. We shall repeat almost word by word the previous proof of the Hensel Lemma. By induction, we construct a sequence

$$b^{(1)}, b^{(2)}, \ldots, b^{(r)}, \ldots, \ (b^{(n)} = (b_1^{(n)}, b_2^{(n)}, \ldots, b_r^{(n)}))$$

of elements in A^r, such as to have

$$(1) \qquad b^{(1)} = a;$$

$$(2) \qquad b^{(n+1)} \equiv b^{(n)} (\text{mod } \underline{m}^{d+n}) \text{ for every } n \geqslant 1;$$

(3) $\qquad P_i(b_1^{(n)}, b_2^{(n)}, \ldots, b_r^{(n)}) \equiv 0 \text{ mod } \underline{m}^{2d+n}$ for every $n \geqslant 1$ and every $i = 1, 2, \ldots, r$.

Suppose that the first n ($n \geqslant 1$) terms $b^{(1)}, b^{(2)}, \ldots, b^{(n)}$ with the required properties have been already constructed. Then one shows in exactly the same manner as in the case of the Hensel Lemma, that

$$J_{b^{(n)}}(P_1, P_2, \ldots, P_r) \not\equiv 0 (\text{mod } \underline{m}^{d+1})$$

hence, in particular,

$$J_{b^{(n)}}(P_1, P_2, \ldots, P_r) \not\equiv 0.$$

Consider the system of r linear equations in the r unknown quantities $x_1, x_2, \ldots, x_r$ with coefficients in A:

$$(2.3.4.3)\quad \begin{cases} \dfrac{\partial P_1}{\partial X_1}(b_1^{(n)}, b_2^{(n)}, \ldots, b_r^{(n)})x_1 + \dfrac{\partial P_1}{\partial X_2}(b_1^{(n)}, b_2^{(n)}, \ldots, b_r^{(n)})x_2 + \ldots \\ \qquad \ldots + \dfrac{\partial P_1}{\partial X_r}(b_1^{(n)}, b_2^{(n)}, \ldots, b_r^{(n)})\, x_r = \\ \qquad = -\dfrac{P_1(b_1^{(n)}, b_2^{(n)}, \ldots, b_r^{(n)})}{\pi^{d+n}} \\ \dfrac{\partial P_2}{\partial X_1}(b_1^{(n)}, b_2^{(n)}, \ldots, b_r^{(n)})x_1 + \dfrac{\partial P_2}{\partial X_2}(b_1^{(n)}, b_2^{(n)}, \ldots, b_r^{(n)})x_2 + \ldots \\ \qquad \ldots + \dfrac{\partial P_2}{\partial X_r}(b_1^{(n)}, b_2^{(n)}, \ldots, b_r^{(n)})\, x_r = \\ \qquad = -\dfrac{P_2(b_1^{(n)}, b_2^{(n)}, \ldots, b_r^{(n)})}{\pi^{d+n}} \\ \cdots\cdots\cdots\cdots\cdots\cdots\cdots\cdots\cdots\cdots \\ \dfrac{\partial P_r}{\partial X_1}(b_1^{(n)}, b_2^{(n)}, \ldots, b_r^{(n)})x_1 + \dfrac{\partial P_r}{\partial X_2}(b_1^{(n)}, b_2^{(n)}, \ldots, b_r^{(n)})x_2 + \ldots \\ \qquad \ldots + \dfrac{\partial P_r}{\partial X_r}(b_1^{(n)}, b_2^{(n)}, \ldots, b_r^{(n)})x_r = \\ \qquad = -\dfrac{P_r(b_1^{(n)}, b_2^{(n)}, \ldots, b_r^{(n)})}{\pi^{d+n}} \end{cases}$$

Using Cramer's rule (and the fact that $J_{b^{(n)}}(P_1, P_2, \ldots, P_r) \neq 0$) and evaluating the valuations of the solutions, it follows exactly as in the case of the Hensel Lemma that the solution of this system, which is *a priori* in K^r ($K =$ the field of quotients of A) actually belongs to A^r. Let this solution be

$$(\xi_1^{(n)}, \xi_2^{(n)}, \ldots, \xi_r^{(n)}).$$

For every $j = 1, 2, \ldots, r$ and every $n \geqslant 1$, we put by definition

$$b_j^{(n+1)} = b_j^{(n)} + \pi^{d+n}\xi_j^{(n)}.$$

In order to verify condition (3), the only one demanding a proof, we use the Taylor formula:

$$P_i(b_1^{(n+1)}, \ldots, b_r^{(n+1)}) = P_i(b_1^{(n)} + \pi^{d+n}\xi_1^{(n)}, \ldots, b_r^{(n)} +$$

$$+ \pi^{d+n}\xi_r^{(n)}) = P_i(b_1^{(n)}, \ldots, b_r^{(n)}) +$$

$$+ \pi^{d+n}\left(\frac{\partial P_i}{\partial X_1}\right)(b_1^{(n)}, \ldots, b_r^{(n)})\, \xi_1^{(n)} + \ldots$$

$$\ldots + \frac{\partial P_i}{\partial X_r}(b_1^{(n)}, \ldots, b_r^{(n)})\xi_r^{(n)} + \pi^{2(d+n)}y_{in},\ y_{in} \in A.$$

In view of the system (2.3.4.3) and of the inequality $2d + 2n \geqslant 2d + n + 1$, this yields exactly relation (3).

As for the uniqueness, it can also be proved almost with the same words as in the case of the Hensel Lemma.

A proposition of the Hensel Lemma type can also be formulated and proved in the case of r polynomials $P_1, P_2, \ldots, P_r \in A[X_1, X_2, \ldots, X_s]$ in s variables, where $r \leqslant s$ and A is a complete discrete valuation ring.

If $M \in \mathcal{M}(r, s; A)$ is an $r \times s$-matrix with elements in A and $\underline{c} \subset A$ is an ideal of the ring A, we denote by $M \bmod \underline{c}$ the $r \times s$-matrix with elements in the quotient ring $A/\underline{c}$, obtained by reducing mod $\underline{c}$ the elements of the matrix M:

$$M \bmod \underline{c} \in \mathcal{M}(r, s; A/\underline{c}).$$

With these notation, for every element $a = (a_1, \ldots, a_r)$ in A^r, we consider the following matrix with elements in the ring A:

$$M_a(P_1, P_2, \ldots, P_r) =$$

$$= \begin{pmatrix} \dfrac{\partial P_1}{\partial X_1}(a_1, \ldots, a_s) & \dfrac{\partial P_1}{\partial X_2}(a_1, \ldots, a_s) \ldots & \dfrac{\partial P_1}{\partial X_s}(a_1, \ldots, a_s) \\ \dfrac{\partial P_2}{\partial X_1}(a_1, \ldots, a_s) & \dfrac{\partial P_2}{\partial X_2}(a_1, \ldots, a_s) \ldots & \dfrac{\partial P_2}{\partial X_s}(a_1, \ldots, a_s) \\ \ldots & \ldots & \ldots \\ \dfrac{\partial P_r}{\partial X_1}(a_1, \ldots, a_s) & \dfrac{\partial P_r}{\partial X_2}(a_1, \ldots, a_s) \ldots & \dfrac{\partial P_r}{\partial X_s}(a_1, \ldots, a_s) \end{pmatrix}.$$

PROPOSITION 2.3.4.4. *Let $P_1, P_2, \ldots, P_r$ be r polynomials in s variables with coefficients in the complete discrete valuation ring A. Assume that $r \leqslant s$ and that an element $a = (a_1, a_2, \ldots, a_s) \in A^s$ and a positive integer $d \geqslant 0$ exist such that the following two conditions be fulfilled:*

(I) $P_i(a_1, a_2, \ldots, a_s) \equiv 0 \,(\text{mod}\, \underline{m}^{2d+1}),\ i = 1, 2, \ldots, r;$

(II) $\text{rang}\,(M_a(P_1, \ldots, P_r) \text{ mod } \underline{m}^{d+1}) = r.$

Then there exists an element $b = (b_1, b_2, \ldots, b_s) \in A^s$ such that:

(i) $P_i(b_1, b_2, \ldots, b_s) = 0,\ i = 1, 2, \ldots, r;$

(ii) $b \equiv a (\text{mod}\, \underline{m}^{d+1}).$

Proof. Renumbering, if necessary, the polynomials $P_1, \ldots, P_r$, we may assume that the following determinant is non-vanishing:

$$\det \begin{pmatrix} \frac{\partial P_1}{\partial X_1}(a_1, \ldots, a_s) & \frac{\partial P_1}{\partial X_2}(a_1, \ldots, a_s) \ldots & \frac{\partial P_1}{\partial X_r}(a_1, \ldots, a_s) \\ \frac{\partial P_2}{\partial X_1}(a_1, \ldots, a_s) & \frac{\partial P_2}{\partial X_2}(a_1, \ldots, a_s) \ldots & \frac{\partial P_2}{\partial X_r}(a_1, \ldots, a_s) \\ \ldots & \ldots & \ldots \\ \frac{\partial P_r}{\partial X_1}(a_1, \ldots, a_s) & \frac{\partial P_r}{\partial X_2}(a_1, \ldots, a_s) \ldots & \frac{\partial P_r}{\partial X_r}(a_1, \ldots, a_s) \end{pmatrix} \text{mod } \underline{m}^{d+1} \neq 0.$$

The r polynomials $P_1, P_2, \ldots, P_r$ can be completed to a system of s polynomials in s variables by adding the polynomials

$$X_{r+1} - a_{r+1}, X_{r+2} - a_{r+2}, \ldots, X_s - a_s.$$

The s polynomials $(P_1, \ldots, P_r, X_{r+1} - a_{r+1}, \ldots, X_s - a_s)$ satisfy the assumptions of Proposition 2.3.4.2, from which the proof of our proposition follows.

DEFINITION 2.3.4.5. Let (A, m) be a local ring and let $k = A/\underline{m}$ be its residual field. We say that the local ring $(A, \underline{m})$ is a *Henselian ring* (cf. Hensel Lemma 2.3.4) if for every polynomial $P \in A[X]$ and every element $a \in A$ satisfying the conditions

1. $P(a) \equiv 0 \bmod \underline{m}$,

2. $P'(a) \not\equiv 0 \bmod \underline{m}$,

there exists at least one element $b \in A$ such that the following conditions be fulfilled:

I. $P(b) = 0$,

II.. $b \equiv a \bmod \underline{m}$.

Example 2.3.4.6. Every complete local rings is a Henselian ring. (For the particular case of discrete valuation rings, see Proposition 2.3.4).

DEFINITION 2.3.4.7. Let $(A, \underline{m})$ be a local ring and let $\hat{A}$ be the complet ring with respect to the m-adic topology. We say that the local ring $(A, \underline{m})$ has the approximation property if for every family $P_1, P_2, \ldots, P_r \in A[X_1, X_2, \ldots, X_s]$, every family $x_1, x_2, \ldots, x_s$ of elements in $\hat{A}$ satisfying the conditions

$$P_1(x_1, x_2, \ldots, x_s) = 0, \ldots, P_r(x_1, \ldots, x_s) = 0$$

and every integer $c \geqslant 0$, there exists at least one system of elements $y_1, \ldots, y_s \in A$ such that the following conditions be fulfilled

$$P_1(y_1, \ldots, y_s) = 0, \ldots, P_s(y_1, \ldots, y_s) = 0,$$

$$x_1 \equiv y_i \bmod \underline{m}^c, \ i = 1, 2, \ldots, s.$$

PROPOSITION 2.3.4.8. *Every local ring $(A, \underline{m})$ satisfying the approximation property (Definition 2.3.4.7) is a Henselian ring.*

Proof. Let $P \in A[X]$ and $a \in A$ be such that

$$P(a) \equiv 0 \bmod \underline{m},$$

$$P'(a) \not\equiv 0 \bmod \underline{m}.$$

Since $\hat{A}$ is a Henselian ring (cf. example 2.3.4.6) and $A \subset \hat{A}$, $A/\underline{m} = \hat{A}/\underline{m}\hat{A}$, it follows that an element $a_1 \in \hat{A}$ exists such that the following conditions be fulfilled:

$$P(a_1) = 0,\ a_1 \equiv a \bmod \underline{m}.$$

In view of the approximation property for $c = 1$, it follows that an element $b \in A$ exists such that

$$P(b) = 0,\ b \equiv a \bmod \underline{m}\,\hat{A}$$

which shows precisely that the local ring $(A, \underline{m})$ is a Henselian ring.

DEFINITION 2.3.4.9. Let $(A, \underline{m})$ be a local ring. We say that the ring A satisfies the strong approximation property if every family of polynomials $P_1, P_2, \ldots, P_r \in A[X_1, \ldots, X_s]$ has a solution in A if and only if for every $n \in \mathbb{N}$, the given family has a solution in the ring $A/\underline{m}^n$.

PROPOSITION 2.3.4.10. *Let $(A, \underline{m})$ be a local ring such that the following conditions be fulfiled:*

(a) *A is complete.*

(b) *For every $n \geqslant 1$, the quotient ring $A/\underline{m}^n$ is finite. Under these assumptions, the ring A has the strong approximation property.*

The proof of this proposition relies on the following

LEMMA 2.3.4.11. *Let $\ldots \to X_n \xrightarrow{\varphi_{n,n-1}} X_{n-1} \to \ldots \to X_1$ be a projective system and let $X = \varprojlim X_n$ be its projective limit. If the sets X_n are finite and non-empty, then X is non-empty.*

Proof of the lemma. If the maps $\varphi_{n,n-1}$ are surjective, then, clearly, $X \neq \emptyset$. We can reduce ourselves to this case in the following way: Consider $X_{n,p} = \operatorname{Im}(X_{n+p} \xrightarrow{\varphi_{n+p,n}} X_n)$. From the relation

$$\varphi_{n+p+1,n} = \varphi_{n+p+1,n+p}\,\varphi_{n+p,n},$$

$$(X_{n+p+1} \xrightarrow{\varphi_{n+p+1,n+p}} X_{n+p} \xrightarrow{\varphi_{n+p,n}} X_n)$$
$$\underbrace{\qquad\qquad\qquad\qquad\qquad}_{\varphi_{n+p+1,n}}$$

we obviously have the inclusions

$$(2.3.4.12)\qquad \ldots \subset X_{n,p+1} \subset X_{n,p} \subset \ldots \subset X_{n,1} \subset X_n.$$

Since the set X_n is finite, the sequence (2.3.4.12) is stationary. Hence there exists a $p(n)$ such that

$$\emptyset \neq Y_n = X_{n,p(n)} = X_{n,p(n)+1} = \ldots = X_{n,p(n)+k} = \ldots$$

Let $i_n\colon Y_n \hookrightarrow X_n$ be the canonical inclusion. For every $n \in \mathbb{N}$, we have $\varphi_{n,n-1}(Y_n) \subset Y_{n-1}$. Denoting by $\Psi_{n,n-1}\colon Y_n \to Y_{n-1}$ the map defined by $\Psi_{n,n-1}(y) = \varphi_{n,n-1}(y)$, we get the projective system

$$\ldots \to Y_n \xrightarrow{\Psi_{n,n-1}} Y_{n-1} \to \ldots \to Y_1$$

The inclusions $(i_n)_{n\geqslant 1}$ define a morphism of the projective system $(Y_n, \Psi_{n+1,n})_{n\geqslant 1}$ into the projective system $(X_n, \varphi_{n+1,n})_{n\geqslant 1}$ which yields a map

$$\varprojlim_n Y_n \to \varprojlim_n X_n.$$

But for every n, the map $\Psi_{n+1,n}\colon Y_{n+1} \to Y_n$ is surjective. Hence $Y \neq \emptyset$ and consequently $X \neq \emptyset$.

Proof of Proposition 2.3.4.10. If $\underline{a} \subset A[X_1, \ldots, X_s]$ is an ideal of the ring $A[X_1, \ldots, X_s]$ and B is an A-algebra, then the set of the zeros of $\underline{a}$ in B can be naturally identified with the set:

$$\mathrm{Hom}_A(A[X_1, \ldots, X_s]/\underline{a}, B).$$

In particular, the set of the zeros of $\underline{a}$ with values in $A/\underline{m}^n$ is identified with the *finite set*

$$\mathrm{Hom}_A(A[X_1, \ldots, X_s]/\underline{a}, A/\underline{m}^n).$$

Since, obviously,

$$\mathrm{Hom}_A(A[X_1, \ldots, X_s]/\underline{a}, A) = $$

$$= \mathrm{Hom}_A(A[X_1, \ldots, X_s]/\underline{a}, \varprojlim_n A/\underline{m}^n) \simeq$$

$$\simeq \varprojlim_n \mathrm{Hom}_A(A[X_1, \ldots, X_s]/\underline{a}, A/\underline{m}^n),$$

the proof of the proposition is a direct consequence of Lemma 2.3.4.11.

Example 2.3.4.13. For every prime integer p the ring of p-adic integers $\mathbb{Z}_p$ has the strong approximation property.

2.3.4.14. *Hensel-Greenberg approximation conditions.* Let as above $(A, \underline{m})$ be a local ring. We introduce the sets $\mathscr{S}_s(A)$, $\mathscr{S}(A)$ defined in the following way:

$$S_s(A) = \{(P_1, P_2, \ldots, P_r | P_i \in A[X_1, \ldots, X_s], r \geqslant 1,$$
$$\text{arbitrary}\},$$

$$\mathscr{S}(A) = \bigcup_{s \geqslant 1} \mathscr{S}_s(A).$$

In other words, an element of the set $\mathscr{S}(A)$ is a finite family of polynomials with coefficients in the ring A.

DEFINITION 2.3.4.15. We say that the local ring $(A, \underline{m})$ satisfies the *Hensel-Geenberg approximation condition* if there exists a function

$$\gamma : \mathscr{S}(A) \times \mathbb{N} \to \mathbb{N}$$

with the following property:

$$\text{If } \pi = (P_1, P_2, \ldots, P_r) \in S_s(A), c \in \mathbb{N}$$

$$\text{and } (y_1, y_2, \ldots y_s) \in A^s \text{ is such that}$$

$$P_i(y_1, \ldots, y_s) \equiv 0 \bmod \underline{m}^{\gamma(\pi, c)}, \qquad i = 1, 2, \ldots, r,$$

then there exists $(x_1, x_2, \ldots, x_s) \in A^s$ such that

$$P_i(x_1, x_2, \ldots, x_s) = 0, \qquad i = 1, 2, \ldots, r,$$

and

$$x_l \equiv y_l \bmod \underline{m}^c \qquad l = 1, 2, \ldots, s.$$

The Hensel-Greenberg approximation condition is inspired by the Hensel Lemma. Indeed, if A is a complete discrete valuation ring and if $r \leqslant s$ and the elements $y_1, \ldots, y_s \in A$ satisfy condition II from

Proposition 2.3.4.3, then according to this proposition, the function γ takes the extremely simple form

$$\gamma(\pi, c) = 2c - 1.$$

PROPOSITION 2.3.4.16 *(Greenberg). Every complete discrete valuation ring has the Hensel-Greenberg approximation property.*

In fact, Greenberg proves the following more precise result:

PROPOSITION 2.3.4.17. *Let $(A, \underline{m})$ be a Henselian discrete valuation ring and $P_1, \ldots, P_r$ be r polynomials in $A[X_1, \ldots, X_s]$.*

Then there exist some integers $M \geqslant 1$, $u \geqslant 1$, $t \geqslant 1$ (depending on $P_1, P_2, \ldots, P_r$) such that, if $v \geqslant M$ and $(y_1, \ldots, y_s) \in A^s$ satisfy the conditions

$$P_i(y_1, \ldots, y_s) \equiv 0 \bmod \underline{m}^v, \qquad i = 1, 2, \ldots, r$$

then there exists $(x_1, x_2, \ldots, x_s) \in A^s$ such that

$$P_i(x_1, \ldots, x_s) = 0, \qquad i = 1, 2, \ldots, r$$

and

$$y_j \equiv x_j (\bmod \underline{m}^{[\frac{v}{u}] - t}).$$

The proof of this proposition can be found in: M. Greenberg, *Rational Points in Henselian Discrete Valuation Ring*, Pub. Math. I.H.E.S., 23 (1964).

2.3.4.18. Besides the approximation conditions already introduced, it is natural to study the following so-called *weak approximation condition.*

DEFINITION 2.3.4.19. We say that the local ring (A, m) satisfies the weak approximation condition if every family of polynomials $P_1, P_2, \ldots$ $\ldots, P_r \in A[X_1, X_2, \ldots, X_s]$ has a solution in A as soon as it has one in A.

For the sake of the discussion which follows, we introduce the following classes of local rings:

$\mathscr{I}_{a,w}$, the class of rings with the weak approximation property;

$\mathscr{I}_a$, the class of rings with the approximation property;

$\mathscr{I}_{a,s}$, the class of rings with the strong approximation property;

$\mathscr{I}_{a,\gamma}$, the class of rings with the Hensel-Greenberg approximation property.

PROPOSITION 2.3.4.20. We have the inclusions

$$\mathscr{I}_{a,w} = \mathscr{I}_a \supset \mathscr{I}_{a,s} \supset \mathscr{I}_{a,\gamma}.$$

Proof. Obviously, only the inclusion

$$\mathscr{I}_{a,w} \subset \mathscr{I}_a$$

raises some problems. Let then, as usual, $(A, \underline{m})$ be a noetherian local ring with the weak approximation property $(y_1, y_2, \ldots, y_s) \in \hat{A}^s$ be a solution of the family of polynomials $P_1, \ldots, P_r \in A[X_1, X_2, \ldots, X_s]$ and let $c \geqslant 0$ be an integer. We have to prove that the family of polynomials $P_1, \ldots, P_r$ has a solution $(x_1, \ldots, x_s) \in A^s$ such that

$$y_j \equiv x_j \bmod \underline{m}^c\hat{A}, \quad j = 1, 2, \ldots, s.$$

Let $(m_1, m_2, \ldots, m_t)$ be a system of generators of the ideal $\underline{m}^c$ and $\eta_1, \eta_2, \ldots, \eta_s$ be elements in A, such that

$$y_j \equiv \eta_j \bmod \underline{m}^c\hat{A}, \ (\underline{m}\hat{A})^c = \underline{m}^c\hat{A}.$$

Such elements exist in view of the canonical isomorphism defined by the inclusion of A in $\hat{A}$:

$$A/\underline{m}^c \to \hat{A}/\underline{m}^c\hat{A}.$$

It follows that for every $j \in \{1, 2, \ldots, s\}$ we have

$$y_j - \eta_j = \sum_{i=1}^{t} \alpha_{ij}m_i, \ \alpha_{ij} \in \hat{A}, \ i = 1, 2, \ldots, t,$$

$$j = 1, 2, \ldots, s.$$

Now, consider the family of polynomials

$$Q_1, Q_2, \ldots, Q_s, \ P_1, \ldots, P_r \in A[X_1, X_2, \ldots, X_s, \xi_{ij}],$$

$$i = 1, 2, \ldots, t, \ j = 1, 2, \ldots, s,$$

where $Q_1, \ldots, Q_s$ are defined by

$$Q_j(X_1, \ldots, X_s, \xi_{ij}) = X_j - \eta_j - \sum_{i=1}^{t} \xi_{ij} m_i.$$

The family of polynomials $Q_1, \ldots, Q_s, P_1, \ldots, P_r$ has a solution in $\hat{A}$, namely $(y_1, y_2, \ldots, y_s, \alpha_{ij}, \alpha_{2j}, \ldots, \alpha_{tj})$.

Since the ring A has the weak approximation property it follows that this family has a solution in A, hence A has the approximation property.

To conclude, for the moment, the discussion of the different approximation properties in local rings, we recall without proof the following two results belonging to M. Artin:

PROPOSITION 2.3.4.21. *Let k be a field or an excellent discrete valuation ring, let R be a commutative algebra of finite type over k, let p be a prime ideal of R and $A = R_p$ be the Henselizing ring of the ring R in the prime ideal p. Then A has the approximation property.*

PROPOSITION 2.3.4.22. *Let k be a field and let $A = k[X_1, \ldots, X_s]$ be the Henselizing ring of $k[X_1, \ldots, X_s]$ in the prime ideal generated by* $(X_1, X_2, \ldots, X_s)$. The local ring A has the Hensel-Greenberg approximation property.

The proof can be found in: M. Artin, *Algebraic Approximation of Structure over Complete Local Rings*, Pub. Math. I.H.E.S., 36 (1969), pp. 23–58.

COROLLARY 2.3.4.23. *Every quadratic form*

$$Q = \sum_{i-1}^{r} \varepsilon_i, \xi_i^2, r \geqslant 3, \tag{2.3.4.24}$$

whose coefficients are p-adic units ($\xi_i \in \mathbb{Z}_p \setminus p\mathbb{Z}_p$) *has at least a non-trivial solution in* $\mathbb{Z}_p$.

Proof. Since the field $\mathbb{F}_p = \mathbb{Z}/_p\mathbb{Z}$ has the C_1 property, it follows that there exists at least a solution of the quadratic form Q mod p. In view of the Hensel Lemma for $\alpha = 0$, this solution can be raised to a non-trivial solution of Q in $\mathbb{Z}_p$.

DEFINITION 2.3.5. Let K be an arbitrary field and let d, i be two positive integers. We say that the field K has the $C_i(d)$ property if every homogeneous polynomial in n variables of degree d with coefficients in K has a non-trivial solution in K as soon as $n > d^i$.

If a field K has the property $C_i(d)$ for every d, we say that it has the property C_i.

2.3.6. *Examples.* (a) K has the property C_0 if and only if it is algebraically closed.

(b) Every finite field has the property C_1 (see J. P. Serre, *Cours d'Arithemtique*).

PROPOSITION 2.3.7. *For every prime number, p, the p-adic number field $\mathbb{Q}_p$, has the property $C_2(2)$.*

The proof of this theorem is a consequence of the following lemmas.

LEMMA 2.3.8. If $p \neq 2$, then the p-adic number field $\mathbb{Q}_p$ has the property $C_2(2)$.

Proof. We have to show that every quadratic form Q in n variables, $n \geqslant 5$, with coefficients in $\mathbb{Q}_p$ has at least a non-trivial zero in $\mathbb{Q}_p$. Consider the quadratic module $(\mathbb{Q}_p^n, Q)$ canonically associated with the form Q. We can assume that this quadratic module is non-degenerate, since otherwise Q obviously has a non-trivial solution in $\mathbb{Q}_p$.

According to Proposition 2.3.2, one can find a basis $e_1, e_2, \ldots, e_n$ of $\mathbb{Q}_p^n$ with respect to which Q writes

$$Q(x) = \varepsilon_1\xi_1^2 + \varepsilon_2\xi_2^2 + \ldots + \varepsilon_r\xi_r^2 + p(\varepsilon_{r+1}\xi_{r+1}^2 + \ldots + \varepsilon_n\xi_n^2). \tag{2.3.8.1}$$

In view of Remark 2.2.3 (replacing if necessary, the form Q by pQ) we may suppose $r \geqslant n - r$, i.e. $r \geqslant 3$. But in this case, the form

$$Q_0(x) = \sum_{i=1}^{r} \varepsilon_i\xi_i^2$$

has a non-trivial solution, since $\mathbb{F}_p$ has the C_1 property and we may apply directly Hensel Lemma with $d = 0$.

LEMMA 2.3.9. *Let Q be a quadratic form in n variables with coefficients in $\mathbb{Q}_2$ written under the form* (2.3.8.1) *and* $a = (a_1, a_2, \ldots, a_n) \in \mathbb{Z}_2^n$ *be an element in* $\mathbb{Z}_2^n$ *such that the following conditions be fulfilled:*

(*i*) $Q(a) \equiv 0 (\mathrm{mod}\ 8)$;

(*ii*) *There exists at least one* $i \in \{1, 2, \ldots, n\}$ *such that* $a_i \notin 2\mathbb{Z}_2$.
Then in $\mathbb{Q}_2$ *there exists at least one non-trivial solution of the quadratic form* Q.

Proof. Straightforward application of the Hensel Lemma together with Remark 2.3.3.

LEMMA 2.3.10. *The field* $\mathbb{Q}_2$ *has the* $C_2(2)$ *property.*

Proof. Using the notation from Proposition 2.3.2, we distinguish the following cases:

1. $r < n$. In this case, it is sufficient to show that the quadratic form

$$Q'(x) = \varepsilon_1\xi_1^2 + \varepsilon_2\xi_2^2 + \varepsilon_3\xi_3^3 + 2\varepsilon_n\xi_n^2$$

has at least one non-trivial solution.

To this aim, we show that the conditions of Lemma 2.3.9 are fulfilled. Indeed, since

$$\varepsilon_1 + \varepsilon_2 = 2\alpha$$

it follows that

$$\varepsilon_1 + \varepsilon_2 + 2\varepsilon_n\alpha^2 \equiv 2\alpha + 2\alpha^2 (\mathrm{mod}\ 4)$$

$$2\alpha + 2\alpha^2 \equiv 0\ (\mathrm{mod}\ 4).$$

Consequently, $\varepsilon_1 + \varepsilon_2 + 2\varepsilon_n\alpha^2 = 4\beta$, $\beta \in \mathbb{Z}_2$.
We show that $(1, 1, 2\beta, \alpha)$ satisfies the conditions of Lemma 2.3.9:

$$\varepsilon_1 \cdot 1 + \varepsilon_2 \cdot 1 + \varepsilon_3(2\beta)^2 + 2\varepsilon_n\alpha^2 = 4\beta + \varepsilon_3(4\beta)^2 \equiv$$

$$\equiv 4\beta + 4\beta^2 (\mathrm{mod}(8) \equiv 0\ \mathrm{mod}(8).$$

2. $r = n$. In this case, it suffices to show that the quadratic form

$$Q''(x) = \varepsilon_1\xi_1^2 + \varepsilon_2\xi_2^2 + \varepsilon_3\xi_3^2 + \varepsilon_4\xi_4^2 + \varepsilon_5\xi_5^2$$

has at least one non-trivial zero in $\mathbb{Q}_2$.
Put

$$\varepsilon_1 + \varepsilon_2 = 2\alpha_{12},\ \varepsilon_3 + \varepsilon_4 = 2\alpha_{34},\ \alpha_{12}, \alpha_{34} \in \mathbb{Z}_2$$

We distinguish two subcases;

(*i*) Neither α_{12}, nor α_{34} is divided by 2.

(*ii*) At least one of the elements α_{12}, α_{34} (say α_{12}) is divided by 2.

In the first subcase, we take $\xi_1 = \xi_2 = \xi_3 = \xi_4 = 1$. In the second subcase, we take $\xi_1 = \xi_2 = 1$, $\xi_3 = \xi_4 = 0$. In both subcases we have

$$\varepsilon_1\xi_1^2 + \varepsilon_2\xi_2^2 + \varepsilon_3\xi_3^2 + \varepsilon_4\xi_4^2 = 4\gamma,\ \gamma \in \mathbb{Z}_2.$$

Taking $\xi_5 = 2\gamma$, we get

$$Q''(x) \equiv 4\gamma + 4\gamma^2 \equiv 0 \pmod 8$$

hence we can apply Lemma 2.3.9.

§ 4. *A formula for the Hilbert "symbol" in the case of p-adic fields*

2.4.1. The aim of this paragraph is to give an explicit formula for the Hilbert "symbol" by means of the Legendre "symbol" and some other elementary invariants.

In the case $p = 2$, between the elementary invariants appearing in this formula two group homomorphisms occur:

$$\text{(2.4.1.1)} \qquad \varepsilon: U/U^{(3)} \to \mathbb{Z}/_2\mathbb{Z}$$

$$\text{(2.4.1.2)} \qquad w: U/U^{(3)} \to \mathbb{Z}/_2\mathbb{Z},$$

defined by the formulae

$$\text{(2.4.1.3)} \qquad \varepsilon(Z) \equiv \frac{z-1}{2} \pmod 2 = \begin{cases} 0 \text{ if } z \equiv 1 \pmod 4, \\ 1 \text{ if } z \equiv -1 \pmod 4. \end{cases}$$

(Obviously, $z \in U \Rightarrow Z \not\equiv 0,2 \bmod 4$),

$$(2.4.1.4) \qquad w(z) \equiv \left\{\frac{z^2-1}{8} \pmod 2\right. = \begin{cases} 0 \text{ if } z \equiv \pm 1 \pmod 8, \\ 1 \text{ if } z \equiv \pm 5 \pmod 8). \end{cases}$$

(Obviously, $z \in U \Rightarrow Z \not\equiv 0,2\,,4,6 \bmod 4$).

Proposition 2.4.2. *If $a, b \in \mathbb{Q}_p$ and $a = p^\alpha u$, $b = p^\beta v$, $\alpha, \beta \in \mathbb{Z}$, $u, v \in U$, then*

$$(2.4.2.1) \qquad (a, b) = (-1)^{\frac{p-1}{2}(\alpha\beta)} \left(\frac{u}{p}\right)^\beta \left(\frac{v}{p}\right)^\alpha \text{ if } p \neq 2,$$

$$(2.4.2.2) \qquad (a, b) = (-1)^{\varepsilon(u)\,\varepsilon(v) + \alpha w(v) + \beta w(u)} \text{ if } p = 2.$$

Proof (Following Serre). The right-hand side of the two formulae depends only on the mod 2 class of α and β. Hence, in view of the symmetry of the Hilbert "symbol" (2.1.5.3), it is sufficient to consider only the following cases: ($\alpha = 0$, $\beta = 0$), ($\alpha = 1$, $\beta = 0$), ($\alpha = 1$, $\beta = 1$). They will be dealt with separately in the case $p \neq 2$ and in the case $p = 2$.

The case $p \neq 2$. (I) $\alpha = 0$, $\beta = 0$. In this case, the formula (2.4.2.1) is equivalent with

$$(2.4.2.3) \qquad (u, v) = 1, \ \forall\, u, v \in U.$$

In other words, it is sufficient to prove that the equation

$$z^2 - ux^2 - vy^2 = 0$$

has a non-trivial solution in $\mathbb{Z}_p$ which follows from Corollary 2.3.4.23.

(II) $\alpha = 1$, $\beta = 0$. In this case, (2.4.2.1) is equivalent with

$$(2.4.2.4) \qquad (pu, v) = \left(\frac{v}{p}\right), \ \forall\, u, v \in U.$$

But (2.4.2.4) is in turn equivalent to

$$(2.4.2.5) \qquad (p, v) = \left(\frac{v}{p}\right).$$

Indeed, since from the analysis of the previous case it follows that $(u, v) = 1$, (2.1.5.5) yields

$$(pu, v) = (p, v).$$

Hence, it is sufficient to prove (2.4.2.5).

We distinguish two subcases:

— $\left(\dfrac{v}{p}\right) = 1$. In this case, v is a square (Proposition 2.2.11) hence $(p, v) = 1$.

— $\left(\dfrac{v}{p}\right) = -1$. In this case we have to show that the equation $z^2 - px^2 - vy^2 = 0$ has only the trivial root in $\mathbb{Q}_p$. Indeed, assuming the contrary and multiplying by a suitable power of p,we get a "primitive" root of this equation, i.e.a solution (ξ, η, ζ) satisfying the following conditions:

$$(i)\ (\xi, \eta, \zeta) \in \mathbb{Z}_p^3.$$

(ii) At least one of the ξ, η, ζ coordinates is a p-adic unit.

It is sufficient to prove that this solution necessarily satisfies the condition $\eta \in U$, which obviously contradicts the hypothesis $\left(\dfrac{v}{p}\right) = -1$.

By *reductio ad absurdum*, assume that $\eta \in p \leqslant \mathbb{Z}_p$. Since, at any rate, $\zeta^2 - v\eta^2 \equiv 0 \pmod p$, $v \not\equiv 0 \pmod p$ and $\eta \in p\mathbb{Z}_p$, it follows that $\zeta \in p\mathbb{Z}_p$. Consequently, $p\xi^2 \in p^2\mathbb{Z}_p$, hence $\xi \in p\mathbb{Z}_p$, contradicting the "primitive" character of the solution (ξ, η, ζ).

(iii) $\alpha = 1$, $\beta = 1$. In this case, the formula (2.4.2.1) is equivalent to

$$(2.4.2.6) \qquad (pu, pv) = (-1)^{\frac{(p-1)}{2}} \left(\frac{u}{p}\right)\left(\frac{v}{p}\right).$$

But, in view of the formulae (2.1.5.6), (2.1.5.4), (2.1.5.5), (2.4.3.4), (2.2.10.1) and of Proposition 2.1.9, we have

$$(pu, pv) = (pu, -p^2uv) = (pu, -uv) = \left(\frac{-uv}{p}\right) =$$

$$= \left(\frac{-1}{p}\right)\left(\frac{u}{p}\right)\left(\frac{v}{p}\right) = (-1)^{\frac{p-1}{2}} \left(\frac{u}{p}\right)\left(\frac{v}{p}\right).$$

The case $p = 2$.(I) $\alpha = 0$, $\beta = 0$. In this case, using (2.4.1.3), the formula (2.4.2.2) is equivalent to

$$(2.4.2.7) \qquad (u, v) = \begin{cases} 1 \text{ if } u \equiv 1 \pmod 4 \text{ sau } v \equiv 1 \pmod 4, \\ -1 \text{ if } u \equiv v \equiv -1 \pmod 4. \end{cases}$$

Assume first that $u \equiv 1 \pmod 4$. Then either $u \equiv 1 \pmod 8$, or $u \equiv 5 \pmod 8$. If $u \equiv 1 \pmod 8$, then u is a square (Proposition 2.2.12), hence $(u, v) = 1$. If $u \equiv 5 \pmod 8$ then

$$u + 4v \equiv 1 + 4 + 4v = 1 + 4(1 + v) \equiv 0 \pmod 8.$$

Consequently (still using Proposition 2.2.12) there exists $w \in U$ such that

$$w^2 = u + 4v.$$

The system $(1, 2, w)$ is a solution of the equation $z^2 - ux^2 - vy^2 = 0$, hence $(u, v) = 1$.

Now, assume that $u \equiv v \equiv -1 \pmod 4$. We show that the equation $z^2 - ux^2 - vy^2 = 0$ has only the trivial root in $\mathbb{Q}_2$. Assuming the contrary, we would find a "primitive" solution (ξ, η, ζ), with

$$\xi^2 + \zeta^2 + \eta^2 \equiv 0 \pmod 4.$$

Since the only squares in $\mathbb{Z}/_4\mathbb{Z}$ are 0 and 1, it follows that $\xi \equiv \eta \equiv \equiv \zeta \bmod 2$, contradicting the primitive character of the solution (ξ, η, ζ).

(II) $\alpha = 1$, $\beta = 0$. In this case, (2.4.2.2) is equivalent to the following formula:

$$(2.4.2.8) \qquad (2u, v) = (-1)^{\varepsilon(u)\,\varepsilon(v) + (w)(v)}.$$

But (2.4.2.8) is equivalent to the conjunction of the following formulae:

$$(2.5.2.9) \qquad (2, v) = (-1)^{w(v)},$$

$$(2.4.2.10) \qquad (2u, v) = (2, v)\,(u, v).$$

Let us prove first (2.4.2.9). In view of (2.4.1.4), the formula (2.4.2.9) is equivalent to

$$(2.4.2.11) \qquad (2, v) = 1 \Leftrightarrow v \equiv \pm 1 \pmod 8.$$

So, assume that $(2, v) = 1$ and let (ξ, η, ζ) be a "primitive" solution of the equation

$$z^2 - 2x^2 - vy^2 = 0.$$

As in the case $p \neq 2$, II, $\left(\frac{v}{p}\right) = -1$, we can assume $\eta \not\equiv 0 \bmod 2$, $\zeta \not\equiv 0 \bmod 2$. On the other hand, as in the case of ordinary integers, the square of every odd 2-adic number is congruent mod 8 with 1. Consequently, we have

$$\zeta^2 \equiv \eta^2 \equiv 1 (\mathrm{mod}\ 8).$$

hence

$$1 - 2\xi^2 - v \equiv 0 (\mathrm{mod}\ 8).$$

The only squares mod 8 being 0, 1 and 4, it follows that $v \equiv \pm 1 (\mathrm{mod}\ 8)$.

Conversely, assume that $v \equiv \pm 1 (\mathrm{mod}\ 8)$. We show that $(2, v) = 1$. If $v \equiv 1 (\mathrm{mod}\ 8)$, then $(2, v) = 1$ since v is a square (Proposition 2.2.12). If $v \equiv -1 (\mathrm{mod}\ 8)$, then the equation $z^2 - 2x^2 - vy^2 = 0$ has the mod 8 solution (1, 1, 1) and applying Lemma 2.3.9, it follows that this equation has a non-trivial solution in $\mathbb{Q}_2$, hence $(2, v) = 1$.

Now, we prove formula (2.4.2.10). We distinguish the following cases:

$(u, v) = 1$. In view of (2.1.5.5), we have

$$(2u, v) = (2, v) = (2, v)\,(u, v).$$

$(2, v) = 1$. Using the same formula, we have

$$(2u, v) = (u, v) = (2, v)\,(u, v).$$

$(u, v) = -1$ and $(2, v) = -1$. In view of what has been already proved in the cases I and II, it follows that $v \equiv 3 (\mathrm{mod}\ 8)$ and $u \equiv 3 (\mathrm{mod}\ 8)$ or $u \equiv -1 (\mathrm{mod}\ 8)$. Hence, it is sufficient to prove that in the cases $v \equiv 3 (\mathrm{mod}\ 8)$, $u \equiv -1 (\mathrm{mod}\ 8)$; $u \equiv 3 (\mathrm{mod}\ 8)$, $v \equiv -5 (\mathrm{mod}\ 8)$ we have $(2u, v) = 1$. But using the fact that in $\mathbb{Z}_2$ one can perform the division by every odd number, we have the implication

$$(x \equiv \xi (\text{mod } 8) \text{ and } \xi \in \mathbb{Z}, \xi \text{ odd}) \Rightarrow (x = \xi\varepsilon, \varepsilon \equiv$$

$$\equiv 1 \text{ mod } 8).$$

On the other hand, if $\varepsilon \equiv 1 (\text{mod } 8)$ it follows (Proposition 2.2.12) that ε is a square, hence, in view of (2.1.5.4) we may assume $u = 3$, $v = 3$; $u = 3$, $v = -5$. But the equations

$$z^2 + 2x^2 - 3y^2 = 0$$

$$z^2 - 6x^2 + 5y^2 = 0$$

have (1, 1, 1) as solution, consequently $(2u, v) = 1$.

(III) $\alpha = 1, \beta = 1$. In this case, (2.4.2.2) becomes

$$(2.4.2.12) \qquad (2u, 2v) = (-1)^{\varepsilon(u)\varepsilon(v) + w(u) + w(v)}.$$

In order to prove this, we use formulae (2.1.5.6), (2.4.2.8):

$$(2u, 2v) = (2u, -4uv) = (2u, -uv) = (-1)^{\varepsilon(u)\varepsilon(-vu) + w(-uv)}$$

which, in view of the relations

$$\varepsilon(-1) = 1,$$

$$w(-1) = 0,$$

$$\varepsilon(u)(1 + \varepsilon(u)) = 0.$$

yield precisely the required formula.

COROLLARY 2.4.3. *The map*

$$(,): \mathbb{Q}_p^* \times \mathbb{Q}_p^* \to \{\pm 1\}$$

defined by the Hilbert "symbol" is bilinear. Hence, it defines a symbol (Definition 2.1.1). (In the following, we shall drop the brackets when referring to the Hilbert "symbol").

For every discrete valuation field K, one can introduce a symbol closely related to the Hilbert symbol.

To this aim, let K be a discrete valuation field and denote by A the ring of this valuation, by $\underline{m}$ the maximal ideal of A, by k the residual field $A/\underline{m}$, by $r: A \to k$ the canonical homomorphism and by π a uniformizing element. For every $a \in K$, let $v(a)$ be its valuation.

DEFINITION 2.4.4. The map

$$(,)_v : K^* \times K^* \to k^*$$

$$(a, b) \to (a, b)_v$$

defined by

$$(a, b)_v = (-1)^{v(a)\, v(b)}\, r\left(\frac{\alpha^{v(b)}}{\beta^{v(a)}}\right), \tag{2.4.4.1}$$

$$(a = \alpha\pi^{v(a)},\ b = \beta\pi^{v(b)},\ \alpha, \beta \in A)$$

is a symbol which will be referred to as the *moderated symbol associated to the valuation* v. (A straightforward calculation shows that condition *i—iii* from Definition 2.1.1 are verified).

Observing that the element $\dfrac{a^{v(b)}}{b^{v(a)}} \in K^*$ actually belongs to A, 2.4.4.1 can also be written as $(a, b)_v = (-1)^{v(a)v(b)}\, r\left(\dfrac{a^{v(b)}}{b^{v(a)}}\right)$

PROPOSITION 2.4.5. *For every prime number* $p \neq 2$, *the Hilbert symbol*

$$(,) : \mathbb{Q}_p^* \times \mathbb{Q}_p^* \to \{\pm 1\}$$

can be factorized through the moderated symbol

$$(,)v_p : \mathbb{Q}_p^* \times \mathbb{Q}_p^* \to \mathbb{F}_p^*$$

associated with the canonical valuation v_p *of the field* $\mathbb{Q}_p$ *of p-adic numbers.*

Proof. In view of formula 2.4.2.1 and of the definition of the Legendre symbol (Definition 2.1.7), we have the following commutative diagram

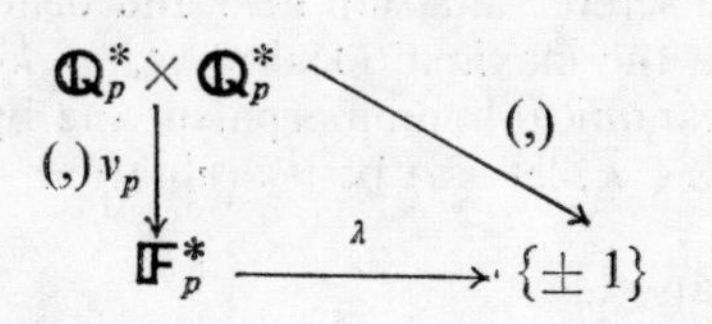

where λ is defined by

$$\lambda(x) = x^{\frac{p-1}{2}}.$$

Definition 2.4.6. The map

$$\mathbb{R}^* \times \mathbb{R}^* \xrightarrow{(,)\infty} \{\pm 1\}$$

defined by

$$(a, b)_\infty = \begin{cases} -1 \text{ if } a < 0,\ b < 0 \\ \ \ 1 \text{ otherwise} \end{cases}$$

represents a symbol.

Proposition 2.4.7 (*Structure of the group $K_2(\mathbb{Q})$*). *The map*

$$\mathbb{Q}^* \times \mathbb{Q}^* \to \{\pm 1\} \times \coprod_{p \geqslant 3} \mathbb{F}_p^*$$

$$(a, b) \to ((a, b)_\infty, (a, b)_3, (a, b)_5, \ldots, (a, b)_p, \ldots)$$

$$((a, b)_p = (a, b)\, v_p),$$

represents a symbol. The corresponding homomorphism

$$K_2(\mathbb{Q}) \xrightarrow{\psi} \{\pm\} \times \coprod_{p \geqslant 3} \mathbb{F}_p^*$$

is an isomorphism.

The proof of the first part of the proposition is straightforward in view of Definition 2.4.4 and of the fact that $(a, b)_p = 1$ for all but a finite number of primes p.

In order to prove the second part of the proposition, namely the fact that ψ is an isomorphism, for every prime number p, consider the subgroups $L(\leqslant p)$ and $L(<p)$ of $K_2(\mathbb{Q}) = \text{Symb}\,(\mathbb{Q})$ defined as follows:

$L(\leqslant p)$ = the subgroup of Symb ($\mathbb{Q}$) generated by elements of the form $\{a, b\}$ where a, b are rational numbers which can be decomposed in products of prime numbers or of inverses of prime numbers *less or equal* to p.

$L(<p)$ = the subgroup of Symb ($\mathbb{Q}$) generated by elements of the form $\{a, b\}$ where a, b are products of prime numbers or of inverses of prime numbers *strictly less* than p.

We shall prove the following lemmas:

LEMMA 2.4.8. *The homomorphism*

$$K_2(\mathbb{Q}) \to \{\pm 1\}$$

defined by the symbol

$$(,)_\infty : \mathbb{Q}^* \times \mathbb{Q}^* \to \{\pm 1\},$$

induces, by restriction to $L(\leqslant 2)$, *an isomorphism*

$$\psi_2 : L(\leqslant 2) \to \{\pm 1\}.$$

Proof. We construct an inverse

$$\chi_2 : \{\pm 1\} \to L(\leqslant 2)$$

for ψ_2 by putting

$$\chi_2(1) = \{1, 1\}$$

$$\chi_2(-1) = \{-1, -1\}.$$

In order to prove that $\chi_2\psi_2 = \mathrm{id}_{L(\leqslant 2)}$ we observe that the elements $\{1, 1\}$, $\{-1, -1\}$ yield a system of generators for $L(\leqslant 2)$. Indeed,

$$\{1, -1\} = \{-1, 1\} = \{-1, -1\} \cdot \{-1, -1\}$$

$$\{2, 1\} = \{2, -1\} \cdot \{2, -1\} = \{2, 1-2\} \cdot \{2, 1-2\} =$$
$$= 1 = \{1, 1\}$$

$$\{2, 2\} = \{-2, 2\} \cdot \{-1, 2\} = \{-1, 2\} =$$
$$= \{1-2, 2\} = \{1, 1\}$$

Now the relation $\chi_2\psi_2 = \mathrm{id}_{L(\leqslant 2)}$ follows as a straightforward consequence.

LEMMA 2.4.9. *The homomorphism*

$$K_2(\mathbb{Q}) \to \mathbb{F}_p^*$$

defined by the symbol

$$(,)_p\colon \mathbb{Q}^* \times \mathbb{Q}^* \to \mathbb{F}_p^*$$

maps the subgroup $L(<p)$ *into the unit element of* $\mathbb{F}_p^*$ *and induces, by restriction to* $L(\leqslant p)$ *and by passing to the quotient with respect to* $L(<p)$, *an isomorphism*

$$\psi_p\colon L(\leqslant p)/L(<p) \to \mathbb{F}_p^*.$$

Proof. The first part of the proposition is obviously verified. In order to prove that ψ_p is an isomorphism, we construct an inverse

$$\chi_p\colon \mathbb{F}_p^* \to L(\leqslant p)/L(<p)$$

for ψ_p, by putting

$$\chi_p(a \bmod p) = \{a, p\} \bmod L(<p),\ a = 1, 2, \ldots, p-1.$$

In order to show that χ_p is a group homomorphism, let a, b, c be integers such that $a, b, c \in \{1, 2, \ldots, p-1\}$; $ab \equiv c \bmod p$.

We show that

$$\{a, p\} \cdot \{b, p\} = \{c, p\} \bmod L(< p).$$

First of all, we have

$$ab = c + rp, \ a \leqslant r < p.$$

(We may obviously assume $r \geqslant 1$). Consequently,

$$1 - \frac{c}{ab} = \frac{r}{ab} p$$

$$1 = \left\{ -\frac{c}{ab}, \frac{c}{ab} \right\} = \left\{ \frac{r}{ab} p, \frac{c}{ab} \right\} =$$

$$= \left\{ \frac{r}{ab}, \frac{c}{ab} \right\} \cdot \left\{ p, \frac{c}{ab} \right\}$$

But since $\left\{ \frac{r}{ab}, \frac{c}{ab} \right\} \in L(< p)$ it follows that

$$\left\{ \frac{c}{ab}, p \right\} = 1 \bmod L(< p).$$

which is the required formula.

In order to show that $\chi_p, \psi_p = id_{L(\leqslant p)/L(p)}$ we observe that the elements of the form $\{a, p\} \bmod L(< p)$, $a = 1, 2, \ldots, p-1$ span the group $L(\leqslant p)/L(< p)$.

Firstly, every element of the form $\{\alpha, p\}$ with $\alpha \in \mathbb{Q}^*$, $\alpha > 0$ and α being a product of prime numbers or of inverses of prime numbers less or equal to p can obviously be expressed (even in the group $L(\leqslant p)$) as a combination of elements of the form $\{a, p\}$ with $a = 1, 2, \ldots$ $\ldots, p-1$ (One applies the multiplicative property of symbols with respect to each variable).

Now suppose that $\alpha < 0$, but α satisfies the same divisibility conditions as above:

$$\{\alpha, p\} = \{-\alpha, p\} \cdot \{-1, p\}.$$

Hence, it is sufficient to prove that $\{-1, p\} \bmod L(<p)$ can be written as a product of elements of the form $\{a, p\} \bmod L(<p)$, $1 \leqslant a < p$ or of inverses of such elements. But we have

$$\{-1, p\} \equiv \{p-1, p\} \bmod L(<p).$$

Indeed,

$$\{-1, p\}\{p-1, p\}^{-1} = \left\{\frac{-1}{p-1}, p\right\} =$$

$$= \left\{\frac{-1+p-p}{p-1}, p\right\} = \left\{1 - \frac{p}{p-1}, p\right\} =$$

$$= \left\{1 - \frac{p}{p-1}, \frac{p}{p-1}(p-1)\right\} =$$

$$= \left\{1 - \frac{p}{p-1}, \frac{p}{p-1}\right\}\left\{1 - \frac{p}{p-1}, p-1\right\} =$$

$$= \left\{1 - \frac{p}{p-1}, p-1\right\} = \left\{\frac{-1}{p-1}, p-1\right\} \in L(<p)$$

i.e.

$$(*) \qquad \{-1, p\} \equiv \{p-1, p\} \bmod L(<p).$$

It remains to show that every element of the form $\{p^r, p^s\} \bmod L(<p)$ can be expressed as a combination of elements of the form $\{a, p\}$ mod $L(<p)$.

First, we have

$$\{p^r, p^s\} = \{p, p\}^{rs}$$

so it is sufficient to consider the case $r = s = 1$.

In view of the equality

$$1 = \{-p, p\} = \{p, p\}\{-1, p\}$$

our assertion follows from (*).

From the fact that the elements of the form $\{a, p\} \bmod L(< p)$, $1 \leqslant a < p - 1$, span $L(\leqslant p) \bmod L(< p)$, we easily get the relation $\chi_p \psi_p = id_{L(\leqslant p)/L(<p)}$, concluding the proof of Lemma 2.4.9.

The proof of Proposition 2.4.7 now follows as a straightforward consequence of Lemmas 2.4.8 and 2.4.9.

§5. *The product formula (Hilbert)*

2.5.0. For every prime number $p \in \mathbb{N}$, let

$$i_p : \mathbb{Q} \to \mathbb{Q}_p$$

be the canonical injection of the rational number field $\mathbb{Q}$ into the p-adic number field $\mathbb{Q}_p$. We denote by i_∞ the canonical injection

$$i_\infty : \mathbb{Q} \to \mathbb{R}.$$

Sometimes, in order to unify notation, we write

$$\mathbb{R} = \mathbb{Q}_\infty.$$

We also denote by V the set obtained by adding the symbol ∞ to the set of all prime numbers p of $\mathbb{N}$.

The fields $\mathbb{Q}_v$ with $v \in V$ yield, up to isomorphisms, all fields which can be obtained from $\mathbb{Q}$ by means of completion with respect to real valuations defined on $\mathbb{Q}$ (theorem of Ostrowski).

For every couple of elements $a, b \in \mathbb{Q}^*$ and for every $v \in V$, we denote by $(a, b)_v$ the Hilbert symbol $(i_v(a), i_v(b))$ of the corresponding images in $\mathbb{Q}_v$.

PROPOSITION 2.5.1 *(Product formula). If $a, b \in \mathbb{Q}^*$, then the symbols $(a, b)_v$ are equal to 1 for almost every $v \in V$, and, in addition,*

$$(2.5.1.1) \qquad \prod_{v \in V} (a,b)_v = 1.$$

Before proving this proposition, we remark that (2.5.1.1) implies the quadratic reciprocity law (Proposition 2.1.11).

Indeed, if l and p are two distinct prime numbers which are not equal to 2, then in view of (2.4.2.1), (2.4.2.2) we have the relations

$$(l, p)_p = \left(\frac{l}{p}\right) \quad (i_p(l) \text{ is a unit in } \mathbb{Q}_p),$$

$$(l, p)_l = \left(\frac{p}{l}\right) \quad (i_l(p) \text{ is a unit in } \mathbb{Q}_l),$$

$$(l, p)_2 = (-1)^{\varepsilon(l)\,\varepsilon(p)} \; (i_2(e), i_2(p) \text{ are units in } \mathbb{Q}_2),$$

$(l, p) = 1$ (l and p are positive numbers, hence squares in $\mathbb{R}$),
$(l, p)_q = 1$ if q is a prime number distinct from 2, p, q.

Applying (2.5.1.1), the previous relations yield

$$\left(\frac{l}{p}\right)\left(\frac{p}{l}\right) = (-1)^{\varepsilon(l)\varepsilon(p)},$$

i.e. the quadratic reciprocity law.

Conversely, we show that the quadratic reciprocity law together with Proposition 2.4.2 essentially implies Proposition 2.5.1.

First, we remark that since the Hilbert symbols are bilinear symmetric maps (2.4.2.1), it is sufficient to prove Proposition 2.4.2 assuming that a and b are equal either to -1 or to a prime number. Consequently, we distinguish the following cases:

(I) $a = b = -1$. In this case, taking into account (2.4.2.1) and (2.4.2.2), we have

$$(-1, -1)_v = \begin{cases} 1 \text{ if } v \neq 2, \infty \\ -1 \text{ if } v = 2 (\varepsilon(-1) = 1), \\ -1 \text{ if } v = \infty. \end{cases}$$

Consequently, $(-1, -1)_v$ is distinct from 1 only for a finite set of elements $v \in V$ (namely for $v = 2$, $v = \infty$) and the product $\prod_{v \in V} (-1, -1)_v$ is equal to 1.

(II) $a = -1$, $b = l$, l prime. We distinguish two subcases:

(α) $l = 2$. Under this assumption, taking into account Proposition 2.4.2, we have $(-1, 2)_v = 1$ for every $v \in V$, yielding the desired conclusion.

(β) $l \neq 2$. Under this assumption, using the same proposition, we get

$$(-1, l)_v = \begin{cases} 1 & \text{if } v \neq 2, l \\ (-1)^{\varepsilon(l)} & \text{if } v = 2, l. \end{cases}$$

In other words, there is only a finite set of elements $v \in V$ for which $(-1, l)_v$ is distinct from 1 ($v = 2$, $v = l$ if $l = -1 \bmod 4$ and the empty set otherwise), so the product $\prod_{v \in V} (-1, l)_v$ is equal to 1.

(III) $a = l$, $b = l'$, l, l' being prime numbers. We distinguish the following three subcases:

(α) $l = l'$. Then (2.1.5.6) and (2.1.5.3) yield

$$(l, l)_v = (-1, l)_v.$$

Indeed, from the above formulae, it follows that

$$(l, l)_v = (l, -l^2)_v = (l, -l)_v = (-1, l)_v$$

hence we are reduced to case II).

(β) $l \neq l'$ and one of them is equal to 2. In view of the symmetry of the Hilbert symbol, we may assume that $l' = 2$. Then, using (2.4.2), we get

$$(l, 2)_v = \begin{cases} 1 & \text{if } v \neq 2, l, \\ (-1)^{w(l)} & \text{if } v = 2, \\ \left(\dfrac{2}{l}\right) = (-1)^{w(l)} & \text{if } v = l \end{cases}$$

Consequently, there is only a finite set of elements $v \in V$ such that $(l, 2)_v \neq 1$ ($v = 2$ and $v = l$ if $l = \pm 5 \bmod 8$ and the empty set otherwise), showing that $\prod_{v \in V} (l, 2)_v = 1$.

(γ) $l \neq l'$ with l and l' distinct from 2. In this case, we have (cf. the comment preceding the proof of Proposition 2.5.1):

$$(l, l')_v = \begin{cases} 1 \text{ if } v \neq 2, l, l', \\ (-1)^{\varepsilon(l)\,\varepsilon(l')} \text{ if } v = 2, \\ \left(\dfrac{l'}{l}\right) \text{ if } v = l \\ \left(\dfrac{l}{l'}\right) \text{ if } v = l' \end{cases}$$

Consequently, in this last case we have also $(l, l') = 1$ for almost every $v \in V$, and in view of the quadratic reciprocity law, we get $\prod_{v \in V} (l, l')_v = 1$.

§ 6. *Applications. Comments.*

LEMMA 2.6.1. *There is a canonical isomorphism*

$$\text{(2.6.1.1)} \qquad \prod_p \mathbb{Z}_p \xrightarrow{\varphi} \varprojlim_n \mathbb{Z}/n\mathbb{Z}$$

such that for every prime number q and every integer i, the following diagram be commutative (the oblique morphisms are defined in a straightforward way):

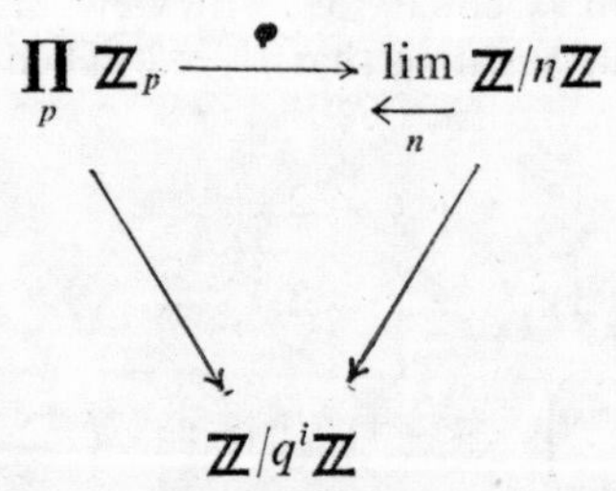

Proof. In order to the morphism φ, is sufficient to define for every integer $n \geqslant 2$ a homomorphism

$$\prod_p \mathbb{Z}_p \xrightarrow{\varphi_n} \mathbb{Z}/n\mathbb{Z}$$

such that the diagram

$$\begin{array}{ccc} \prod_p \mathbb{Z}_p & \xrightarrow{\varphi_n} & \mathbb{Z}/n\mathbb{Z} \\ & \searrow^{\varphi_m} & \downarrow \\ & & \mathbb{Z}/m\mathbb{Z} \end{array}$$

be commutative as soon as n is divisible by m. But according to the Bézout theorem, we have

$$\mathbb{Z}/n\mathbb{Z} \xrightarrow{\sim} \mathbb{Z}/p_1^{s_1}\mathbb{Z} \times \mathbb{Z}/p_2^{s_2}\mathbb{Z} \times \ldots \times \mathbb{Z}/p_r^{s_r}\mathbb{Z},$$

where $n = p_1^{s_1}p_2^{s_2} \ldots p_r^{s_r}$ is the prime factors decomposition of n. Hence, it is sufficient to define for every prime number q and for every integer i a morphism

$$\prod_p \mathbb{Z}_p \to \mathbb{Z}/q^i\mathbb{Z}.$$

But this can be done by composing the obvious morphisms

$$\prod_p \mathbb{Z}_p \xrightarrow{pr\,\mathbb{Z}_q} \mathbb{Z}_q \to \mathbb{Z}/q^i\mathbb{Z}.$$

LEMMA 2.6.2. *There exists an isomorphism*

$$\varprojlim U(\mathbb{Z}/n\mathbb{Z}) \xrightarrow{\sim} \prod_p U(\mathbb{Z}_p)$$

where by $U(A)$ we denote the group of the units of the ring A.

The proof is an immediate consequence of Lemma 2.6.1 and of the fact that the functor U commutes with projective limits, since

$$U(A) = \mathrm{Hom}_{\mathrm{Ann}}(\mathbb{Z}[X, X^{-1}], A).$$

PROPOSITION 2.6.3. *Consider $\mathbb{Q} \subset \mathbb{Q}^{ab}$, $\mathbb{Q}^{ab} \subset \mathbb{C}$ being the subfield of $\mathbb{C}$ consisting from all finite Galois extensions of $\mathbb{Q}$ which are contained in $\mathbb{C}$ and whose Galois group is Abelian. Then we have*

$$\mathrm{Gal}(\mathbb{Q}^{ab}/\mathbb{Q}) = \prod_p U(\mathbb{Z}_p).$$

Proof. According to a famous and deep theorem of Weber, the field $\mathbb{Q}^{ab}$ is generated by the fields of the form $\mathbb{Q}(w)$, where w is a root of the unit. It follows that

$$\mathrm{Gal}(\mathbb{Q}^{ab}/\mathbb{Q}) = \varprojlim_n \mathrm{Gal}(\mathbb{Q}(w_n)/\mathbb{Q}),$$

where w_n is a $n^{-\mathrm{th}}$ order primitive root of the unit. But

$$\mathrm{Gal}(\mathbb{Q}(w_n)/\mathbb{Q}) = U(\mathbb{Z}/n\mathbb{Z})$$

hence, using Lemma 2.6.2, we have

$$\mathrm{Gal}(\mathbb{Q}^{ab}/\mathbb{Q}) = \prod_p U(\mathbb{Z}_p).$$

DEFINITION 2.6.4. Let K be a field, $a, b \in K^*$, $G = \mathrm{Gal}(K\sqrt{a})/K)$, let $\varphi_a: G \to \mathbb{Z}/2\mathbb{Z}$ be the canonical isomorphism and let $\chi_a: G \to \mathbb{Q}/\mathbb{Z}$ be the character of G, defined by

$$\chi_a(g) = \frac{1}{2}\varphi_a(g) \bmod \mathbb{Z}.$$

We can introduce the following elements:

$\chi_a \in H^1(G, \mathbb{Q}/\mathbb{Z})$, where $\mathbb{Q}/\mathbb{Z}$ has the trivial G-module structure;

$b \in H^0(G, K(\sqrt{a})^*)$, where $K(\sqrt{a})^*$ has an obvious G-module structure.

Consider the exact sequence of trivial G-modules:

$$0 \to \mathbb{Z} \to \mathbb{Q} \to \mathbb{Q}/\mathbb{Z} \to 0.$$

From this we get the morphism

$$\delta: H^1(G, \mathbb{Q}/\mathbb{Z}) \to H^2(G, \mathbb{Z}).$$

Using the cup-product

$$H^2(G, \mathbb{Z}) \otimes H^0(G, K(\sqrt{a})^*) \to H^2(G, K(\sqrt{a})^*) = Br_K$$

we obtain the element $(\chi_a, b) \in Br_K$.

PROPOSITION 2.6.5. *If K is a discrete valuation field, then we have an isomorphism:*

$$\text{inv}: Br_K \to \mathbb{Q}/\mathbb{Z}$$

With the above notation, we have

$$\text{inv}\,(\chi_a, b) = (a, b).$$

§ 7. *The theorem of Minkowski-Hasse*

2.7.0. In this paragraph, we intend to prove one of the most remarkable propositions of modern algebra, known as the Minkowski-Hasse theorem.

Let (E, Q) be a quadratic module over the field k and let $K \supset k$ be an extension of the field k. A quadratic module (E_k, Q_k) over K can be defined as follows:

$$E_K = E \otimes_k K;$$

$$Q_K(x \otimes \lambda) = \lambda^2 Q(x) \text{ for every } x \in E.$$

This quadratic module will be called the quadratic module obtained from (E, Q) by extending the coefficients from k to K.

If $e_1, e_2, \ldots, e_n$ is an orthogonal basis of the quadratic module (E, Q), then $e_1 \otimes 1, e_2 \otimes 1, \ldots, e_n \otimes 1$ is obviously an orthogonal basis of the quadratic module (E_K, Q_K).

The quadratic module (E, Q) is non-degenerate, if and only if (E_K, Q_K) is non-degenerate.

If the quadratic space (E, Q) has a non-vanishing isotropic vector, then obviously (E_K, Q_K) also has a non-vanishing isotropic vector, but the converse is not true.

We remark also that even the concept of Witt decomposition is not invariant with respect to extensions of coefficients.

Assume that the quadratic module (E, Q) is defined over the field $\mathbb{Q}$ of rational numbers.

Using the notation from 2.5.0, we denote by V the set obtained by adding the symbol ∞ to the set of all prime numbers $p \in \mathbb{N}$. Using the field extensions

$$\mathbb{Q} \subset \mathbb{Q}_v$$

we obtain by coefficients extension the quadratic modules (E_v, Q_v), where, for every $v \in V$, we have set

$$E_v = E \otimes_{\mathbb{Q}} \mathbb{Q}_v,$$

$$Q_v = Q_{\mathbb{Q}_v}.$$

(Here Q_v is a quadratic form defined on a vector space over the field $\mathbb{Q}_v$). In particular, (E_∞, Q_∞) is a quadratic module over the field $\mathbb{R}$ of real numbers.

PROPOSITION 2.7.1. (Theorem of Minkowski-Hasse). *Let (E, Q) be a non-degenerate quadratic module over $\mathbb{Q}$. (E, Q) has a non-vanishing isotropic vector if and only if for every $v \in V$, the quadratic module (E_v, Q_v) has a non-vanishing isotropic vector.*

(In other words, the non-degenerate quadratic form Q with rational coefficients has a non-trivial rational zero if and only if it has a non-trivial zero in the field of real numbers and in the field of p-adic numbers for every prime number p).

Proof. The necessity is trivial. In order to prove the sufficiency, assume $\dim_{\mathbb{Q}} E = n$ and let $e_1, e_2, \ldots, e_n$ be an orthogonal basis of (E, Q). With respect to this basis, Q can be written

$$Q(x) = a_1\xi_1^2 + a_2\xi_2^2 + \ldots + a_n\xi_n^2,$$

where

$$x = \xi_1 e_1 + \xi_2 e_2 + \ldots + \xi_n e_n,$$

$$\xi_i, a_i \in \mathbb{Q}^*.$$

Replacing, if necessary, Q by $a_1 Q$, we may assume $a_1 = 1$. In the case $n = 1$, the sufficiency of the condition is trivial. In what follows, we consider separately the cases $n = 2$, $n = 3$, $n = 4$ and $n \geqslant 5$. The case $n = 3$ has been proved for the first time by Legendre.

(*i*) *The case* $n = 2$. In view of the previous remarks, we may assume the existence of an orthogonal basis e_1, e_2 of the quadratic module (E, Q), such that

$$Q(x) = \xi_1^2 - a\xi_2^2, \ a \in \mathbb{Q}^*.$$

Since the quadratic module (E_∞, Q_∞) has a non-trivial isotropic vector, we can find two real numbers x_1, x_2 satisfying the conditions

$$(x_1, x_2) \neq (0, 0),$$

$$x_1^2 - ax_2^2 = 0.$$

Consequently, $a > 0$. On the other hand, writing the rational number a as

$$a = \prod_p p^{v_p(a)}$$

from the fact that (E_p, Q_p) has a non-trivial isotropic vector for every prime number p it follows that $v_p(a)$ is an even integer. Hence a is a square in $\mathbb{Q}$, consequently the equation

$$\xi_1^2 - a\xi_2^2 = 0$$

has non-trivial solutions in $\mathbb{Q}$.

(*ii*) *The case* $n = 3$. In this case, the quadratic module (E, Q) has an orthogonal basis e_1, e_2, e_3 such that

$$Q(x) = \xi_1^2 - a\xi_2^2 - b\xi_3^2, \ a, b \in \mathbb{Q}^*.$$

Replacing, if necessary, the basis e_1, e_2, e_3, we may assume that a, b are integers: $a, b \in \mathbb{Z}$. Indeed, if $a = \dfrac{\alpha_1}{\alpha_2} = \dfrac{\alpha_1\alpha_2}{\alpha_2^2}$, $b = \dfrac{\beta_1}{\beta_2} = \dfrac{\beta_1\beta_2}{\beta_2^2}$, it is sufficient to consider the new orthogonal basis

$$e_1' = e_1, \ e_2' = \alpha_2 e_2, \ e_3' = \beta_2 e_3$$

We may also assume that each of the numbers a, b is free of quadratic factors. Consequently, for every prime number p, we have

$$v_p(a) = 0 \text{ or } 1;$$

$$v_p(b) = 0 \text{ or } 1.$$

We may assume that $|a| \leqslant |b|$ and we argue by induction on $m = |a| + |b|$. If $m = 2$, then in the orthogonal basis e_1, e_2, e_3, Q writes

$$Q(x) = \xi_1^2 \pm \xi_2^2 \pm \xi_3^2.$$

Obviously, the case

$$Q(x) = \xi_1^2 + \xi_2^2 + \xi_3^2$$

cannot appear, since the quadratic module (E_∞, Q_∞) has at least one non-trivial isotropic vector. In the other cases, the quadratic module (E, Q) obviously has at least one non-trivial isotropic vector.

Now assume that $m > 2$, hence $|b| \geqslant 2$. Since b is free of quadratic factors, it can be written as

$$b = \pm p_1 p_2 \ldots p_k,$$

where p_i are distinct prime numbers. Let p be one of the p_i; we show that a is square mod p. We distinguish two cases: $a \equiv 0 (\operatorname{mod} p)$, $a \not\equiv 0 (\operatorname{mod} p)$.

In the first case, a is obviously a square mod p. Consequently, assume that $a \not\equiv 0 \operatorname{mod} p$.

In this case, the image of a in $\mathbb{Q}_p$ is obviously a p-adic unit. By assumption, there exists $(\xi, \eta, \zeta) \in \mathbb{Q}_p^3$ such that

$$\zeta^2 - a\xi^2 - b\eta^2 = 0.$$

We may obviously suppose that (ξ, η, ζ) is a primitive solution of this equation (2.4.2).

We show that $\xi \not\equiv 0 (\operatorname{mod} p)$. Indeed, since $\zeta^2 - a\xi^2 \equiv 0 (\operatorname{mod} p)$, the contrary assumption would imply $b\eta \equiv 0 (\operatorname{mod} p)$, hence $b\eta$ would be divisible by p, and taking into account the relation $v_p(b = 1$, we would get $\eta \equiv 0 \operatorname{mod} p$, contradicting the primitive character of the solution (ξ, η, ζ). We have shown that $\xi \not\equiv 0 (\operatorname{mod} p)$; since $\zeta^2 - a\xi^2 \equiv 0 (\operatorname{mod} p)$, this implies that a is a mod p-square.

In view of the isomorphism (Chinese lemma)

$$\mathbb{Z}/b\mathbb{Z} \xrightarrow{\sim} \mathbb{Z}/p_1\mathbb{Z} \times \mathbb{Z}/p_2\mathbb{Z} \times \ldots \times \mathbb{Z}/p_n\mathbb{Z}$$

it follows that a is a mod b-square. Consequently, there exists an integer t with the following properties:

(2.7.1.1) $t^2 = a + bb'$,

(2.7.1.2) $|t| \leqslant \frac{|b|}{2}$.

(Condition (2.7.1.2) can be realized taking into account the fact that every segment of length b contains b points with integer coordinates).

(2.7.1.1) shows that the element bb' is the norm (2.1.6) of an element of an extension

$$k(\sqrt{a}) \supset k, \ k = \mathbb{Q} \text{ or } k = \mathbb{Q}_v, \ v \in V.$$

Indeed,

$$N(t + \sqrt{a}) = t^2 - a = bb'.$$

Consequently, using Proposition 2.1.6, we have

(2.7.1.3) $(a, bb') = 1, \ a, b, b' \in \mathbb{Q}$,

(2.7.1.4) $(a, bb')_v = 1$.

But in view of the bilinearity of the Hilbert symbol (Corollary 2.4.3), from (2.7.1.3) and (2.7.1.4) it follows:

$$(a, b)\,(a, b') = 1,$$

$$(a, b)_v(a, b')_v = 1, \ v \in V.$$

In other words,

(2.7.1.5) $(a, b) = (a, b')$

(2.7.1.6) $(a, b)_v = (a, b')_v, \ v \in V$

Now consider the quadratic form Q' which with respect to the basis e_1, e_2, e_3 of E writes

$$Q'(x) = \xi_1^2 - a\xi_2^2 - b'\xi_3^2.$$

From the definition of the Hilbert symbol and relations (2.7.1.5), (2.7.1.6) it follows that if the quadratic module (E, Q) (or (E_v, Q_v), $v \in V$) has a non-trivial isotropic vector, then the quadratic module (E, Q') (or (E_v, Q'_v), $v \in V$) also has a non-trivial isotropic vector. In particular, the hypotheses imply $(a, b)_v = 1$, hence (2.7.1.6)' yields $(a, b')_v = 1$ and the quadratic module (E_v, Q'_v) has a non-trivial isotropic vector for every $v \in V$. But since $|b| \geqslant 2$, (2.7.1.1), (2.7.1.2) imply

$$|b'| = \left|\frac{t^2 - a}{b}\right| \leqslant \left|\frac{t^2}{b}\right| + \left|\frac{a}{b}\right| \leqslant \frac{|b|}{4} + 1 < |b|.$$

If b' contains quadratic factors, i.e. if

$$b' = b''u^2,$$

where b'' does not contain quadratic factors, then there exists an orthogonal basis (e_1, e_2, e_3) of the quadratic module (E, Q') such that

$$Q'(x) = \xi_1'^2 - a\xi_2'^2 - b''\xi_1'^2.$$

Since $|b''| \leqslant |b'| < |b|$, the induction assumption shows that (E, Q') has a non-trivial isotropic vector, hence (E, Q) also has a non-trivial isotropic vector.

(iii) The case $n = 4$. The proof of Proposition 2.7.1 in this case makes use of the following results:

PROPOSITION 2.7.1.7. *Let K be an infinite field, let (F, R) stand for a non-degenerate quadratic module over K and let $f_1, f_2, \ldots, f_m$ be an orthogonal basis of (F, R). If (F, R) has a non-trivial isotropic vector, then there exists an isotropic vector*

$$y = \eta_1 f_1 + \eta_2 f_2 + \ldots + \eta_m f_m$$

such that $\eta_i \neq 0$ for $i = 1, 2, \ldots, m$.

For the proof, see Shafarevich I. R., Borevich Z. U., *Theory of Numbers*, Academic Press, New York, 1966, Algebraic Appendix, § 1, 3, Theorem 8.

PROPOSITION 2.7.1.8. *(Theorem of Dirichlet). If a and m are mutually prime integers, and $a, m \geqslant 1$, then there exists an infinity of prime numbers p such that $p \equiv a \pmod{m}$.*

For the proof, see Jean-Pierre Serre, *Cours d'Arithmétique*, P.U.F., 1970, Chap. VI.

Now, let e_1, e_2, e_3, e_4 be an orthogonal basis of the quadratic module (E, Q) such that

$$Q(x) = a_1\xi_1^2 + a_2\xi_2^2 + a_3\xi_3^2 + a_4\xi_4^2,$$

$$a_i \in \mathbb{Z},\ i = 1, 2, 3, 4,$$

$$a_1 > 0,\ a_4 > 0,$$

a_i are free of quadratic factors.

Let also (E_1, Q_1), (E_2, Q_2) be quadratic modules over $\mathbb{Q}$ with the following properties:

E_1 has a basis f_1, f_2 such that, for every $y_1 \in E_1$, we have

$$Q_1(y_1) = a_1\eta_1^2 + a_2\eta_2^2,\ y_1 = \eta_1 f_1 + \eta_2 f_2. \tag{2.7.1.9}$$

E_2 has a basis f_3, f_4 such that, for every $y_2 \in E_2$, we have

$$Q_2(y_2) = -a_3\eta_3^2 - a_4\eta_4^2,\ y_2 = \eta_2 f_2 + \eta_3 f_4. \tag{2.7.1.10}$$

In order to prove the Minkowski-Hasse theorem in the case $n = 4$, we show that there exists a rational number $a \neq 0$ and two vectors x_1, x_2 such that

$$x_1 \in E_1,\ x_2 \in E_2,$$

$$a = Q_1(x_1) = Q_2(x_2).$$

To this aim, let $p_1, p_2, \ldots, p_s$ be all the odd prime numbers dividing one of the integers a_1, a_2, a_3, a_4 and consider the set of prime numbers

$$P = \{2, p_1, p_2, \ldots, p_s\}.$$

For every $p \in P$, there exist some elements $b_p \in \mathbb{Z}_p$ and $(\eta_1, \eta_2, \eta_3, \eta_4) \in \mathbb{Q}_p^4$ such that one of the following conditions be fulfilled:

(α) $b_p = a_1\eta_1^2 + a_2\eta_2^2 = -a_3\eta_3^2 - a_4\eta_4^2$;

(β) $b_p \neq 0$;

(γ) $\eta_i \neq 0$ for every $i = 1, 2, 3, 4$;

(δ) $b_p \not\equiv 0 \pmod{p^2}$.

Indeed, from the assumption of Proposition 2.7.1 and from Proposition 2.7.1.7 it follows that there exist $b_p \in \mathbb{Q}_p$ and $\eta_1, \eta_2, \eta_3, \eta_4$ such that

$$\eta_i \neq 0, \; i = 1, 2, ,3 ,4,$$

$$b_p = a_1\eta_1^2 + a_2\eta_2^2 = - a_3\eta_3^2 - a_4\eta_4^2.$$

If $b_p = 0$, then the quadratic modules (E_p, Q_{1p}), (E_p, Q_{2p}) are hyperbolic planes (Proposition 1.4.5) and in view of Corollary 1.4.13, for every $a \in \mathbb{Q}$ there exists $(\eta_1, \eta_2, \eta_3, \eta_4) \in \mathbb{Q}_p^4$ such that

$$a = a_1\eta_1^2 + a_2\eta_2^2 = - a_3\eta_3^2 - a_4\eta_4^2. \tag{2.7.1.11}$$

Applying Proposition 2.7.1.7 to the quadratic form

$$a\eta_0^2 - a_1\eta_1^2 - a_2\eta_2^2, \; a\eta_0^2 + a_3\eta_3^2 + a_4\eta_4^2$$

we see that the elements $\eta_1, \eta_2, \eta_3, \eta_4$ in (2.7.1.11) may be supposed to be non-zero.

In this way we have proved the existence of $b_p \in \mathbb{Q}_p$ and $(\eta_1, \eta_2, \eta_3, \eta_4) \in \mathbb{Q}_p^4$ satisfying the above conditions α, β, γ. If $b_p \notin \mathbb{Z}_p$, then it has the form

$$b_p = \frac{\beta'_p}{\beta''_p} = \frac{\beta'_p\beta''_p}{\beta''^2_p}, \; \beta'_p, \beta''_p \in \mathbb{Z}_p$$

and replacing b_p by $\beta'_p\beta''_p$ and η_i by $\eta_i\beta''_p$, we may obviously realize conditions α, β, γ and $b_p \in \mathbb{Z}_p$.

If $b_p = b'_p p^{2\rho}$, then replacing b_p by $b'_p/p^{2\rho}$ and η_i by $\eta_i p^\rho$ we obviously realize all the required conditions.

Now consider the system of congruences *

$$\begin{aligned} X &\equiv b_2 (\text{mod } 16) \\ X &\equiv b_{p_1} (\text{mod } p_1^2) \\ &\cdots\cdots\cdots \\ X &\equiv b_{p_s} (\text{mod } p_s^2). \end{aligned} \tag{2.7.1.12}$$

* Of course, in order to choose the elements $b_i \in \mathbb{Z}$, we use the isomorphisms

$$\mathbb{Z}_p/p^n\mathbb{Z}_p \simeq \mathbb{Z}/p^n\mathbb{Z}.$$

In view of the isomorphism (Chinese lemma):

$$\mathbb{Z}/16\mathbb{Z} \times \mathbb{Z}/p_1^2\mathbb{Z} \times \ldots \times \mathbb{Z}/p_s^2\mathbb{Z} \rightleftarrows \mathbb{Z}/m\mathbb{Z},\ m =$$

$$= 16p_1^2p_2^2 \ldots p_s^2,$$

it follows that the system (2.7.1.12) has (obviously integer) solutions. We show that there exists a solution $a \in \mathbb{N}$ of this system, having at most one prime factor q distinct from the elements of $P = \{2, p_1, p_2, \ldots, p_s\}$.

Let, indeed, a^* be an arbitrary natural number satisfying (2.7.1.12), consider $m = 16p_1^2p_2^2 \ldots p_s^2$ and denote by d the greatest common divisor of a^* and m. Since $\frac{a^*}{d}$, $\frac{m}{d}$ are mutually prime integers, from the Theorem of Dirichlet (Proposition 2.7.1.8) it follows that there exist $k \geqslant 0$ and a prime number q such that we have:

$$\frac{a^*}{d} + k\frac{m}{d} = q.$$

Consequently, putting

$$a = a^* + km,$$

we get

$$a \equiv a^* \operatorname{mod}(16), \operatorname{mod}(p_1^2), \ldots, \operatorname{mod}(p_s^2)$$

$$a = dq$$

i.e. a satisfies the system (2.7.1.12) and, in addition, the only possible divisor of a lying outside P is q.

Now, remark that we have

$$\text{(2.7.1.13)} \qquad \frac{b_{p_i}}{a} \equiv 1(\operatorname{mod} p_i), \quad i = 1, 2, \ldots, s.$$

$$\text{(2.7.1.14)} \qquad \frac{b_2}{a} \equiv 1(\operatorname{mod} 8).$$

These relations follow from δ) and from (2.7.1.12). In view of Propositions 2.2.11 and 2.2.12, the element $\frac{b_{p_i}}{a}\left(\text{or } \frac{b_2}{a}\right)$ is a square in $\mathbb{Q}_{p_i}$ (or in $\mathbb{Q}_2$).

Now consider the quadratic modules (E_1', Q_1'), (E_2', Q_2') over $\mathbb{Q}$, satisfying the following conditions:

E_1' has a basis f_0', f_1', f_2' such that, for every $x_1' \in E_1'$, we have

$$Q_1'(x_1') = -a\eta_0'^2 + a_1\eta_1'^2 + a_2\eta_2'^2 (x_1' = \eta_0' f_0' + \eta_1' f_1' + \\ + \eta_2' f_2')$$

E_2' has a basis f_0'', f_3', f_4' such that, for every $x_2' \in E_2'$, we have

$$Q_2'(x_2') = -a\eta_0'^2 - a_3\eta_3'^2 - a_4\eta_4'^2 (x_1' = \eta_0'' f_0'' + \\ + \eta_3' f_3' + \eta_4' f_4').$$

Assuming $a > 0$ (a condition which obviously can always be realized), the following assertions hold:

— Each of the quadratic modules $(E_{1\infty}', Q_{1\infty}')$, $(E_{2\infty}', Q_{2\infty}')$ over $\mathbb{R}$ has a non-trivial isotropic vector. This follows from the relations $a > 0$, $a_1 < 0$ and $a > 0$, $a_4 < 0$ respectively.

— For each prime number p, the quadratic modules (E_{1p}', Q_{1p}') (E_{2p}', Q_{2p}') over $\mathbb{Q}_p$ have non-trivial isotropic vectors.

In order to see this, we distinguish the following three cases:

(A) $p \in P = \{2, p_1, p_2, \ldots, p_s\}$. In this case, the assertion follows from condition α and relations

$$b_p = a\xi_0^2, \; \xi_0 \in \mathbb{Q}_p, \; (p \in P),$$

(B) $p \neq q$. In this case, we may obviously assume $p \notin P$, hence p does not divide any one of the integers a, a_1, a_2, a_3, a_4. Consequently, each of them is a p-adic unit. Replacing, if necessary, Q_1', Q_2' by aQ_1', aQ_2', we may assume $a = 1$. Hence, it is sufficient to show that

$$(a_3, a_4)_p = 1, \; (-a_1, -a_2)_p = 1$$

which follows from Proposition 2.4.2.

(C) $p = q$. We may as above assume $a = 1$. Hence, we have to prove the relations:

(2.7.1.15) $\quad (a_3, a_4)_q = 1, \; (-a_1, -a_2)_q = 1.$

But from the above discussion it follows that

$$(a_3, a_4)_v = 1, \; (-a_1, -a_2)_v = 1 \quad \text{for every } v \neq q.$$

In view of the product formula (Proposition 2.5.1) we get relations (2.7.1.15). Now we may conclude the proof of the case $n = 4$ as follows.

Using the Minkowski-Hasse theorem in the case $n = 3$, it follows that the quadratic modules (E_1', Q_1'), (E_2', Q_2') have non-trivial isotropic vectors. Applying if so once more Proposition 2.7.1.7, it follows that there exists $(\xi_1, \xi_2, \xi_3, \xi_4) \in \mathbb{Q}^4$ such that

$$a = a_1\xi_1^2 + a_2\xi_2^2 = -a_3\xi_3^2 - a_4\xi_4^2.$$

This yields the proof of the theorem in the case $n = 4$.

(*iv*) *The case* $n \geqslant 5$. From Proposition 2.3.7 (which states that every p-adic field $\mathbb{Q}_p$ has the property $C_2(2)$), it follows that Minkowski-Hasse theorem is in this case equivalent to the following statement:

Let (E, Q) be a non-degenerate quadratic module over $\mathbb{Q}$, of dimension 5. In order for (E, Q) to have a non-trivial isotropic vector, it is necessary and sufficient for the quadratic module (E_∞, Q_∞) over $\mathbb{R}$ to have a non-trivial isotropic vector.

Using this equivalent setting, we show that in order to get the result in the case $n \geqslant 5$, it is sufficient to consider only the case $n = 5$.

Assume, indeed, that the Minkowski-Hasse theorem has been already proved in the case $n = 5$, and let (E, Q) be a non-degenerate quadratic space over $\mathbb{Q}$, such that

(2.7.1.16) $\quad \dim_{\mathbb{Q}} E > 5,$

(2.7.1.17) $\quad (E_\infty, Q_\infty)$ has a non-trivial isotropic vector.

Let $e_1, e_2, \ldots, e_n$ be an orthogonal basis of (E, Q). Consequently, we have

$$Q(x) = a_1\xi_1^2 + a_2\xi_2^2 + \ldots + a_n\xi_n^2, \; a_i \in \mathbb{Q}^*$$

and from (2.7.1.17) it follows that we may assume $a_1 > 0$, $a_2 < 0$.

Consider a non-degenerate quadratic module (E', Q') with a basis $e_1', e_2', e_3', e_4', e_5'$ such that

$$Q'(x') = a_1\xi_1'^2 + a_2\xi_2'^2 + a_3\xi_3'^2 + a_4\xi_4'^2 + a_5\xi_5'^2,$$

$$x' = \xi_1'e_1 + \xi_2'e_2 + \xi_3'e_3 + \xi_4'e_4 + \xi_5'e_5.$$

(E', Q') satisfies the hypotheses of the Minkowski-Hasse theorem and, in addition, $\dim_{\mathbb{Q}} E' = 5$. Hence, if the Minkowski-Hasse theorem is true for $n = 5$, it is also true for $n > 5$.

Now, let (E, Q) be a quadratic module of dimension 5, satisfying the condition of Proposition 2.7.1, and let e_1, e_2, e_3, e_4, e_5 be an orthogonal basis of (E, Q). We have:

$$Q(x) = a_1\xi_1^2 + a_2\xi_2^2 + a_3\xi_3^2 + a_4\xi_4^2 + a_5\xi_5^2,$$

$$a_i \in \mathbb{Z},\ a_i \neq 0,$$

$$a_1 > 0,\ a_5 < 0,$$

a_i is free of quadratic factors.

As in the case $n = 4$, we consider the quadratic modules (E_1, Q_1), (E_2, Q_2), defined over $\mathbb{Q}$ with the following properties:

E_1 has a basis f_1, f_2 such that, for every $y_1 \in E_1$,

$$Q_1(y_1) = a_1\eta_1^2 + a_2\eta_2^2,\ y_1 = \eta_1 f_1 + \eta_2 f_2. \tag{2.7.1.18}$$

E_2 has a basis f_3, f_4, f_5 such that, for every $y_2 \in E_2$,

$$Q_2(y_2) = -a_3\eta_3^2 - a_4\eta_4^2 - a_5\eta_5^2, \tag{2.7.1.19}$$

$$y_2 = \eta_3 f_3 + \eta_4 f_4 + \eta_5 f_5.$$

As in the case $n = 4$, we show that there exists a rational number $a \neq 0$ and two vectors x_1, x_2 such that

$$x_1 \in E_1,\ x_2 \in E_2,$$

$$a = Q_1(x_1) = Q_2(x_2).$$

The argument goes on as in the case $n = 4$, apart from the following modifications:

In cases B ($p \neq q$) and C ($p = q$), the existence of elements $\eta_3, \eta_4, \eta_5 \in \mathbb{Q}_p$ such that

$$a = - a_3\eta_3^2 - a_4\eta_4^2 - a_5\eta_5^2 \tag{2.7.1.20}$$

is proved as follows:

Since by assumption p is not a divisor of a_3, a_4, a_5, these are p-adic units. Hence the equation

$$- a_3\xi_3^2 - a_4\xi_4^2 - a_5\xi_5^2 = 0$$

has a non-trivial solution in $\mathbb{Q}_p$ (Corollary 2.3.4.23). Consequently, (Proposition 1.4.3) every element $a \in \mathbb{Q}_p$ can be represented under the form (2.7.1.20).

The proof is ended as in the case $n = 4$.

Remark 2.7.1.21. The Minkowski-Hasse theorem does not hold for homogeneous forms of degree strictly greater than 5. A counter-example can be obtained in the following way (cf. Borevič-Safarevič, *Theory of Numbers*, chap. I, § 7, 6).

Consider the homogeneous form of degree 4 in $2n$ variables

$$(\xi_1^2 + \xi_2^2 + \ldots + \xi_n^2)^2 - 2(\eta_1^2 + \eta_2^2 + \ldots + \eta_n^2)^2,$$

where $n \geqslant 5$. This form obviously has a non-trivial root in $\mathbb{R}$. On the other hand, since every non-degenerate quadratic form in at least 5 variables always has a non-trivial root in the p-adic number field for every p (Proposition 2.3.7), it follows that the above form has a non-trivial root in the p-adic number field for every p. But, since $\sqrt{2}$ is irrational, this form has in $\mathbb{Q}$ only the trivial root.

Recently, Yu. Manin has shown that an obstruction to the validity of the Minkowski-Hasse theorem can be introduced (Cf. Yu. I. Manin, *Le groupe de Brauer-Grothendieck en Géométrie Diophantienne*, Actes, Congrès intern. Math., 1970, I, p. 401–411).

2.8.0. In the following, we intend to prove the following result (cf. James Ax, Simon Kochen, *Diophantine Problems over Local Fields*, I, Amer. J. Math., LXXXVII, 3, 1965, pp. 605–630):

PROPOSITION 2.8.1. *Given* $P_1, P_2, \ldots, P_r \in \mathbb{Z}[X_1, X_2, \ldots, X_s]$, *there exists a finite set* $U(P_1, \ldots, P_r)$ *of prime numbers such that for every prime number p which does not belong to* $U(P_1, P_2, \ldots, P_r)$, *the folowing condition be fulfilled:*

If $(n_1, n_2, \ldots, n_s) \in \mathbb{Z}^s$ is a system of integers with

$$P_i(n_1, n_2, \ldots, n_s) \equiv 0 \bmod p, \quad i = 1, 2, \ldots, r,$$

then there exist $(\xi_1, \xi_2, \ldots, \xi_s) \in \mathbb{Z}_p^s$ such that

$$P_i(\xi_1, \xi_2, \ldots, \xi_s) = 0, \quad i = 1, 2, \ldots, r,$$

$$\xi_j \equiv \eta_j \bmod p, \qquad j = 1, 2, \ldots, s.$$

The proof makes use of the following preliminary concepts and results:

PROPOSITION 2.8.2. *Every local Henselian ring with a residual field of zero characteristic has a field of representatives.*

Proof. Let $(A, \underline{m})$ be a local Henselian ring, let $k = A/\underline{m}$ be its residual field and let $\pi: A \to k$ be the canonical homomorphism. We have to show that, assuming $\operatorname{char}(k) = 0$, there exists a ring homomorphism

$$\sigma: k \to A$$

such that

$$\pi \cdot \sigma = \mathrm{id}_k.$$

Since $\operatorname{char}(k) = 0$, $\mathbb{Q}$ can be canonically embedded in A, so there exist subfields of A. Let S be a maximal subfield of A, the existence of which is ensured by the Zorn lemma; we show that $\pi(S) = k$. To see this, we first prove that k is algebraic over $\pi(S)$. Indeed, if we assume the existence of an element $a \in A$ such that $\pi(a) = \bar{a}$ be transcedental over $\bar{S} = \pi(S)$, then from the ring homomorphisms

$$S[a] \to \overline{S[a]}$$

$$\xi \to \pi(\xi)$$

$$S[X] \simeq \bar{S}[X] \simeq \bar{S}[\bar{a}] \to S[a]$$

$$X \to a$$

it would follow that $S[a]$ is isomorphic to $S[X]$. On the other hand, we obviously have $S[a] \cap \underline{m} = \{0\}$. Consequently, the field of quotients of $S[a]$ would be a subfield of A, strictly including S, so contradicting the maximality character of S.

Now we are able to prove that $\pi(S) = k$. Consider an element $\lambda \in k$ and a polynomial $P \in A[X]$ such that $\bar{P} \in \bar{S}[X]$ be the minimal polynomial of λ with respect to S. Since k is a field of zero characteristic, it follows that λ is a simple root of the polynomial $\bar{P}$. As, by assumption, A is henselian it follows that there exists an element $\xi \in A$ such that $P(\xi) = 0$. Consequently, $S[\xi]$ is isomorphic to $\bar{S}(\lambda)$, hence $\xi \in S$ and $\lambda \in \bar{S}$.

PROPOSITION 2.8.3. *Let $\mathscr{P}$ be the set of prime positive integers and let $\mathscr{U}$ be a non-principal ultrafilter* on $\mathscr{P}$. Then the ultraproduct $\pi_{\mathscr{U}}\mathbb{F}_p$ is a field of zero characteristic.*

Proof. First, we remark that a non-principal ultrafilter on $\mathscr{P}$ cannot contain any finite subset of $\mathscr{P}$.

Now, consider $1 \in \pi_{\mathscr{U}}\mathbb{F}_p$ and let n be an arbitrary positive integer; we show that $n \cdot 1 \neq 0$.

The element 1 is defined by the map

$$\mathscr{P} \to \bigcup_{p \in P} \mathbb{F}_p,$$

$$p \to 1 \bmod p \in \mathbb{F}_p.$$

The assumption $n \cdot 1 = 0$ leads to a contradiction, since it implies

$$\{p | n \equiv 0 \bmod p\} \in \mathscr{U}.$$

which is impossible, $\{p|n \equiv 0 \bmod p\}$ being a finite set.

DEFINITION 2.8.4. Let G be a totally ordered Abelian group and let K be a commutative field. A *G-valuation* on K is a map

$$v\colon K - \{0\} \to G$$

* For the concepts of ultrafilter and ultraproduct used in this section, see paragraph 4.8.2 9.2 from Chapter II.

satisfying the following conditions:

(*i*) v is onto:

(*ii*) $v(xy) = v(x) + v(y)$, $x, y \in K - \{0\}$;

(*iii*) $x, y, x + y \in K - \{0\} \Rightarrow v(x + y) \geqslant \min\{v(x), v(y)\}$.

If K is a G-valuated field, then the local subring

$$\{x | x \in K,\ v(x) \geqslant 0\} \cup \{0\}$$

is called the valuation ring of the G-valuated field K. A G-valuated field is called *Henselian* if its associated valuation ring is Henselian. The residual field of a G-valuated field is by definition the residual field of its associated valuation ring.

PROPOSITION 2.8.5. *Let I be an infinite set, let $\mathscr{U}$ be an ultrafilter on I, and let $(K_i)_{i \in I}$ be a family of valuated fields. Denote by G_i the valuation group of the field K_i and let k_i stand for its residual field.*

The ultraproduct $\prod_{\mathscr{U}} K_i$ is a $\prod_{\mathscr{U}} G_i$-valuated field, and its residual field is $\prod_{\mathscr{U}} k_i$.

If every field of the family $(K_i)_{i \in I}$ is Henselian, then the valuated field $\prod_{\mathscr{U}} K_i$ is Henselian too.

The proof is left to the reader.

Proof of Proposition 2.8.1. Consider again the polynomials $P_1, P_2, \ldots, P_r$. By *reductio ad absurdum*, suppose that the set $U(P_1, P_2, \ldots, P_r)$ of all prime integers p, such that there exist mod-p solutions which cannot be lifted to p-adic solutions, is infinite. Obviously, a non-principal ultrafilter $\mathscr{U}$ on the set $\mathscr{P}$ of prime integers exists *, such that

$$\mathscr{U} \ni U(P_1, P_2, \ldots, P_r).$$

For every $p \in U(P_1, P_2, \ldots, P_r)$, we denote by $(n_1(p), \ldots, n_s(p))$ the mod p-solution of the polynomials $P_1, \ldots, P_r$. We also introduce the notation

$$v_i(p) = n_i(p) \bmod p, \quad i = 1, 2, \ldots, s.$$

$p^{\bar{P}_i}$ = mod p-reduction of the polynomial P_i

$$p_i^{\bar{P}} \in \mathbb{F}_p[X_1, \ldots, X_s], \quad i = 1, 2, \ldots, r.$$

* Such an ultrafilter can be defined for instance in the following way: First consider an arbitrary non-principal ultrafilter $\mathscr{F}$ on $U(P_1, \ldots, P_r)$. The set of all subsets of $\mathscr{P}$ cutting on $U(P_1, \ldots, P_r)$ two subsets belonging to $\mathscr{F}$ is an ultrafilter on $\mathscr{P}$ which obviously contains the subset $U(P_1, \ldots, P_r)$.

For every $p \in U(P_1, \ldots, P_r)$ we obviously have

$$p_i^{\overline{P}}(v_1(p), v_2(p), \ldots, v_s(p)) = 0, \quad i = 1, 2, \ldots, r.$$

Now, let

$$\mu_i: \mathscr{P} \to \bigcup_{p \in \mathscr{P}} \mathbb{F}_p, \quad i = 1, 2, \ldots, s$$

be the maps defined by

$$\mu_i(p) = \begin{cases} v_i(p) & \text{if } p \in U(P_1, \ldots, P_r), \\ 0 & \text{otherwise.} \end{cases}$$

The mappings μ_i define elements $\mu_1 \bmod \mathscr{U}$, $\mu_2 \bmod \mathscr{U}$, $\ldots$, $\mu_s \bmod \mathscr{U}$ in the ultraproduct $\mathbb{F} = \prod_{\mathscr{U}} \mathbb{F}_p$.

In other words, we have:

$$(\mu_1 \bmod \mathscr{U}, \mu_2 \bmod \mathscr{U}, \ldots, \mu_s \bmod \mathscr{U}) \in \mathbb{F}^s.$$

From the definition of the ultrafilter $\mathscr{U}$, it follows that the system

$$(\mu_1 \bmod \mathscr{U}, \mu_2 \bmod \mathscr{U}, \ldots, \mu_s \bmod \mathscr{U})$$

is a solution in $\mathbb{F}$ of the polynomial system $P_1, \ldots, P_r$. But the field $\mathbb{F}$ is the residual field (Proposition 2.8.5) of the valuated field $\prod_{\mathscr{U}} \mathbb{Q}_p$. This being a henselian field and its residual field having zero characteristic, Proposition 2.8.5 implies the existence of a system of representatives

$$\sigma: \mathbb{F} \to \prod_{\mathscr{U}} \mathbb{Q}_p.$$

consequently, the system

$$(\sigma(\mu_1 \bmod \mathscr{U}), \sigma(\mu_2 \bmod \mathscr{U}), \ldots, \sigma(\mu_s \bmod \mathscr{U}))$$

is a solution of the system $P_1, \ldots, P_r$ in the field $\prod_{\mathscr{U}} \mathbb{Q}_p$.

For every $i = 1, 2, \ldots, s$, let

$$\rho_i: \mathscr{P} \to \bigcup_{p \in \mathscr{P}} \mathbb{Z}_p$$

be a map such that

$$\rho_i \bmod \mathscr{U} = \sigma(\mu_i \bmod \mathscr{U}), \quad i = 1, 2, \ldots, s.$$

The existence of these maps follows from the definition of the valuation of the field $\prod_{\mathscr{U}} \mathbb{Q}_p$.

We conclude that the subset $T = \{p \in \mathscr{P} | (\rho_1(p)_1, \ldots, \rho_s(p))$ is a solution of the system $(P_1, \ldots, P_r)\}$ belongs to $\mathscr{U}$.

Consider $p_0 \in T \cap U(P_1, \ldots, P_r)$. The given set of polynomials admit in $\mathbb{Z}_{p_0}$ the solution $(\rho_1(p_0), \ldots, \rho_s(p_0))$ yielding by mod p reduction the solution $(\mu_1(p), \mu_2(p), \ldots, \mu_s(p))$ which by assumption cannot be lifted to a solution in $\mathbb{Z}_{p_0}$, an obvious contradiction.

Chapter II

Methods of logic and algorithmic methods in algebra

§ 1. *Countable sets*

1.1. The set $\mathbb{N}$ of natural numbers plays a fundamental role for the algorithmic study of mathematical problems. This set can be characterized as follows:

Let $(\mathbb{N}, 0, s)$ be a triplet consisting of: a set $\mathbb{N}$, an element $0 \in \mathbb{N}$ and an endomorphism $s: \mathbb{N} \to \mathbb{N}$ of $\mathbb{N}$ such that the following conditions be fulfilled:

(*i*) s is an injection

(*ii*) $s(\mathbb{N}) = \mathbb{N} - \{0\}$;

(*iii*) If $P \subseteq \mathbb{N}$ is such that

α) $0 \in P$,

β) $s(P) \subseteq P$,

then $P = \mathbb{N}$.

If (M, m_0, t) is another triplet satisfying conditions *i–iii*, then there exists a unique bijection $\mu: \mathbb{N} \to M$ such that

(I) $\mu(0) = m_0$;

(II) the diagram

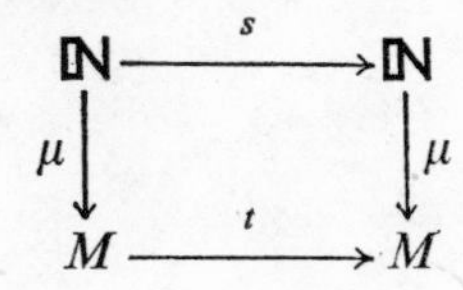

commutes*.

* These axioms have been first formulated by G. Peano. His main works — *Arithmeticae principia, novo methodo exposita* (Rivista di mat. Turin, 1, 1981), *Formulaire de Mathématiques*, 5 volumes, Turin, 1895–1906 — are aimed towards constructing mathematics by means of a small number of logical principles. Their spirit is closely related to that of the German mathematician G. Frege, who was in fact his contemporary.

The triplet can also be characterized by the following property: For every triplet (M, m_0, t) where $m_0 \in M$ is an element of M and $t: M \to M$ is an endomorphism of M, there exists a unique map

$$\mu: \mathbb{N} \to M,$$

such that conditions I, II be fulfilled *.

The possibility of using the set $\mathbb{N}$ in the algorithmic treatment of various mathematical problems follows from the fact that some remarkable sets considered in mathematics (for instance sets of polyhedra, polynomials, groups, propositions, etc.) can be naturally imbedded in $\mathbb{N}$.

PROPOSITION 1.2. *For every infinite subset A of $\mathbb{N}$, there exists a natural bijection $\nu: \mathbb{N} \to A$.*

Proof: Let S_n be the "segment" determined by n, i.e.

$$S_n = \{i \in \mathbb{N} | i \leqslant n\}.$$

Consider the pairs of the form (S_n, σ), where $\sigma: S_n \to A$ is such that

$$\sigma(0) = \inf(A),$$

$$\sigma(r) = \inf(A - \{\sigma(0), \sigma(1), \ldots, \sigma(r-1)\}).$$

If (S_n, σ), (S_m, τ) are such that $S_n \subset S_m$, then $\tau / S_n = \sigma$ whence the existence of the required bijection.

PROPOSITION 1.3. *There exists a natural bijection*

$$\mu: \bigcup_{i \geqslant 1} \mathbb{N}^i \to \mathbb{N},$$

where

$$\mathbb{N}^i = \underbrace{\mathbb{N} \times \mathbb{N} \times \ldots \times \mathbb{N}}_{i \text{ times}}$$

* These axioms have been introduced by the American mathematician W. F. Lawvere. Their advantage consists in their formulation merely in general terms of category theory.

Proof. It will be sufficient to prove the existence of an injection $\mu_1 \colon \bigcup_{i \geq 1} \mathbb{N}^i \to \mathbb{N}$ (cf. Proposition 1.2). In fact, we prove the following

LEMMA 1.4. If $\theta \colon A \to \mathbb{N} - \{0\}$ *is an injection, then there exists an injection*

$$\mu_1 \colon \bigcup_{i \geq 1} \mathbb{N}^i \to \mathbb{N}.$$

Proof. Let p_n be the n-th prime number from the sequence of natural numbers (e.g. $p_1 = 2$, $p_2 = 3$, $p_3 = 5, \ldots$). The conclusion of the lemma follows by setting

$$\mu_1(a_1, a_2, \ldots, a_n) = p_1^{\theta(a_1)} \cdot p_2^{\theta(a_2)} \ldots p_n^{\theta(a_n)}.$$

For the proof of the proposition it is sufficient to take $\theta(n) = n + 1$.

COROLLARY 1.5. *There exists a natural bijection*

$$\nu_i \colon \mathbb{N}^i \to \mathbb{N}.$$

COROLLARY 1.6. *We have the natural bijections*

$$\rho \colon \mathbb{Z}[X] \to \mathbb{N},\ \rho_n \colon \mathbb{Z}[X_1, X_2, \ldots, X_n] \to \mathbb{N},$$

$$\sigma \colon \mathbb{Q}[X] \to \mathbb{N},\ \sigma_n \colon \mathbb{Q}[X_1, X_2, \ldots, X_n] \to \mathbb{N}.$$

If A is an arbitrary set, we denote by $P_f(A)$ the set of finite subsets of A.

COROLLARY 1.7. *We have the natural bijections*

$$\varkappa \colon P_j(\mathbb{N}) \to \mathbb{N}$$

$$\lambda \colon P_f\Big(\bigcup_{i \geq 1} \mathbb{N}^i\Big) \to \mathbb{N}$$

$$\tau \colon \bigcup_{i \geq 1} A^i \to \mathbb{N} \text{ where } A = \bigcup_{i \geq 1} \mathbb{N}^i.$$

Enumeration of finite polyhedra.

DEFINITION 1.8. We say that a finite complex $\mathscr{K}$ is given if a *finite* set V is given together with a set $\mathscr{S}$ of non-void subsets of V, such that the following conditions be fulfilled:

(a) For every element $x \in V$, $\{x\} \in \mathscr{S}$;

(b) If $S \in \mathscr{S}$ and $S' \subset S$, then $S' \in \mathscr{S}$. By definition, the elements of the set V are the vertices of the complex $\mathscr{K}$ and the elements of $\mathscr{S}$ are the simplexes of $\mathscr{K}$.

In the following, we suppose that $V \subset \mathbb{N}$, which implies no restriction due to the straighforward notion of complex isomorphism.

PROPOSITION 1.9. *Let* $\mathscr{K}_f$ *be the set of finite complexes. There exists a natural bijection*

$$\gamma : \mathscr{K}_f \to \mathbb{N}.$$

Proof. In fact, every finite complex $(V, \mathscr{S})$ points out an element of $\bigcup_{i \geq 1} A^i$,

$$(V, S_1^1, S_2^1, \ldots, S_{r_1}^1, S_1^2, \ldots, S_{r_2}^2, \ldots, S_1^i, S_2^i, \ldots, S_{r_i}^i),$$

$$(V \subset \mathbb{N}^i, \; S_i^r \in \mathbb{N}^*)$$

where S_j^i is a simplex with i vertices of the complex $(V, \mathscr{S})$, the ordering of its vertices being given by the natural order of $\mathbb{N}$.

Enumerations of groups

DEFINITION 1.10. Let M be an arbitrary set. We say that the group G is a free group of basis M if there exists a map $i: M \to G$ such that for every map $f: M \to G'$, where G' is an arbitrary group, there exists a unique group homomorphism

$$F: G \to G'$$

making commutative the diagram

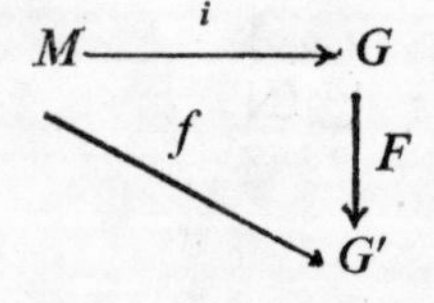

It can be proved that for every set M, there exists a free group $\mathscr{L}(M)$ having the basis M. Clearly, if $(M, i: M \to G)$ is a free group of basis M, then i is an injection. Moreover, if $(M, i: M \to G_1)$, $(M, i_2: M \to G_2)$ are free groups with the same basis, then there exists a unique group isomorphism $G_1 \xrightarrow{\alpha} G_2$, making commutative the diagram

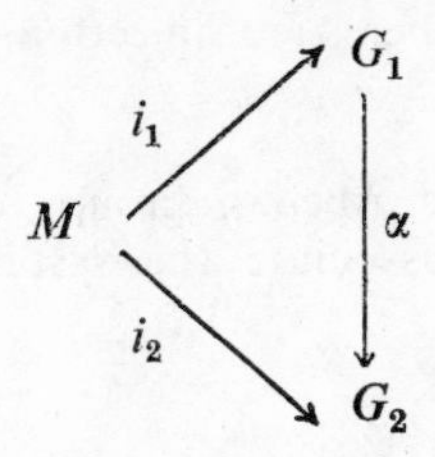

The set $\mathscr{L}(M)$ is built-up from all sequences of the form

$$a_1^{\varepsilon_1} a_2^{\varepsilon_2} \ldots a_r^{\varepsilon_r},$$

$a_i \in M$, $\varepsilon_i = \pm 1$, and $\varepsilon_i \varepsilon_{i+1} = 1$ if $a_i = a_{i+1}$, plus the void sequence. The composition rule of $\mathscr{L}(M)$ is induced by the juxtaposition of sequences, the void sequence being obviously the unit element of $\mathscr{L}(M)$.

PROPOSITION 1.11. *To every injection*

$$\alpha: M \to \mathbb{N}$$

one can associate an injection

$$\gamma_\alpha: \mathscr{L}(M) \to \mathbb{N}$$

defined by

$$\gamma_\alpha(a_1^{\varepsilon_1} a_2^{\varepsilon_2} \ldots a_r^{\varepsilon_r}) = p_1^{\alpha_{\varepsilon_1}(a_1)} p_2^{\alpha_{\varepsilon_2}(a_2)} \ldots p_r^{\alpha_{\varepsilon_r}(a_r)},$$

where

$$\alpha_\varepsilon(a) = \begin{cases} 2\alpha(a) + 1 & \text{if } \varepsilon = 1, \\ 2\alpha(a) + 2 & \text{if } \varepsilon = -1. \end{cases}$$

(Obviously, we set $\gamma_\alpha(\ ) = 1$*).*

DEFINITION 1.12. *A finitely-generated group presentation* is a couple $(M, (r_1, r_2, \ldots, r_s))$ consisting of a subset M of $\mathbb{N}$ and of a sequence $r_1, r_2, \ldots, r_s$ of elements of the free group $\mathscr{L}(M)$ having the basis M. Let $\mathscr{T}$ be the set of finitely-generated group presentations.

PROPOSITION 1.13. *There exist injections of* $\mathscr{T}$ *in* $\mathbb{N}$.

For some categories of groups, there are injections arising in a natural way.

Example 1.14. Let F be the set of Abelian groups of finite type. To every such group G, one can associate the system of natural numbers

$$(n, \theta_1, \theta_2, \ldots, \theta_r),$$

where: n is the rank of G and $(\theta_1, \theta_2, \ldots, \theta_r)$ are the elementary divisors of the group G.

The group G is isomorphic to the group

$$\mathbb{Z} \oplus \mathbb{Z} \oplus \ldots \oplus \mathbb{Z} \oplus \mathbb{Z}/\theta_1\mathbb{Z} \oplus \mathbb{Z}/\theta_2\mathbb{Z} \oplus \ldots \oplus \mathbb{Z}/\theta_n\mathbb{Z}.$$

In this way we get an injection of the set of isomorphism classes of finitely generated Abelian groups into the set $\bigcup_{i \geq 1} \mathbb{N}_i$ and consequently (Proposition 1.3) an injection into $\mathbb{N}$.

§ 2. *The notion of lattice*

DEFINITION 2.1. A partially ordered set M is a lattice if for every couple of elements $a, b \in M$, there exist $\min(a, b)$ and $\max(a, b)$.

Clearly, if M is a lattice, then its dual partially ordered set M^* is a lattice too, i.e. the notion of lattice is autodual. Consequently, the usual duality principle in lattice theory holds. If M is a lattice, we put

$$\min(a, b) = a \cap b,$$

$$\max(a, b) = a \cup b.$$

2.2. The standard examples of lattices occur when studying the set of subobjects or of quotient objects of a suitable category:

In the category of groups: the set of the subgroups of a group forms a lattice. Clearly, if A, B are subgroups of a group, then

$$\min(A, B) = A \cap B,$$

$\max(A, B) =$ the subgroup generated by A and B.

In the category of modules: the set of submodules of a module forms a lattice. In particular, the set of left (right, bilateral) ideals of a ring forms a lattice.

In the category of rings: the set of subrings of a ring forms a lattice.

In an Abelian category: the set of subobjects (respectively of quotient objects) of an object forms a lattice (See I. Bucur, A. Deleanu, *Introduction to the Theory of Categories and Functors*, John Wiley, 1968).

PROPOSITION 2.3. *If M is a lattice, the following relations hold for every elements $a, b, c \in M$.*

(I) $$a \cap a = a, a \cup a = a,$$

(II) $$a \cap b = b \cap a, \quad a \cup b = b \cup a,$$

(III) $$(a \cap b) \cap c = a \cap (b \cap c), \ (a \cup b) \cup c = a \cup (b \cup c),$$

(IV) $$a \cap (a \cup b) = a, \ a \cup (a \cap b) = a.$$

The proof, which is very simple, is left to the reader. We merely remark that, in view of the general duality principle, it is sufficient to prove a single equality for each of the relations I—IV.

PROPOSITION 2.4. *Let S be a set endowed with two internal composition rules*

$$S \times S \to S, \ S \times S \to S,$$

$$(a, b) \to a \cup b, \ (a, b) \to a \cap b,$$

fulfilling conditions I—IV from Proposition 2.3. Then there exists a lattice structure on S, such that for every couple of elements $a, b \in S$ we have

$$\min(a, b) = a \cap b,$$

$$\max(a, b) = a \cup b.$$

Proof. We introduce on S a partial order relation setting by definition:

$a \leqslant b$ if and only if one of the following equivalent relations is fulfilled:

$$a \cap b = a,$$

$$a \cup b = b.$$

One can easily show that this is a partial order relation such that for every couple of elements c, d in S, we have:

$$\min(c, d) = c \cap d,$$

$$\max(c, d) = c \cup d.$$

DEFINITION 2.5. *Distributive lattices.* If M is a lattice, the following statements are equivalent:

D_1. For every triplet a, b, c of elements in M, we have

$$a \cap (b \cup c) = (a \cap b) \cup (a \cap c).$$

D_2. For every triplet a, b, c of elements in M, we have

$$a \cup (b \cap c) = (a \cup b) \cap (a \cup c),$$

Obviously, for every lattice M, we have

$$(2.5.1) \qquad (a \cap b) \cup (a \cap c) \leqslant a \cap (b \cup c), \; a \cup (b \cap c) \leqslant$$

$$\leqslant (a \cup b) \cap (a \cup c), \; \forall\, a, b, c \in M.$$

Hence, in order to prove the implication $D_1 \Rightarrow D_2$, we have to show that $D_1 \Rightarrow (a \cup b) \cap (a \cup c) = a \cup (b \cap c)$ for every $a, b, c \in M$. But in view of D_1, we have

$$(a \cup b) \cap (a \cup c) = [(a \cup b) \cap a] \cup [(a \cup b) \cap c] =$$

$$= a \cup [c \cap (a \cup b)] = a \cup (c \cap a) \cup (c \cap b) = a \cup (b \cap c).$$

In order to prove the implication $D_2 \Rightarrow D_1$, we have to show that $D_2 \Rightarrow a \cap (b \cup c) \leqslant (a \cap b) \cup (a \cap c)\ \forall\, a, b, c \in M$. But in view of D_2, we have

$$(a \cap b) \cup (a \cap c) = ((a \cup b) \cup a) \cap ((a \cap b) \cup c) =$$

$$= a \cap (c \cup (b \cap a)) = a \cap ((c \cup b) \cap (c \cup a))$$

whence the required relation follows.

We say that a lattice M is *distributive* if one of the equivalent conditions D_1, D_2 is fulfilled.

Example. The lattice defined by the diagram

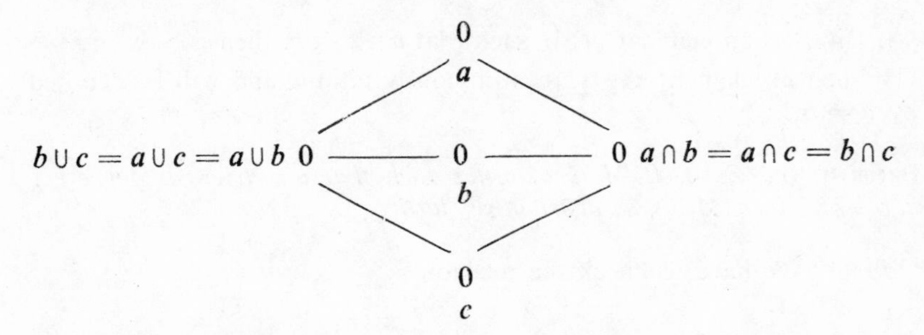

is not distributive:

$$a \cap (b \cup c) = a, \ (a \cap b) \cup (a \cap c) = a \cap b.$$

DEFINITON 2.6. *(Modular (Dedekindian) lattices)*. The lattice M is called *modular* if for every triplet a, b, c of elements in M we have the implication

$$a \geqslant b \Rightarrow a \cap (b \cup c) = b \cup (a \cap c).$$

Clearly, every distributive lattice is modular.

The importance of modular lattices is shown in particular by the following two propositions:

PROPOSITION 2.7. *The lattice of normal subgroups of a group is modular.*

Proof. See Kurosh, *The theory of groups*, Chelsea, New York, 1960.

PROPOSITION 2.8. *The lattice of subobjects of an object of an Abelian category is modular.*

Proof. See Bucur-Deleanu, *Introduction to the Theory of Categories and Functors.*

DEFINITION 2.9. The relative pseudo-complement of an element (cf. Rosiowa-Sikorski, *The Mathematics of Metamathematics*). Let M be a lattice and let a, b be a couple of elements in M. An element $c \in M$ will be called the *pseudocomplement of a with respect to b* if the following conditions are fulfilled:

$$(i)\ a \cap c \leqslant b;$$

(*ii*) if x is an element of M such that $a \cap x \leqslant b$, then $c \geqslant x$.

If such an element exists, it is obviously unique and will be denoted by $a \Rightarrow b$.

PROPOSITION 2.10. *If M is a lattice such that $a \Rightarrow b$ exists for every $a, b \in M$, then M is a distributive lattice.*

Proof. We have to check the relation

$$(a \cap b) \cup (a \cap c) \geqslant a \cap (b \cup c)$$

We put $d = (a \cap b) \cup (a \cap c)$. In order to show that $a \cap (b \cup c) \leqslant d$, it is sufficient to show (see Definition 2.9) that

$$b \cup c \leqslant (a \Rightarrow d).$$

This is equivalent to the relations

$$b \leqslant (a \Rightarrow b).$$
$$c \leqslant (a \Rightarrow d),$$

i.e.

$$a \cap b \leqslant d, a \cap c \leqslant d$$

which are obviously true.

DEFINITION 2.11 *(The concept of Heyting algebra)*. We say that the set M has a structure of *Heyting algebra* if an order relation has been given on M such that the following conditions be fulfilled:

(*i*) M has a first element.

(*ii*) M has a last element.

(*iii*) For every couple (a, b) of elements in M, $a \Rightarrow b$ exists, i.e. there exists an element c satisfying the conditions:

(α) $a \cap c \leqslant b$,

(β) if $a \cap x \leqslant b$, then $x \leqslant c$.

Example 2.12. The importance of the concept of Heyting algebra follows on one hand from the fact that the set of propositions of intuitionistic logic forms a Heyting algebra, and on the other hand, from the proposition:

If $\mathscr{C}$ is a topos (in the sense of Grothendieck or of Lawvere-Tierney) and if A is an arbitrary object of $\mathscr{C}$, then the set M of its subobjects forms a Heyting algebra (see *Séminaire Lawvere-Tierney*, 1970—1971).

Equivalent definition of the concept of Heyting algebra. Let A be a lattice. A is in particular an ordered set. As it is well-known, with the ordered set A one can associate a category $\underline{A}$:

$$\mathrm{Ob}(\underline{A}) = A,$$

$$\mathrm{Hom}_{\underline{A}}(a, b) = \begin{cases} \{(a, b)\} & \text{if} \quad a \leqslant b \\ \emptyset & \text{otherwise} \end{cases}$$

$\mathrm{Hom}_{\underline{A}}(a, b) \times \mathrm{Hom}_{\underline{A}}(b, c) \Rightarrow \mathrm{Hom}_{\underline{A}}(a, c)$ is the canonical map.

Assuming that A is a lattice, it follows that for every couple of objects $a, b \in A$, there exists the direct product in the category $\underline{A}$:

$$a\pi b = a \cap b.$$

For every $b \in A$, one can consider the functor

$$\pi b \colon \underline{A} \to \underline{A}$$

defined by

$$(\pi b)(a) = a\pi b.$$

We say that A is a Heyting algebra if for every $b \in A$, the functor πb admits a right adjoint:

$$b \Rightarrow : \underline{A} \to \underline{A}.$$

In other words, for every $a, b, c \in A$, we have

$$\mathrm{Hom}_{\underline{A}}(a \cap b, c) = \mathrm{Hom}_{\underline{A}}(a, b \Rightarrow c)$$

or, using the definition of morphisms in the category $\underline{A}$:

$$a \cap b \leqslant c \text{ is equivalent to } a \leqslant (b \Rightarrow c).$$

Setting $a = (b \Rightarrow c)$ in the adjonction formula, we obtain:

$$(b \Rightarrow c) \cap b \leqslant c$$

i.e. the latticial form of the "modus ponens" rule.

Remark. The definition given above to the concept of Heyting algebra coincides in fact with one of the conditions imposed to a category with products in order to be a topos in the sense of Lawvere-Tierney.

DEFINITION 2.13. *The concept of Brouwer algebra.* An ordered set M will be called a Brouwer algebra if the following conditions are fulfilled:

(α) For every family $(a_i)_{i \in I}$ of objects in M, $\min_{i \in I}(a_i)$ exists. We put

$$\bigcap_{i \in I} a_i = \min_{i \in I}(a_i).$$

(β) For every family $(a_i)_{i \in I}$ of objects in M, $\max_{i \in I}(a_i)$ exists. We put

$$\bigcup_{i \in I} a_i = \max_{i \in I}(a_i).$$

(γ) For every object $a \in M$ and every family $(b_j)_{j \in J}$ of objects in M we have

$$a \cap (\bigcup_{j \in J} b_j) = \bigcup_{j \in J} (a \cap b_j)$$

Every Brouwer algebra becomes a Heyting algebra if for every couple $(a, b) \in M \times M$, we put

$$(a \Rightarrow b) = \bigcup_{a \cap x = b} (x).$$

LEMMA 2.14. *Let M be a distributive lattice having a first element, denoted by 0, and a last element, denoted by 1. We assume that to the element a there corresponds at least an element c such that the following conditions be fulfilled:*

(2.14.1) $\qquad a \cap c = 0, \ a \cup c = 1.$

Then, for every $b \in M$, the element $a \Rightarrow b$ exists and we have

(2.14.2) $\qquad (a \Rightarrow b) = c \cup b.$

Proof. First, we show that

$$c \cup b \leqslant (a \Rightarrow b).$$

It will be sufficient to prove that

$$a \cap (c \cup b) \leqslant b.$$

But in view of the distributivity property (Definition 2.5), we have

$$a \cap (c \cup b) = (a \cap c) \cup (a \cap b) = a \cap b \leqslant b.$$

Now, let x be an element of M such that $a \cap x \leqslant b$.

In order to prove (2.14.2), it is sufficient to show that the previous relation implies $x \leqslant c \cup b$. But we have

$$c \cup (a \cap x) \leqslant c \cup b,$$

$$(c \cup a) \cap (c \cup x) = c \cup x \leqslant c \cup b,$$

whence $x \leqslant c \cup b$.

COROLLARY 2.15. *With the notation and under the assumptions of Lemma 2.14. for every element* $a \in M$, *there exists at most an element* c *satisfying* (2.14.1).

Indeed, Lemma 2.14 shows that $c = (0 \Rightarrow a)$.

DEFINITION 2.16. *The concept of Boole algebra.* A distributive lattice M, having a first and a last element, is called a *Boole algebra* if for every $a \in M$, there exists an element $c \in M$ such that

$$a \cap c = 0, \; a \cup c = 1.$$

The element c is unique by Corollary 2.15, and it will be denoted by $\neg a$. Obviously, $\neg \neg a = a$.

COROLLARY 2.17. *Every Boole algebra is a Heyting algebra:*

$$(a \Rightarrow b) = (\neg a) \cup b.$$

REMARK 2.18. The concept of Boole algebra is autodual. Consequently, the duality principle holds in the theory of Boole algebras.

PROPOSITION 2.19. *In a Boole algebra, the following relations hold:*

$$\left.\begin{aligned} \neg(a \cup b) &= (\neg a) \cap (\neg b) \\ \neg(a \cap b) &= (\neg a) \cup (\neg b) \end{aligned}\right\} \quad \text{The rules of De Morgan.} \tag{2.19.1}$$

$$(a \Rightarrow \neg b) = (b \Rightarrow \neg a) \qquad \text{The contraposition rule.}$$

The proof is straightforward.

DEFINITION 2.20. *The concept of Boole ring.* A ring A having a unit element is called a *Boole ring* if every element of A is idempotent:

$$a^2 = a, \quad \forall a \in A.$$

PROPOSITION 2.21. *Every Boole ring is commutative and has the characteristic* 2.

Proof. Consider $a, b \in A$. We have

$$\begin{aligned} a + b = (a + b)^2 &= a^2 + ab + ba + b^2 = \\ &= a + b + ab + ba. \end{aligned}$$

Consequently,

(2.21.1) $\qquad ab + ba = 0.$

Taking $a = b$ and using the hypothesis, we get

$$2a = 0,$$

i.e. the ring A has the characteristic 2. In particular, $a = -a$, hence (2.21.1) becomes

$$ab = ba.$$

PROPOSITION 2.22. *To every Boole algebra M one can associate a Boole ring with the same underlying set, by putting:*

$$a + b = (a \cap \neg b) \cup (\neg a \cap b), \; ab = a \cap b.$$

Conversely, to every Boole ring A, one can associate a Boole algebra with the same underlying set, by putting

$$a \cup b = a + b - ab, \; a \cap b = ab.$$

DEFINITION 2.23. *Morphisms of Boole algebras.* Let M and N be two Boole algebras. A morphism of Boole algebras is a map

$$f\colon M \to N$$

such that the following conditions be fulfilled:

$$f(\neg x) = \neg f(x), \; \forall x \in M,$$

$$f(x \cup y) = f(x) \cup f(y), \quad \forall x, y \in M.$$

PROPOSITION 2.24. *If $f\colon M \to N$ is a morphism of Boole algebras, then*

$$f(x \cap y) = f(x) \cap f(y),$$

$$f(1) = 1, \; f(0) = 0.$$

Proof. $f(x \cap y) = f(\neg(\neg x \cup \neg y)) = \neg f(\neg x \cup \neg y) =$

$$= \neg c(f(\neg x) \cup f(\neg y)) = \neg(\neg f(x) \cup \neg f(y)) =$$

$$= \neg(\neg(f(x) \cap f(y))) = f(x) \cap f(y),$$

$$f(0) = f(x \cap \neg x) = f(x) \cap \neg f(x) = 0,$$

$$f(1) = f(x \cup \neg x) = f(x) \cup \neg f(x) = 1.$$

PROPOSITION 2.25. *There exists an isomorphism*

$$F: \mathscr{B} \to \mathscr{A}$$

between the category $\mathscr{B}$ of Boole algebras (the morphisms being morphisms of Boole algebras) and the category $\mathscr{A}$ of Boole rings.

The proof is straightforward; it is sufficient to associate to every Boole algebra M the Boole ring $F(M)$ with the same underlying set, defined by Proposition 2.22.

DEFINITION 2.26. *Free Boole algebras.* Let E be a set, let M be a Boole algebra and let $\gamma: E \to M$ be a map. We say that M is a free Boole algebra of basis M and of canonical map γ if for every Boole algebra N and every map $u: E \to N$, there exists a unique morphism of Boole algebras $v: M \to N$, such that the following diagram commutes:

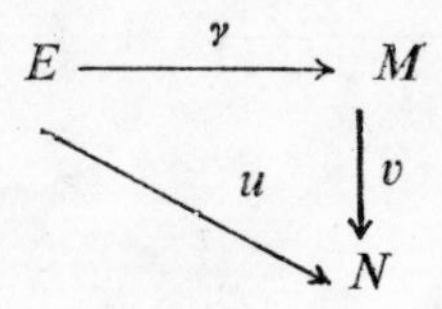

PROPOSITION 2.27. *For every set E, there exists a free Boole algebra with the basis E.*

Proof. In view of Proposition 2.25, it is sufficient to show the existence of a Boole ring $\mathscr{L}(E)$ with the basis E. This fact can be seen as follows:

Let $\mathcal{M}(E)$ be the free monoid with the basis E, and let $\mathbb{Z}[\mathcal{M}(E)]$ be the ring associated with this monoid. Consider the bilateral ideal I generated in $\mathbb{Z}[\mathcal{M}(E)]$ by the elements of the form x^2-x, $x \in \mathbb{Z}[\mathcal{M}(E)]$. We put

$$\mathcal{L}(E) = \mathbb{Z}[\mathcal{M}(E)]/\mathrm{I}$$

$$(E \xrightarrow{\gamma} \mathcal{L}(E)) = (E \to \mathcal{M}(E)) \to \mathbb{Z}[\mathcal{M}(E)] \to \mathbb{Z}[\mathcal{M}(E)]/\mathrm{I}$$

The system $(E, \mathcal{L}(E), \gamma)$ obviously defines a free Boole ring with the basis E.

Variant. Instead of $\mathbb{Z}[\mathcal{M}(E)]$ one can use the tensor algebra $T(f(E))$, where $f(E)$ is the free Abelian group with the basis E.

PROPOSITION 2.28. *For every injection $\lambda: E \to N$, there exists a canonical injection of the Boole algebra $L(E)$ into N.*

§ 3. *The algebra of propositions*

DEFINITION 3.1. Let E be a set. The free Boole algebra $L(E)$ having the basis E is called an *algebra of propositions.* Its elements will be called propositions. If $i: E \to L(E)$ is the canonical inclusion, the elements of the form $i(x)$ will be called elementary propositions.

DEFINITION 3.2. *The formal system of logic.* Given a set E, we consider the set

$$E' = E \cup \{\neg, \vee, (,)\}$$

and the free monoid $\mathcal{M}(E')$ with the basis E'.

A sequence

$$A_1, A_2, \ldots, A_n$$

of elements of $\mathcal{M}(E')$ will be called a formative construction (cf. Bourbaki, Livre 1, Chap. 1, § 1.3) if for every $i \in \{1, 2, \ldots, n\}$ one of the following three conditions is fulfilled:

(a) $A_i \in E$;
(b) there exists a $j < i$ such that $A_i = \neg(A_j)$;

(c) there exist $j, k < i$ such that $A_i = (A_j) \vee (A_k)$ (the indices j, k are not necessarily distinct).

An element of $\mathscr{M}(E')$ which can be inserted in a formative construction is called a *logical formula*.

Example of logical formula:

$$(x \vee (\neg(\neg y \vee z))) \vee (u \vee v), \quad (x, y, z, u, v \in E).$$

We denote by $\mathscr{F}\ell(E)$ the set of logical formulae defined by the set E. We introduce the notation:

$$(x \Rightarrow y) = (\neg x) \vee y,$$

$$\neg((\neg x) \vee (\neg y)) = x \wedge y,$$

$$(x \Leftrightarrow y) = ((x \Rightarrow y) \wedge (y \Rightarrow x)).$$

PROPOSITION 3.3. *There exists a unique map*

$$\varphi : \mathscr{F}\ell(E) \to L(E)$$

such that the following conditions be fulfilled:

(α) $\varphi(x) = \gamma(x)$, *for every* $x \in E$;

(β) $\varphi(A \vee B) = \varphi(A) \vee \varphi(B)$ for every logical formulae $A, B \in \mathscr{F}\ell(E)$;

(γ) $\varphi(\neg A) = \neg \varphi(A)$ for every $A \in \mathscr{F}\ell(E)$.

DEFINITION 3.4. *The axioms of propositional calculus.* The following logical formulae will be called *axioms of propositional calculus:*

I

1. $x \Rightarrow (y \Rightarrow x), \qquad x, y \in \mathscr{F}\ell(E)$

2. $(x \Rightarrow (y \Rightarrow z)) \Rightarrow ((x \Rightarrow y) \Rightarrow (x \Rightarrow z)), \quad x, y, z \in \mathscr{F}\ell(E)$

II

1. $(x \wedge y) \Rightarrow x, \qquad x, y \in \mathcal{F}\ell(E),$

2. $(x \wedge y) \Rightarrow y, \qquad x, y \in \mathcal{F}\ell(E).$

3. $(x \Rightarrow y) \Rightarrow ((x \Rightarrow z) \Rightarrow (x \Rightarrow y \wedge z)), \quad x, y, z \in \mathcal{F}\ell(E).$

III

1. $x \Rightarrow x \vee y, \qquad x, y \in \mathcal{F}\ell(E),$

2. $y \Rightarrow x \vee y \qquad x, y \in \mathcal{F}\ell(E),$

3. $(x \Rightarrow z) \Rightarrow ((y \Rightarrow z) \Rightarrow (x \vee y \Rightarrow z)), \quad x, y, z \in \mathcal{F}\ell(E),$

IV

1. $(x \Rightarrow y) \Rightarrow (\neg y \Rightarrow \neg x), \qquad x, y \in \mathcal{F}\ell(E),$

2. $x \Rightarrow \neg(\neg x), \qquad x \in \mathcal{F}\ell(E),$

3. $\neg(\neg x) \Rightarrow x, \qquad x \in \mathcal{F}\ell(E),$

DEFINITION 3.5. *Demonstrative text.* A sequence of logical formulae

$$P_1, P_2, \ldots, P_n$$

is said to form a demonstrative text if for every $i \in \{1, 2, \ldots, n\}$ one of the following two conditions is fulfilled:

(a) P_i is an axiom of propositional calculus (Definition 3.4);

(b) there exist j, k such that

$$j < k < i \text{ and } P_k = (P_j \Rightarrow P_i).$$

Every logical formula which can be inserted in a demonstrative text is called a theorem of propositional calculus.

Examples of theorems of propositional calculus:

— Every axiom P of propositional calculus is a theorem.

The demonstrative text is reduced to P.
— If P and $P \Rightarrow Q$ are theorems, then Q is a theorem.
If

$$P_1, P_2, \ldots, P_n$$

is a demonstrative text for P and

$$Q_1, Q_2, \ldots, Q_n$$

is a demonstrative text for $P \Rightarrow Q$, then

$$P_1, P_2, \ldots, P_n, Q_1, Q_2, \ldots, Q_n, Q$$

is a demonstrative text for Q.

PROPOSITION 3.6. *The necessary and sufficient condition for the formula P to be a theorem of propositional calculus is that* $\varphi(P) = 1$ (cf. Proposition 3.3)

PROPOSITION 3.7. *Let P be a logical formula. There exist logical formulae D_p, C_p of the form*

$$D_P = (e_{11} \wedge e_{12} \wedge \ldots \wedge e_{1i_1}) \vee (e_{21} \wedge e_{22} \wedge \ldots \wedge e_{2i_2}) \vee \ldots \tag{3.7.1}$$

$$\ldots \vee (e_{p_1} \wedge e_{p_2} \wedge \ldots \wedge e_{p\,i_p}),$$

$$C_P = (f_{11} \vee f_{12} \vee \ldots \vee f_{2j_1}) \wedge (f_{21} \vee f_{22} \vee \ldots \vee f_{2j_2}) \wedge \ldots \tag{3.7.2}$$

$$\ldots \wedge (f_{q1} \vee f_{q2} \vee \ldots \vee f_{qi_q}),$$

where e_{rs} (respectively f_{tu}) is either an element of the set E or an element of the form $\neg u$, with $u \in E$, such that the formulae

$$P \Leftrightarrow D_P, \tag{3.7.3}$$

$$P \Leftrightarrow C_P. \tag{3.7.4}$$

be theorems of propositional calculus.

DEFINITION 3.8. A logical formula of the form D_p (or C_p) satisfying the conditions stated in Proposition 3.7 is called a *normal disjunctive* (or *conjunctive*) *form* for the formula P.

§ 4. *Formal systems*

4.1. One of the characteristics of the mathematical science consists in the great diversity of its applications. In working at them, the mathematician often uses the intuitive meaning of mathematical concepts. The thoughtless use of this method and the ply toward conferring some qualities or properties to general mathematical concepts starting from their intuitive verification or verisimilitude may lead, as we are going to show, to the arising of paradoxes.

Example 4.2. *Russel's paradox*. Using the intuitive concept of a set, we can obviously consider the set:

$$R = \{a |\ a \notin a\}$$

in other words, the set of all sets a, such that a is not an element of a. By means of the set R, we can form the following set-theoretic proposition: $P = (R \in R)$.
Obviously, we have

$$\neg P = (R \notin R).$$

On the other hand, we also have

$$\neg P \Rightarrow P$$

$$P \Rightarrow \neg P$$

which shows that both P and $\neg P$ are set-theoretic theorems. But this implies the fact that every set-theoretic proposition is a theorem, which is obviously absurd.

The following example shows that a particular attention must be paid not only to the use of concepts, in example 4.2— the concept of a set, but also to the way in which propositions using or regarding such concepts are formed.

Example 4.3. Let $\mathbb{N}$ be the set of natural numbers. Denote by M the subset of $\mathbb{N}$ consisting of the natural numbers which can be defined by concrete, specific English propositions. For instance, $1 \in M$, for it can be defined by the following specific proposition:

"The number one is the only natural number which is invertible with respect to multiplication".

In this acception, two distinct natural numbers must be obviously defined by distinct propositions. In particular, it follows that the set B of all numbers in N which can be defined by English propositions involving at most 1000 letters has a non-empty complement: there are only a finite number ($\leqslant 24^{1000}$) of English propositions which can be formed using at most 1000 letters. Let A be the complement of B, i.e. the set of all natural numbers which cannot be defined by means of English propositions involving at most 1000 letters.

Consider $n_0 = \inf \{n \in \mathbb{N} \mid n \in A\}$.

Let P be the arithmetic proposition $(n_0 \in A)$. Clearly, P is a theorem, since A has a first element. On the other hand, the proposition

$$\neg P = (n_0 \notin A) = (n_0 \in B).$$

is also true.

Indeed, n_0 can be defined by the specific English proposition:

"n_0 is the first number which cannot be defined by a specific English proposition involving at most 1000 letters".

As in example 4.2, we obviously get a paradox.

In order to avoid the arising of paradoxes of the above types or of some other type (see for instance St. Kleene, *Introduction to Metamathematics*, chap. III, § 11), one introduces the concept of a formal system or of a formal mathematical theory.

4.4. *The symbols of a formal system.* The symbols of a formal system are given if the following sets are given:

(a) A set V, whose elements are called *variables*, more precisely, *free variables*, and which are usually denoted by $x, y, z, x_1, x_2, \ldots$ $\ldots, y_1, y_2, \ldots$ The set V is usually supposed to be infinite;

(b) A set U, whose elements are called *bound variables*, and which are usually denoted by $\xi, \eta, \zeta, \xi_1, \xi_2, \ldots, \eta_1, \eta_2, \ldots$ The set U is usually supposed to be infinite;

(c) The set of logical symbols $L = (\neg, \vee, \wedge, (,), \exists, \forall)$;

(d) A family Φ_m of sets. The elements of Φ_m are called operations of rank m. The elements of Φ_0 are also called individual constants.

(e) A family $(P_m)_{m \in \mathbb{N}}$ of sets. The elements of P_m are called predicates in m variables *.

In this acception, the specific symbols of a formal system obviously are those from points d, e.

4.4.1. *The symbols of the formal system of arithmetics:*

$$\Phi_0 = \{0\},\ \Phi_1 = \{,\},\ \Phi_2 = \{+, .\},\ \Phi_i = \emptyset,\ i \geqslant 3,$$

$$P_0 = P_1 = \emptyset,\ P_2 = \{=\},\ P_i = \emptyset,\ i \geqslant 3.$$

4.4.2. *The symbols of the formal system of set theory:*

$$\Phi_i = \emptyset, \qquad i = 0, 1, 2, \ldots$$

$$P_0 = P_1 = \emptyset,\ P_2 = \{=, \in\}, \quad P_i = \emptyset,\ i \geqslant 3, \ldots$$

4.4.3. *The symbols of the formal system of theory of topologic spaces:*

$$\Phi_0 = \emptyset,\ \Phi_1 = (-),\ \Phi_i = \emptyset,\ i \geqslant 2$$

$$P_0 = P_1 = \emptyset,\ P_2 = (=, \subseteq),\ P_i = \emptyset,\ i \geqslant 3.$$

4.4.4. *The symbols of the formal system of the theory of partially ordered sets:*

$$\Phi_i = \emptyset, \qquad i = 0, 1, 2, \ldots$$

$$P_0 = P_1 = \emptyset,\ P_2 = (=, \leqslant),\ P_i = \emptyset,\ i \geqslant 3.$$

4.4.5. *The symbols of the formal system of group theory:*

$$\Phi_0 = \Phi_1 = \emptyset,\ \Phi_2 = (.),\ \Phi_n = \emptyset \text{ for } n \geqslant 3,$$

$$P_0 = P_1 = \emptyset,\ P_2 = (=).$$

* It is strange to observe that some authors suppose $P_0 = \emptyset$, which excludes for instance the very important system of the logic of propositions.

4.4.6. *The symbols of the formal system of logic:*

$$V = U = \emptyset, \Phi_m = \emptyset, P_0 = E, P_m = \emptyset \text{ for } m \geqslant 1.$$

4.4.7. *The symbols of the formal system of predicate calculus:*

$$V \neq \emptyset, U \neq \emptyset, \Phi_0, \Phi_m = \emptyset, (i \geqslant 1),$$

$$P_0 = E_0, P_1 = E_1, P_2 = E_2, \ldots, P_m = E_m, \ldots$$

The Aristotelian logic is obtained in the particular case when $E_2 = E_3 = \ldots = \emptyset$, i.e. when only predicates in one variable are considered.

4.5. *The elementary terms of a formal system.* Let

$$\{V, U, L, (\Phi_m)_{m \in \mathbb{N}}\}$$

be the symbols of a formal system.

Consider the free monoid $\mathscr{M}(E)$, where $E = V \cup (\bigcup_{m \in \mathbb{N}} \Phi_m) \cup \{,\}$.

An *elementary term* of the formal system is by definition an element of $\mathscr{M}(E)$, which can be inserted in a sequence of elements $T_1, T_2, \ldots, T_n \in \mathscr{M}(E)$, satisfying for every $i \in \{1, 2, \ldots, n\}$ one of the following conditions:

(a) $T_i \in V$;

(b) $T_i \in \Phi_0$;

(c) $T_i = \varphi(T_{r_1}, T_{r_2}, \ldots, T_{r_m})$, where $\varphi \in \Phi_m$, $m \geqslant 1$ and $r_1 \leqslant i$, $r_2 \leqslant i, \ldots, r_m \leqslant i$.

It is easy to verify that if τ is an elementary term containing the free variables $x_1, x_2, \ldots, x_m$ and if $\tau_1, \tau_2, \ldots, \tau_m$ are elementary terms, then the expression obtained by substituting $\tau_1, \tau_2, \ldots, \tau_m$ for $x_1, x_2, \ldots, x_m$ is also an elementary term. The set of elementary terms of the formal system $\mathscr{S}$ is denoted by $\mathscr{T}(\mathscr{S})$ or simply by $\mathscr{T}$.

If $\Phi_m = \emptyset$ for every $m \geqslant 0$, then the set of the terms is equal to V: $\mathscr{T}(\mathscr{S}) = V$.

If $\Phi_m = \emptyset$ for every $m \geqslant 1$, then the set of elementary terms is equal to $V \cup \Phi_0$: $\mathscr{T}(\mathscr{S}) = V \cup \Phi_0$.

4.6. *The elementary relations of a formal system.*
Consider the free monoid $\mathscr{M}(F)$, where $F = V \cup U \cup L \cup (\bigcup_{m \in \mathbb{N}} P_m) \cup \cup (\bigcup_{m \in \mathbb{N}} \Phi_m)$. An *elementary relation* is by definition an element of the least subset $\mathscr{R}$ of $\mathscr{M}(F)$ satisfying the following conditions:

(a) $P(x_1, x_2, \ldots, x_m) \in \mathscr{R}$ for every $P \in P_m$ and $x_i \in V$, $i = 1, 2, \ldots, m$. (If $m = 0$ and $P \in P_0$, we shall obviously understand that $P \in \mathscr{R}$);

(b) If $R(x_1, x_2, \ldots, x_m) \in \mathscr{R}$ contains the free variables $x_1, x_2, \ldots, x_m$ and if $\tau_1, \tau_2, \ldots, \tau_m$ are elementary terms, then $R(\tau_1, \tau_2, \ldots, \tau_m) \in \mathscr{R}$;

(c) If $R \in \mathscr{R}$, then $\neg R \in \mathscr{R}$;

(d) If $R \in \mathscr{R}$ and $S \in \mathscr{R}$, then $R \vee S \in \mathscr{R}$;

(e) If x is a free variable, $R(x) \in \mathscr{R}$ contains the variable x and $\xi \in U$ is a bound variable which does not occur in $R(x)$, then $\exists\, \xi R(\xi) \in \mathscr{R}$ and $\forall\, \xi R(\xi) \in \mathscr{R}$, where $R(\xi)$ is the expression obtained by substituting ξ for all occurrences of x in R.

An elementary relation is *closed* if it contains no free variable.

4.6.1. *Examples.* (a) Consider the formal system of logic (propositional calculus). Obviously, the set of elementary relations coincides in this case with the set of *logical formulae* which are generated by the set of elementary propositions $E = P_0$: $\mathscr{R} = \mathscr{F}\ell(E)$ (cf. Definition 3.2).

(b) Now consider the symbols $V, U, L, \Phi_0, (P_m)_{m \in \mathbb{N}}$ of the formal system of predicate calculus. The following elements of $\mathscr{M}(F)$ (cf. 4.6) are elementary relations of the formal system of predicate calculus:

$$(\forall\, \xi A(\xi)) \Rightarrow (A(c)),\ A \in P_1,\ \xi \in U,\ c \in \Phi_0 \text{ (specialization rule)};$$

$$(A(c) \Rightarrow B) \Rightarrow (\exists\, \xi A(\xi) \Rightarrow B).$$

PROPOSITION 4.6.2. *If V, U, P_m, Φ_m are countable sets, then the set $\mathscr{R}$ of elementary relations is also countable.*

4.7. *The axioms of a formal system* are given if a family $\mathscr{A}$ of elementary relations of this system is given.

4.7.1. *The theorems of a formal system.* An elementary relation R is called a theorem if it can be inserted in a demonstrative text, i.e. in a sequence $R_1, R_2, \ldots, R_n$ of elementary relations satisfying, for every i, at least one of the following conditions:

(α) R_i is an axiom;

(β) there exist j, h such that $j < k < i$ and $R_k = (R_j \Rightarrow R_i)$;

(γ) there exist $j < i$ and terms $\tau_1, \tau_2, \ldots, \tau_m$ such that R_j contains the free variables $x_1, x_2, \ldots, x_m$ and R_i is obtained by substituting the terms $\tau_1, \tau_2, \ldots, \tau_m$ for the variables $x_1, \ldots, x_m$.

4.7.2. *Examples.* (a) *Logical systems.* A formal system is a logical system if the family of relations obtained from the axioms of propositional calculus (Definition 3.4) by taking for $\mathscr{F}\ell(E)$ the set $\mathscr{R}$ is a subfamily of the family $\mathscr{A}$ of axioms of the formal system.

(b) *Quantified systems.* A formal system is quantified if the set of its axioms includes the following families of relations:

(σ) $(\forall\, \xi R(\xi)) \Rightarrow (R(c))$, $R \in \mathscr{R}$ contains the free variable x,

(τ) $(R(c) \Rightarrow S) \Rightarrow (\exists\, \xi R(\xi) \Rightarrow S)$, $R \in \mathscr{R}$, R contains the free variable x, and S contains neither the constant c, nor the variable x.

(μ) $(\neg(\forall\, \xi(R(\xi))) \Rightarrow \exists \xi \neg(R(\xi)))$ for every relation R containing the free variable x;

$\nu((\forall\, \xi R(\xi)) \wedge (B)) \Leftrightarrow (\forall\, \xi((R(\xi)) \wedge (B)))$ and $((\exists\, \xi R(\xi)) \wedge (B)) \leftrightarrow$

$\leftrightarrow (\exists \xi((R(\xi)) \wedge (B)))$ for every relations R, B such that the free variable x occurs in R but not in B.

(c) *Equalitary systems.* A formal system is *equalitary* if the set P_2 contains the symbol $=$ and if the following relations

$$x = x, \; x \in V,$$

$$(x = y) \Leftrightarrow (y = x), \; x, y \in V,$$

$$((x = y) \wedge (y = z) \Rightarrow (x = z), \; x, y, z \in V.$$

are axioms.

If x is a free variable occurring in the relation $R = R(x)$ and if τ, μ are terms of the formal system, then the relation $(\tau = \mu) \Rightarrow R(\tau) = R(\mu)$ is an axiom.

(d) *The axioms of the formal system of arithmetics.* The formal system of arithmetics, whose symbols have been described in 4.4.1 is by definition a logical quantified equalitary system and has in addition the following axioms (In writing these axioms, we use the abbreviation $\exists!\, \xi A(\xi) = ((\exists \xi A(\xi)) \wedge (\forall \eta\, A(\eta))) \Rightarrow (\xi = \eta))$.

1. $\forall \xi\, \forall \eta\, \exists!\, \tau(\xi + \eta = \zeta)$;

2. $\forall \xi \forall\, \eta\, \exists!\, \tau(\xi\eta = \zeta)$;

3. $\forall \xi((\xi + 0 = \xi) \wedge (\xi \cdot 1 = \xi))$. (We have put $1 = 0'$).

4. $\forall \xi \forall \eta(\xi + \eta') = (\xi + \eta')$;

5. $\forall \xi \forall\, \eta(\xi\eta = \xi\eta + \xi)$;

6. $\forall \xi \forall\, \eta(\xi' = \eta') \Rightarrow (\xi = \eta)$;

7. $\forall \xi(\neg(\xi' = 0))$;

8. $(A(0) \wedge \forall\, \xi(A(\xi) \Rightarrow A(\xi'))) \Rightarrow (\forall\, \xi A(\xi))$.

The last axiom has obviously to be understood in the following way: for every relation A of the formal system of arithmetics, relation 8 is an axiom.

(e) *The formal theory of field theory.* The symbols of this system are defined as follows:

$$\Phi_0 = \{0, 1\},\ \Phi_1 = \emptyset,\ \Phi_2 = \{+, .\},\ \Phi_i = \emptyset,\ i \geqslant 3$$

$$P_0 = P_1 = \emptyset,\ P_2 = \{=\},\ P_i = \emptyset,\ i \geqslant 3.$$

With the usual notation

$$= (x, y) = (x = y),\ + (x, y) = x + y,\ \cdot(x, y) = x \cdot y,$$

the axioms of the formal system of field theory can be written as follows:

$$\forall \xi \, \forall \eta((\xi + \eta) = (\eta + \xi));$$

$$\forall \xi(\xi + 0 = \xi);$$

$$\forall \xi \, \exists \xi'((\xi + \xi') = 0);$$

$$\forall \xi_1 \, \forall \xi_2 \, \forall \xi_3((\xi_1 + \xi_2) + \xi_3 = \xi_1 + (\xi_2 + \xi_3));$$

$$\forall \xi_1 \, \forall \xi_2 \, \forall \xi_3(\xi_1(\xi_2 + \xi_3) = (\xi_1 \xi_2 + \xi_1 \xi_3));$$

$$\forall \xi_1 \, \forall \xi_2 \, \forall \xi_3((\xi_2 + \xi_3)\xi_1 = (\xi_2 \xi_1 + \xi_3 \xi_1));$$

$$\forall \xi(\xi, 1) = 1\xi = \xi;$$

$$\forall \xi(\neg(\xi = 0) \Rightarrow \exists \eta(\xi \eta = \eta \xi = 1));$$

$$\forall \xi_1 \, \forall \xi_2 \, \forall \xi_3(\xi_1(\xi_2 \xi_3) = (\xi_1 \xi_2)\xi_3).$$

It is easily seen that every relation in the formal system of field theory is equivalent to a relation obtained in the following way:

Let $\mathbb{Z}[X_1, \ldots, X_n]$ be the ring of polynomials in n variables with integer coefficients. If $P \in \mathbb{Z}[X_1, X_2, \ldots, X_n]$ and $x_1, x_2, \ldots, x_n$ are free variables, a field-theoretic relation is obtained by substituting $x_1, x_2, \ldots, x_n$ for $X_1, X_2, \ldots, X_n$ in the expression $P(X_1, \ldots, X_n) = 0$.

The interpretation of $n \in \mathbb{Z}$ is clear. Obviously, a polynomial P may generate several such relations.

Let R_n be the set of the relations obtained in this way. Consider the formal system of predicate calculus, having the symbols:

$$\Phi_m = \emptyset, \; m \geqslant 0, \; P_0 = \{n = 0 | \, n \in \mathbb{Z}, \; P_i = R_i, \; i \geqslant 1\}^*.$$

Any relation of this formal system obviously defines a unique field-theoretic relation and any relation of the formal system of field theory is equivalent to a relation obtained in this way. More precisely we can

* In any formal system of the predicate calculus, one supposes $\Phi_0 \neq \emptyset$.

say that there exists an essentially surjective mapping from the set of these relations into a subset of the set of field-theoretic relations.

(f) *The formal system of ordered fields theory.* The symbols of field theory will be modified as follows:

$$P_2 = \{=, <\}.$$

The axioms include the axioms of field theory and in addition the following ones:

$$\forall \xi \, \forall \eta ((\xi = \eta) \vee (\xi < \eta) \vee (\eta < \xi));$$

$$\forall \xi \, \forall \eta (\neg ((\xi < \eta) \wedge (\eta < \xi));$$

$$\forall \xi \, \forall \eta ((\xi > 0) \wedge (\eta > 0) \Rightarrow ((\xi + \eta) > 0) \wedge (\xi \cdot \eta > 0)).$$

As in the case (e), a formal system of predicate calculus and an essentially surjective mapping of the set of the relations of this formal system into the set of relations of the formal system of ordered field theory can be defined.

Indeed, for every $n > 0$, consider the set R_n of relations of the form $\{P(x_1, x_2, \ldots, x_n) = 0, \; P(x_1, \ldots, x_n) < 0|, \; P \in \mathbb{Z}[X_1, \ldots, X_n], \; x_i \in V\}$ while $R_0 = \{n < 0, \, n = 0| \, n \in \mathbb{Z}\}$.

DEFINITION 4.7.3. A relation is *closed* if it contains no free variable. A relation is *open* if it contains only free variables.

DEFINITION 4.7.4. A *prenex relation* is a relation of the form

$$Q_1\xi_1 Q_2\xi_2 \ldots Q_n\xi_n(P(\xi_1, \xi_2, \ldots, \xi_n)), \text{ where}$$

– $Q_1, Q_2, \ldots, Q_n$ is one of the quantifiers $\exists$, $\forall$;
– $P(x_1, x_2, \ldots, x_n)$ is an open relation (possibly containing some other free variables besides $x_1, x_2, \ldots, x_n$).

PROPOSITION 4.7.5. *Let* $(\mathscr{S}, \mathscr{A})$ *be a logical and quantified formal system and let* R *be one of its relations. Then there exists an open relation* $R_P(x_1, x_2, \ldots, x_n)$ *such that the following relation be a theorem:*

$$R \Leftrightarrow Q_1\xi_1 Q_2\xi_2 \ldots Q_n\xi_n(R_P(\xi_1, \xi_2, \ldots, \xi_n)),$$

where, as in Definition 4.7.4, $Q_1, Q_2, \ldots, Q_n$ designate one of the quantifiers $\exists$, $\forall$. In other words, in a logical and quantified system, every relation is equivalent to a prenex relation.

4.8. *The concept of model of a formal system*

4.8.1. *Preliminary lemmas and concepts.* Recall that a lattice L is called *complete* if for every family $(x_i)_{i \in I}$ of elements in L, there exists $\min_{i \in I}(x_i)$, $\max_{i \in I}(x_i)$. Obviously, every complete lattice has a first and a last element:

$0 =$ the maximum of the family of elements in L defined by the canonical map $\emptyset \hookrightarrow L$;

$1 =$ the minimum of the family of elements in L defined by the canonical map $\emptyset \hookrightarrow L$.

Also, recall that to every lattice one can associate a category, whose elements coincide with the elements of L, and whose morphisms are defined by the formula

$$\mathrm{Hom}_L(x, y) = \begin{cases} (x, y) & \text{if } x \leqslant y; \\ \emptyset & \text{otherwise,} \end{cases}$$

the composition of morphisms being defined in a straightforward way.

If X is any set and if L is a complete lattice, then set L^X of all maps of X into L has a natural structure of complete lattice. If L is a Boole lattice, then L^X is a Boole lattice too. Moreover, if $X \xrightarrow{f} Y$ is any map, then the map

$$L^f \colon L^Y \to L^X$$

$$\mu \to \mu \cdot f$$

is a morphism of lattices (a monotone map).

Let

$$\exists_f \colon L^X \to L^Y$$

$$\forall f \colon L^Y \to L^Y$$

be the morphisms of lattices defined as follows:

$$[(\exists_f)(x)](y) = \max_{\xi \in f^{-1}(y)} \{\chi(\xi)\}$$

$$[(\forall_f)(x)](y) = \min_{\xi \in f^{-1}(y)} \{\chi(\xi)\}$$

The map L^f, $\exists_f$, $\forall_f$ define functors between the corresponding associated categories. These functors will be denoted by the same symbols.

LEMMA 4.8.2. *The functor* $\exists_f: L^X \to L^Y$ (*or* $\forall_f: L^X \to L^Y$) *is a left (or right) adjoint of the functor* L^f.

Proof. It is sufficient to prove, for every $\chi \in L^X$, $\mu \in L^X$, the following equivalences:

(4.8.2.1) $[(\exists_f(\chi))(y) \leqslant \mu(y)$ for every $y \in Y] \Leftrightarrow$

$[\chi(x) \leqslant (\mu \circ f)(x)]$ for every $x \in X$

(4.8.2.2) respectively $[\mu(y) \leqslant (\forall_f(\chi))(y)$ for every $y \in Y] \Leftrightarrow$

$\Leftrightarrow [(\mu_0 f)(x) \leqslant \chi(x)$ for every $x \in X]$.

We shall content ourselves with proving the equivalence 4.8.2.1. First assume that for every $x \in X$, we have $\chi(x) \leqslant (\mu \circ f)(x)$ and consider $y \in Y$. Two cases can occur:

(a) $f^{-1}(x) = \emptyset$;

(b) $f^{-1}(y) \neq \emptyset$.

In the first case, we have

$$(\exists_f(\chi))(y) = 0 \leqslant \mu(y).$$

In the second case, we have

$$(\exists_f(\chi))(y) = \max_{\xi \in f^{-1}(y)} \{\chi(\xi)\} \leqslant \mu(y).$$

Now assume that for every $y \in Y$, we have

$$(\exists_f(\chi))(y) \leqslant \mu(y)$$

and consider an element $x \in X$. Writing $y = f(x)$, we have

$$\chi(x) \leqslant \sup_{\xi \in f^{-1}(y)} \{\chi(\xi)\} = (\exists_f(\chi))(y) \leqslant \mu(y) = (\mu \circ f)(x).$$

4.8.3. *A particular case.* Let L be the complete lattice consisting of two elements 0, 1. In this case, for any set X, the lattice $(0, 1)^X$, which we abbreviate as 2^X, is isomorphic to the lattice $\mathscr{P}(X)$ of all subsets of X, in which the order is defined by inclusion. This isomorphism is realized as follows:

$$\mathscr{P}(X) \to 2^X,$$

$$A \to \chi_A,$$

where χ_A is the characteristic function associated with the subset A of X:

$$\chi_A(x) = \begin{cases} 1 \text{ if } x \in A; \\ 0 \text{ if } x \notin A. \end{cases}$$

If $f: X \to Y$ is a map, we have

(4.8.3.1) $\exists_f \chi_A = \chi_{f(A)}$,

(4.8.3.2) $\forall_f \chi_A = \chi\{y \in Y | f^{-1}(y) \subset A\}$.

4.8.4. *The case of topoi.* For details, see: F. W. Lawvere, *Quantifiers and Sheaves (Actes du Congrès international des mathématiciens*, Nice, 1970); *Séminaire Benabou*, 1970–71; *Colloquium of Category Theory, Intuitionistic Logic and Algebraic Geometry*, Halifax, 1971, Springer, Lecture Notes, 274.

For every topos $\mathscr{C}$, there exists an object Ω, which "classifies" subobjects in the following sense. Denoting the final element by 1, there exists a morphism

$$t: 1 \Rightarrow \Omega$$

such that for every subobject

$$i: F' \hookrightarrow F$$

of F, there exists a unique morphism

$$\varphi: F \to \Omega$$

for which the following diagram is cartesian:

$$\begin{array}{ccc} F' & \to & 1 \\ i \downarrow & \varphi & \downarrow t \\ F & \to & \Omega \end{array}$$

i.e. such that $(F', i, F' \to 1)$ is the fiber product associated to the morphisms φ, t.

The morphism φ is called the characteristic function of the subobject $F' \hookrightarrow F$.

Example. If $\mathscr{C} = Ens$, then $\Omega = \{0, 1\}$ and one finds again the classical concept of characteristic function of a subset.

The object Ω is a Heyting algebra in the category $\mathscr{C}$ in the sense that for every object X in $\mathscr{C}$ the set $\mathrm{Hom}_{\mathscr{C}}(X, \Omega)$ has a natural structure of Heyting algebra (see Definitions 2.11, 2.12). Moreover, this Heyting algebra is complete in the sense that for every morphism $f: X \to Y$ in $\mathscr{C}$, the functor

$$T^f: \Omega^X \to \Omega^X$$

defined by

$$T^f(\mu) = \mu \circ f$$

has a left adjoint

$$\exists_f: \Omega^X \to \Omega^Y$$

and a right adjoint

$$\forall_f: \Omega^X \to \Omega^Y.$$

4.8.5. *Intuitive interpretation.* In order to get the intuitive interpretation of the functors $\exists_f$ and $\forall_f$, it is sufficient to consider the particular case

$$X \times Y \xrightarrow{\text{pr}_Y} Y$$

and the predicate $P(x, y)$ defined by a subset A of $X \times Y$:

$$P(x, y) \Leftrightarrow ((x, y) \in A).$$

In this case, the maps $\exists_{\text{pr}_Y}(\chi_A)$, $\forall_{\text{pr}_Y}(\chi_A)$ define two subsets of Y which are equivalent to the predicates

$$\exists \xi P(\xi, y), \quad \forall \xi P(\xi, y).$$

Indeed, we have:

$$\exists \xi P(\xi, y) \Leftrightarrow (y \in pr_Y(A)) \Leftrightarrow \exists \xi((\xi, y) \in A)$$

$$\forall \xi (P(\xi, y) \Leftrightarrow (y \in \{y \in Y \mid (\xi, y) \in A \text{ for every } \xi \in X) \Leftrightarrow$$

$$\Leftrightarrow \forall \xi ((\xi, y) \in A).$$

4.8.6. *The concept of model of a formal system.* Let $\mathcal{S}$ be a formal system, let V be the set of its free variables, let $\mathcal{R}$ be the set of its elementary relations, let $\mathcal{A}$ be the set of its axioms and consider a Boole algebra B.

A premodel of the formal system $\mathcal{S}$ in the Boole algebra B is a map $\mu: \mathcal{R} \to B$ satisfying the following conditions:

(a) $\mu(\neg R) = \neg \mu(R)$ for every $R \in \mathcal{R}$;

(b) $\mu(R_1 \vee R_2) = \mu(R_1) \vee \mu(R_2)$ for every $R_1, R_2 \in \mathcal{R}$;

(c) for every $R \in \mathcal{R}$ containing the free variable x, $\max\limits_{a \in V} \mu(R(x/a))$ exists and we have $\mu(\exists \xi R(x/\xi)) = \max\limits_{a \in V} \mu(R(x/a))$;

(d) for every $R \in \mathcal{R}$ containing the free variable x, $\min\limits_{a \in V} \mu(R(x/a))$ exists and we have $\mu(\forall \xi R(x/\xi)) = \min\limits_{a \in V} \mu(R(x/a))$.

DEFINITION 4.8.7. *We say that an elementary relation R of the system $\mathcal{S}$ is valid in the premodel (B, μ) if $\mu(R) = 1$. We say that a premodel (B, μ) for $\mathcal{S}$ is a model if every axiom of $\mathcal{S}$ is valid in this premodel.*

PROPOSITION 4.8.8. *Any theorem* R *of the formal system* $\mathscr{S}$ *is valid in every model* (B, μ) *of* $\mathscr{S}$.

The proof is straightforward.

Example 4.8.9. (a) Let $\mathscr{S}$ be the formal system of logic: $\Gamma_0 = E$ (cf. 4.4.6). In this case, we have $\mathscr{R} = \mathscr{F}\ell(E)$., (cf. 4.6.1, a)) Consequently, a model for $\mathscr{S}$ is a map $\mu: \mathscr{F}\ell(E) \to B$ satisfying conditions a, b, in fact a map $E \to B$.

(b) Let B_0 be the Boole algebra with a single element: $0 = 1$. The (unique) map $\mu: \mathscr{R} \Rightarrow B_0$ yields a model for $\mathscr{S}$.

DEFINITION 4.8.10. Let (B_1, μ_1), (B_2, μ_2) be two models of the same formal system $\mathscr{S}$. We say that these models are *elementarily equivalent* if the set of closed elementary relations of $\mathscr{S}$ which are valid in the model (B_1, μ_1) coincides with the set of elementary closed relations of $\mathscr{S}$ which are valid in the system (B_2, μ_2).

Example. Let $(\mathbb{N}, 0, s)$ be the standard model of arithmetics and let $(\mathbb{N}^*, 0, s^*)$ be the non-standard model of arithmetics in the sense of Robinson. These two models of arithmetics are elementarily equivalent.

DEFINITION 4.8.11. *The concept of universal model.* Let $\mathscr{B}c$ be the full subcategory of the category of Boole algebras, generated by the complete Boole algebras (cf. Definition 2.2.3) and let

$$\mathfrak{M}_f: \mathscr{B}c \to \text{Ens}$$

be the functor which associates with every Boole algebra B the set of the models (B, μ) of $\mathscr{S}$ in B.

If (B, μ) is a representation couple for the functor $\mathfrak{M}_f$, we say that the model (B, μ) is a universal model for the formal system $\mathscr{S}$. In other words if (B, μ) is a universal model for the formal system $\mathscr{S}$ then for every model (C, ν) of $\mathscr{S}$, there exists a unique morphism of Boole algebras $v: B \to C$, such that $\nu = v\mu$.

PROPOSITION 4.8.12. *For every formal system, there exists a universal model.*

For the proof, see N. Bourbaki, *Théorie des Ensembles,* IV.

Example 4.8.13: For the formal system of logic (see 4.8.9, b) a universal model can be obtained as follows: $B =$ the free Boole algebra generated by E and $\mu: \mathscr{F}\ell(E) \to B$ is the canonical map induced by the inclusion $E \hookrightarrow B$.

DEFINITION 4.8.14. *The concept of strict model.* Let V and J be two sets, $J \neq \emptyset$ and let A be a Boole algebra. Obviously, the set

$$A^{J^V} = \{f: J^V \to A\}$$

has a natural structure of Boole algebra. If A is complete, then A^{J^V} is also complete.

A strict premodel of the formal system $\mathscr{S}$ in the set J and in the complete Boole algebra A is a premodel (A^{J^V}, μ), satisfying the following condition:

If R is an elementary relation of the formal system $\mathscr{S}$ and $X = \{x_1, x_2, \ldots, x_n\}$ is the set of free variables occurring in R, there exists a map

$$\varphi: J^X \to A$$

such that $\mu(R) = \varphi \circ J^i$, where $i: X \hookrightarrow V$ is the canonical inclusion of the subset X in V.

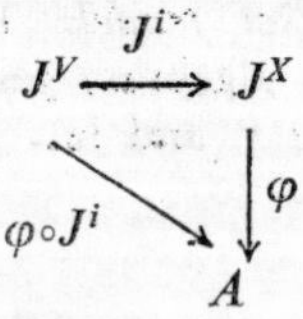

$$\begin{array}{ccc} J^V & \xrightarrow{J^i} & J^X \\ & \underset{\varphi \circ J^i}{\searrow} & \downarrow \varphi \\ & & A \end{array}$$

PROPOSITION 4.8.15. *(Interpretation of the elementary terms of a formal system). Let*

$$\mathscr{S} = (V, U, L, (\Phi_m)_{m \in \mathbb{N}}, (P_m)_{m \in \mathbb{N}})$$

be the symbols of a formal system and let J be an arbitrary set. Assume that for every operation $\varphi \in \Phi_m$, $m \geqslant 0$, a map

$$J_\varphi : \underbrace{J \times J \times \ldots \times J}_{m \text{ times}} \to J$$

is given.
(*For* $m = 0$, *this means a map*

$$\Phi_0 \to J(a \mapsto (\operatorname{Im} J^{\varnothing} = * \xrightarrow{J_a} J))),$$

Then there exists a unique map associating with every term τ *of the formal system* $\mathscr{S}$, *containing the free variables* $x_1, x_2, \ldots, x_n$, *a map*

$$J_\tau \colon J^{\{x_1, x_2, \ldots, x_n\}} \to J$$

such that the following conditions be fulfilled:

(a) *If* $\tau = \varphi(x_1, x_2, \ldots, x_n)$, $x_i \in V$, $\varphi \in \Phi_m$, *then* $J_\tau = J_\varphi$.

(b) *If* $\tau = \varphi(\tau_1, \tau_2, \ldots, \tau_r)$ $\varphi \in \Phi_r$ *and if* J_{τ_i}, $i = 1, 2, \ldots, r$, *is already defined*

$$J_{\tau_i} \colon J^{n_i} \to J,$$

then $J_\tau = J_\varphi \circ (J_{\tau_1} \times J_{\tau_2} \times \ldots \times J_{\tau_r})$, $(J_\tau \colon J^{n_1} \times J^{n_2} \times \ldots \times J^{n_r} \to J)$.

The proof of this proposition is obtained by induction of the length of the term τ and is left to the reader. We merely remark that for every free variable x, we have $J_x = \mathrm{id}_J$.

Proposition 4.8.16. *With the notation and hypotheses of Proposition* 4.8.15, *assume in addition that a Boole algebra* A *is given together with a map*

$$J_P \colon \underbrace{J \times J \times \ldots \times J}_{m \text{ times}} \to A.$$

defined for every predicate $P \in P_m$, $m \geqslant 0$.

Then there exists a unique map associating with every relation R *of the formal system* $\mathscr{S}$ *containing the free variables* $x_1, x_2, \ldots, x_n$ *a map*

$$J_R \colon J^{\{x_1, x_2, \ldots, x_n\}} \to A$$

such that the following conditions be fulfilled:

(a) *If* $R = P(x_1, x_2, \ldots, x_n)$, $x_i \in V$; $P \in P_m$, *then* $J_R = J_P$.

(b) *If* $\tau_1, \tau_2, \ldots, \tau_n$ *are terms of the formal system* $\mathscr{S}$, $R = P(\tau_1, \tau_2, \ldots, \tau_n)$, $P \in P_r$, *then*

$$J_R = J_P \times (J_{\tau_1} \times J_{\tau_2} \times \ldots \times J_{\tau_n}) \colon J^{n_1} \times J^{n_2} \ldots J^{n_r} \to A$$

(c) *If* $S = R_1 \vee R_2$ *and if* J_{R_1}, J_{R_2} *are defined, then* J_S *is defined as follows:*

Let X_1 *(respectively* X_2*) be the set of free variables occurring in* R_1 *(respectively* R_2*). Then the set of free variables occurring in* S *is* $X = X_1 \cup X_2$. *Then*

$$J_S: J^X \to A,$$

$$J_S(v) = J_{R_1}(v_1) \vee J_{R_2}(v_2), \text{ where } v_i = v| \, x_i, \; i = 1, 2.$$

(d) *If* $S = \neg R$ *and if* J_R *is defined, then* $J_S = \neg J_R$.

(e) *If* $S = \exists \xi R(x_1 | \, \xi, x_2, \ldots, x_n)$ *and if* J_R *is defined, then*

$$J_S(v(x_2), v(x_3), \ldots, v(x_n)) = \max_{j \in J} J_R(j, v(x_2), \ldots, v(x_n)) \text{ for every}$$
$$v: \{x_2, \ldots x_n\} \to \mathcal{J}.$$

(f) If $S = \forall \xi R(x_1 | \, \xi, x_2, \ldots, x_n)$ and J_R is defined, then

$$J_S(v(x_2), \ldots v(x_n)) \min_{j \in J} J_R(j, v(x_2), \ldots, v(x_n)) \text{ for every } v: \{x_2, \ldots, x_n\} \to J.$$

COROLLARY 4.8.17. *Under the assumptions of Proposition* 4.8.16, *there exists a strict premodel* (A^{J^V}, μ) *of the formal system* $\mathcal{S}$ *in the set* J *and the Boole algebra* A, *such that*

$$\mu(R) = J_R \circ J^i,$$

where $i: X \hookrightarrow V$ *is the canonical inclusion of the set of the free variables occurring in* R.

A premodel which is obtained in this way will be called *normal.*

DEFINITION 4.8.18. *Semantic model.* A strict model (A^{J^V}, μ) is called *semantic* if A is the Boole algebra consisting of two distinct elements $\{0, 1\}$, i.e. the Boole algebra of the subsets of a set consisting of a single element.

In other words, if a semantic model is given, with every relation $R(x_1, x_2, \ldots, x_n)$ we can associate a subset of $J^{\{x_1, x_2, \ldots, x_n\}}$ namely the set of functions

$$\{f: \{x_1, x_2, \ldots, x_n\} \to J | \, \varphi(f) = 1\}.$$

If we identify the set $J^{\{x_1, x_2, \ldots, x_n\}}$ with the cartesian product

$$\underbrace{J \times J \times \ldots \times J}_{n \text{ times}},$$

φ associates with $R(x_1, x_2, \ldots, x_n)$ the subset of the above cartesian product, consisting of the systems $(\xi_1, \xi_2, \ldots, \xi_n) \in J \times J \times \ldots \times J$ for which the relation R is valid in the interpretation of the given model.

In particular if $n = 0$, i.e. if the relation is *closed*, then through

$$\varphi : J = \{*\} \to \{0, 1\}$$

the model under consideration associates with the given relation the element 0 or 1. In the first case, we say that R is false in this model, in the second case we say that R is true (or valid) in this model.

Given a formal system $(\mathcal{S}, \mathcal{A})$, with

$$\mathcal{S} = (V, U, L, (\Phi_m)_{m \geqslant 0}, (P_m)_{m \geqslant 0})$$

it follows that a normal semantic premodel is given by the data system

$$\mathfrak{M} = (J, (J_\varphi)_{\varphi \in \underset{m \geqslant 0}{\cup} \Phi_m}, \quad (J_P)_{P \in \underset{m \geqslant 0}{\cup} P_m}),$$

where:

$$J_\varphi : \underbrace{J \times J \times \ldots \times J}_{m \text{ times}} \to J, \text{ if } \varphi \in \Phi_m;$$

in particular, if $\varphi \in \Phi_0$, then

$$J_\varphi : * \to J$$

is defined by an element of J.

$$J_P : \underbrace{J \times J \times \ldots \times J}_{m \text{ times}} \to \{0, 1\} \text{ if } P \in P_m;$$

in particular, if $P \in P_0$, then $J_P : * \to \{0,1\}$.

The concept of normal semantic premodel can be given in a more general form, considering objects in a topos $\mathscr{C}$ instead of objects in the category Ens.

In this line, let Ω be a classifying object for the subobjects of the category $\mathscr{C}$ and let

$$1 \xrightarrow{t} \Omega$$

be the "truth" function.

From the definition of the object Ω and of the morphism t, it follows that there exists a bijection

$$\mathscr{P}(1) \to \mathrm{Hom}_{\mathscr{C}}(1, \Omega)$$

$$(X' \overset{\xi}{\hookrightarrow} 1) \mapsto \chi_\xi : 1 \to \Omega$$

$$\left(\begin{array}{ccc} X' & \longrightarrow & 1 \\ \xi \downarrow & & \downarrow t \\ 1 & \xrightarrow{\chi_\xi} & \Omega \end{array}\right)$$

between the set $\mathscr{P}(1)$ of subobjects of the final object 1 of the category $\mathscr{C}$ and the set Hom $(1, \Omega)$.

Now let $\tau = (\mathscr{S}, \mathscr{A})$ be a formal system given by its symbols:

$$\mathscr{S} = \{V, U, L, (P_n)_{n\geqslant 0}, (\Phi_n)_{n\geqslant 0}\}$$

and by its axioms $\mathscr{A}$.

Let M be an object of a topos $\mathscr{C}$. Assume that a map $\mathscr{I}$ is given which assigns to every predicate $p_n \in P_n$ a subobject $\mathscr{I}(p_n) \subseteq M^n$, and to every operation $\varphi_n \in \Phi_n$ a morphism

$$\mathscr{I}(\varphi_n) : M^n \to M$$

In particular, for every $p_0 \in P_0$, $\mathscr{I}(p_0)$ is a subobject of the final object 1 of $\mathscr{C}$, and for every constant $\varphi_0 \in \Phi_0$, $\mathscr{I}(\varphi_0)$ is a morphism

$$1 \xrightarrow{\mathscr{I}(\varphi_0)} M$$

Since for every subobject A of an object X and for every morphism

$$f: X \to Y$$

we can define the subobjects $\exists_f(A)$, $\forall_f(A)$ of Y:

$$\exists_f(A) = f(A)$$

$$\forall_f(A) = \sup\{B \mid B \subset Y, f^{-1}(B) \subset A\},$$

With every relation R containing n free variables ($n \geqslant 0$), defined by the system of symbols $\mathscr{S}$, we can associate a subobject $\mathscr{I}(R)$ of M^n.
For instance if $p_n \in P_n$, $\varphi_{i_1} \in \Phi_{i_1}$, $\varphi_{i_2} \in \Phi_{i_2}$, $\ldots$, $\varphi_{i_n} \in \Phi_{i_n}$, then the subobject $\mathscr{I}(p_n(\varphi_{i_1}(x_{11}, \ldots, x_{1i_1}), \varphi_{i_2}(x_2, \ldots, x_{2i_2}), \ldots, \varphi_{i_n}(x_{n1}, \ldots, x_{ni_n}))$ is by definition the fiber product $(\mathscr{I}(\varphi_{i_2}) \times \mathscr{I}(\varphi_{i_n}) \times \ldots \times \mathscr{I}(\varphi_{i_n})) \times \times \mathscr{I}(p_n)$ associated to the diagram

$$\begin{array}{ccc} \mathscr{I}(\varphi_{i_1}) \times \mathscr{I}(\varphi_{i_2}) \times \ldots \times \mathscr{I}(\varphi_{i_n}) & \longrightarrow & i(p_n) \\ \downarrow & & \downarrow \\ M^{i_1 + \ldots i_n} = M^{i_1} \times \ldots \times M^{i_n} & \xrightarrow{i(\varphi i_1) i(\varphi i_2) \ldots i(\varphi i_n)} & M^n \end{array}$$

In particular, for every closed relation R of the system $(\mathscr{S}, \mathscr{A})$, $\mathscr{I}(R)$ is a subobject of 1.
The map $\mathscr{I}$ defines a model of the formal system $(\mathscr{S}, \mathscr{A})$ if for every axiom R which is a closed relation, $\mathscr{I}(R)$ coincides with the subobject $t: 1 \hookrightarrow \Omega$.

DEFINITION 4.8.19. *Non-contradictory formal system.* A formal system $(\mathscr{S}, \mathscr{A})$ is said to be *contradictory* if a relation R of this system exists, such that both R and $\neg R$ are theorems of the system $(\mathscr{S}, \mathscr{A})$. A formal system which is not contradictory is said to be *non-contradictory*.

Problem 4.8.20. (*Hilbert's second problem*). Is the formal system of arithmetics (4.7.2, d)) contradictory?
This second problem, from the twenty-three celebrated problems posed by Hilbert at the Mathematical Congress from 1900 in Paris is still unsolved, in spite of the impressive efforts made in this direction.

DEFINITION 4.8.21. *Consistent formal system.* A formal system is said to be *consistent* if it has at least one semantic model.

PROPOSITION 4.8.22. *If the formal system of set theory is non-contradictory, then every consistent formal system is non-contradictory.*

Proof. Let (A^{J^V}, μ) be a semantic model of the formal system $(\mathscr{S}, \mathscr{A})$ and assume by *reductio ad absurdum* that the formal system $(\mathscr{S}, \mathscr{A})$ is contradictory. Then a relation R of $(\mathscr{S}, \mathscr{A})$ would exist, such that both R and $\neg R$ be theorems of $(\mathscr{S}, \mathscr{A})$. In other words, $\mu(R) = \mu(\neg R) = \neg \mu(R)$, i.e. $\emptyset = A^{J^V} \neq \emptyset$, which is absurd. Proposition 4.8.22 has a remarkable converse:

PROPOSITION 4.8.23. *(Gödel's theorem). Every formal non-contradictory system is consistent.*

For the proof, see for instance Paul J. Cohen, *Set Theory and the Continuum Hypothesis*, Benjamin, 1966, Chap. I, § 4, Theorem 2.

DEFINITION 4.8.24. *The concept of mathematical structure.* Let Ens be the category of sets and consider the groupoid BijEns associated with the category of sets, i.e. the category whose objects coincide with the objects of Ens and whose morphisms are the bijections in Ens. A structure species is by definition a functor

$$F\colon \mathrm{BijEns} \to \mathrm{Ens}.$$

If J is a set and $\sigma \in F(J)$, we say that σ is a structure of the species F defined on the set J (see for details I. Bucur, A. Deleanu, *Introduction to the Theory of Categories and Functors*, Chap. I).

If $u\colon J_1 \to J_2$ is a bijection and $\sigma \in F(J_1)$ (or $\tau \in F(J_2)$) is a structure of the species F defined on the set J_1 (or on the set J_2) and if $\tau = F(u)(\sigma)$, then we say that the structure τ has been obtained from the structure σ through the structure transport defined by the bijection u.

PROPOSITION 4.8.25. *To every formal system τ one can associate a structure species*

$$F_\tau\colon \mathrm{BijEns} \to \mathrm{Ens}$$

defined by

$F_\tau(J)$ = the set of strict models (A^{J^V}, μ) of τ, where A is a fixed complete Boole algebra.

DEFINITION 4.8.26. A structure species

$$F\colon \text{BijEns} \to \text{Ens}$$

is said to be univalent if for every $\sigma \in F(J_1)$ and $\tau \in F(J_2)$, there exists a unique injection

$$u\colon J_1 \to J_2$$

such that $F(u)(\sigma) = \tau$.

A formal system is said to be *univalent* if the associated structure species (Definition 4.8.25) is univalent.

Example. The arithmetic structure defined by the Peano axioms defines a univalent structure species.

On the other hand, one has to be warned that the formal system of arithmetics (4.7.2, d) does not define a univalent structure species.

As already mentioned, "non-standard" models of arithmetics do exist. In fact, assuming that the formal system of arithmetics is non-contradictory, it follows from Gödel's theorem that it has models of any cardinal.

The explanation of this strangeness consists in the fact that the mathematical structure defined by the Peano axiom (for every subset of $\mathbb{N}$) cannot be considered a structure associated to a formal system. Roughly speaking, for any model of the formal system of arithmetics, the Peano axioms apply only to the subsets which define arithmetical predicates (relations). But it is easy to see that the set of arithmetical predicates is countable, while the set of subsets of $\mathbb{N}$ has the power of the continuum.

Exercise 4.8.27. Explain why the structure species of infinite cyclic Abelian group cannot be considered as a structure species associated with a formal system. Hint: one cannot formalize the statement: there exists x_0, such that for every x, there exists $n \in \mathbb{N}$, such that $x = nx_0$.

Exercise 4.8.28. Explain why the structure species of Archimedean ordered fields cannot be considered as a structure species associated to a formal system.

Comment 4.8.29. Hilbert's mathematical school intended to construct he whole edifice of mathematics starting from the concept of (possibly generalized) formal system. Gödel's result showing in particular that the mathematical univalent structures having infinite models cannot be formalized, i.e. cannot be considered as a structure species associated to a formal system, is discouraging in this direction.

4.8.29.1. Let

$$\mathfrak{M} = (J, (J_\varphi)_{\varphi \in \bigcup_{m \geq 0} \Phi_m}, (J_P)_{P \in \bigcup_{m \geq 0} P_m})$$

and

$$\mathfrak{N} = (L, (L_\varphi)_{\varphi \in \bigcup_{m \geq 0} \Phi_m}, (L_P)_{P \in \bigcup_{m \geq 0} P_m})$$

be two normal semantic premodels of the same formal system $(\mathcal{S}, \mathcal{A})$.

A new normal semantic premodel of the same formal system, the rartesian product $\mathfrak{M} \times \mathfrak{N}$ of the given premodels, can be defined as follows:

$$\mathfrak{M} \times \mathfrak{N} = (J \times L, (J_\varphi \times L_\varphi)_\varphi (J_P \times L_P)_P,$$

where

$$J_\varphi \times L_\varphi : (J \times L) \times (J \times L) \times \ldots \times (J \times L) \to J \times L$$

$$((j_1, l_1), (j_2, l_2), \ldots, (j_m, l_m) \mapsto (J_\varphi(j_1, \ldots, j_m), L_\varphi(l_1, \ldots, l_m))$$

$$J_P \times L_P = ((j_1, l_1), (j_2, l_2), \ldots, (j_m, l_m))/(j_1, \ldots, j_m) \in J_P,$$

$$(j_1, \ldots, j_m) \in L_P$$

Unfortunately, the product of two normal semantic systems is not necessarily a model. A classical counterexample is the product of two integrity domains which is never an integrity domain. In order to avoid such a difficulty, one introduces the concept of ultraproduct of premodels.

4.8.29.2. We recall that an ultrafilter on a set X is a maximal element of all partially ordered filters on X, with respect to the order relation of filters (cf. N. Bourbaki, *Topologie générale)*.

If $\mathcal{U} \in \mathcal{P}(\mathcal{P}(X))$ is an ultrafilter on the set X, then the following conditions are fulfilled:

(a) $\emptyset \notin \mathcal{U}$;

(b) if U, V are subsets of X such that $U \in \mathscr{U}$ and $V \in \mathscr{U}$, then $U \cap V \in \mathscr{U}$;

(c) if $U \in \mathscr{U}$ and $U' \supset U$, then $U' \in \mathscr{U}$;

(d) if W is a subset of X, then one (and only one) of the subsets W, $X \setminus W$ belongs to the ultrafilter $\mathscr{U}$.

Example 4.8.29.3. The set of all subsets containing an element x of X forms an ultrafilter.

4.8.29.4. Let $(M_i)_{i \in X}$ be a family of sets and let $\mathscr{F}$ be a filter on the set X. On the product set $\prod_{i \in X} M_i$ we define the equivalence relation $\mathscr{R}_{\mathscr{F}}$ as follows:

$f, g \in \prod_{i \in X} M_i \overset{\text{Def}}{\Leftrightarrow} f \sim g \bmod \mathscr{R}_{\mathscr{F}}$ if and only if the subset of X

$$\{i \in X \mid f(i) = g(i)\}$$

is an element of $\mathscr{F}$

We denote by $\prod_{\mathscr{F}} M_i$ the quotient set obtained from the product $\prod_{i \in I} M_i$ by the equivalence relation $\mathscr{R}_{\mathscr{F}}$.

In the case when $\mathscr{F}$ is an ultrafilter, the set $\prod_{\mathscr{F}} M_i$ is called the ultraproduct of the family of sets $(M_i)_{i \in X}$ with respect to the ultrafilter $\mathscr{F}$.

4.8.29.5. Consider the family $(\mathfrak{M}_i)_{i \in X}$ of normal semantic models of the formal system $(\mathscr{S}, \mathscr{A})$:

$$\mathfrak{M}_i = (J(i), J(i)_\varphi, J(i)_P).$$

If $\mathscr{U}$ is an ultrafilter on X, a new normal semantic model $\prod_{\mathscr{U}} \mathfrak{M}_i$, called the ultraproduct of the family of models $(\mathfrak{M}_i)_{i \in X}$ with respect to the ultrafilter $\mathscr{U}$, can be defined as follows:

$$\prod_{\mathscr{U}} \mathfrak{M}_i = (\prod_{\mathscr{U}} J(i), \prod_{\mathscr{U}} J(i), \prod_{\mathscr{U}} J(i)_P),$$

where the map

$$\prod_{\mathscr{U}} J(i)_\varphi : \underbrace{\prod_{\mathscr{U}} J(i) \times (\prod_{\mathscr{U}} J(i)) \times \ldots \times (\prod_{\mathscr{U}} J(i))}_{m \text{ times}} \to \prod_{\mathscr{U}} J(i)$$

and the subset

$$\prod_{\mathcal{U}} J(i)_P \subset \underbrace{(\prod_{\mathcal{U}} J(i)) \times (\prod_{\mathcal{U}} J(i)) \times \ldots \times (\prod_{\mathcal{U}} J(i))}_{m \text{ times}}, P \in P_m$$

are defined in the following way:

— Let $f_1, f_2, \ldots, f_m$ be elements of the product $\prod_{i \in X} J(i)$ and let $(f_1 \bmod \mathcal{R}_{\mathcal{U}}, f_2 \bmod \mathcal{R}_{\mathcal{U}}, \ldots, f_m \bmod \mathcal{R}_{\mathcal{U}})$ be the associated element in the product $(\prod_{\mathcal{U}} J(i)) \times (\prod_{\mathcal{U}} J(i)) \times \ldots \times (\prod_{\mathcal{U}} J(i))$. In order to define the element $(\prod_{\mathcal{U}} J(i))_\varphi((f_1 \bmod \mathcal{R}_{\mathcal{U}}, f_2 \bmod \mathcal{R}_{\mathcal{U}}, \ldots, f_m \bmod \mathcal{R}_{\mathcal{U}})$ we first consider the map

$$J_\varphi(f_1, f_2, \ldots, f_m): X \to \bigcup_{i \in X} J(i)$$

In other words, $J_\varphi(f_1, f_2, \ldots, f_m)$ is an element of the product $\prod_{i \in X} J(i)$. Now we put

$$\prod_{\mathcal{U}} (J(i)_\varphi)(f_1 \bmod \mathcal{R}_{\mathcal{U}}, f_2 \bmod \mathcal{R}_{\mathcal{U}}, \ldots, f_m \bmod \mathcal{R}_{\mathcal{U}}) =$$

$$= J_\varphi(f_1, \ldots, f_m) \bmod \mathcal{R}_{\mathcal{U}}$$

The subset $\prod_{\mathcal{U}} J(i)_P$ is defined by the formula:

$$(f_1 \bmod \mathcal{R}_{\mathcal{U}}, f_2 \bmod \mathcal{R}_{\mathcal{U}}, \ldots, f_m \bmod \mathcal{R}_{\mathcal{U}}) /$$

$$\{j \in X \mid (f_1(j), f_2(j), \ldots, f_m(j)) \in J(j)_P\} \in \mathcal{U}.$$

In the particular case when $\mathfrak{M}_i = \mathfrak{M}$ for every $i \in X$, the ultraproduct $\prod_{\mathcal{U}} \mathfrak{M}_i$ is called the ultrapower of the model $\mathfrak{M}$ with respect to the ultrafilter $\mathcal{U}$ and it is denoted by $\prod_{\mathcal{U}} \mathfrak{M}$.

Example 4.8.29.6. (*Non-standard Analysis*, cf. A. Robinson, *Non-standard Analysis*, Amsterdam, 1966, North Holland Publ. Co.) Let $\mathbb{R}$ be the field of real numbers considered as model for the formal system of really-closed ordered fields, and let $\mathcal{U}$ be an ultrafilter on the set $\mathbb{N}$ of natural numbers, containing the Cauchy filter.

The ultrapower $\prod_{\mathcal{U}} \mathbb{R}$ is a really-closed ordered field in which an adequate "infinitesimal" calculus can be carried out.

PROPOSITION 4.8.29.7. (Keisler) *Two normal semantic models* $\mathfrak{M}$, $\mathfrak{N}$ *of the same formal system are elementarily equivalent if and only if an ultrafilter* $\mathcal{U}$ *exists such that the ultrapowers* $\prod_{\mathcal{U}}\mathfrak{M}$, $\prod_{\mathcal{U}}\mathfrak{N}$ *be isomorphic.*

(The concept of isomorphism of normal semantic models is obviously the one resulting from the concept of structure isomorphism).

The proof of Proposition 4.8.29.7 will not be given here. Nevertheless, we show how this proposition can be applied in order to obtain a new proof of Hilbert's "Nullstellensatz".

4.8.29.8. *A new proof of Hilbert's Nullstellensatz using the concept of ultrapower.* Let $\underline{a} \subset k[X_1, \ldots, X_n]$ be a proper ideal of the ring $k[X_1, \ldots, X_n]$. We prove that in the algebraic closure $\bar{k}$ of the field k there exists at least a common zero of the polynomials in $\underline{a}$. Since $\underline{a}$ is finitely generated, the field $\bar{k}$ can obviously be assumed to be countable.

Let L be an algebraically closed field extending k and containing a common zero of the polynomials in $\underline{a}$. The field L can be supposed to be countable.

Now consider an ultrafilter $\mathcal{U}$ on the set $\mathbb{N}$ of natural numbers, containing the Cauchy filter. Using a direct argument or applying Proposition 4.8.29.7, we see that the two extensions $\prod_{\mathcal{U}} L$, $\prod_{\mathcal{U}} \bar{k}$ of the field k are algebraically closed.

On the other hand, we have

$$\text{Card}\,(\textstyle\prod_{\mathcal{U}} L) = \text{Card}\,(\textstyle\prod_{\mathcal{U}} \bar{k}) = 2^{\aleph_0} = c.$$

Applying Steinitz's theorem, we get an isomorphism

$$\textstyle\prod_{\mathcal{U}} L \simeq \prod_{\mathcal{U}} \bar{k}.$$

Again by Proposition 4.8.29.7 or by a direct argument, it follows that the field $\prod_{\mathcal{U}} L$, hence also the field $\prod_{\mathcal{U}} \bar{k}$, contains a zero of $\underline{a}$. From this, the existence in $\bar{k}$ of a zero of the ideal $\underline{a}$ follows directly or applying once more Proposition 4.8.29.7.

DEFINITION 4.8.30. *Complete formal systems.* Let $\tau = (\mathcal{S}, \mathcal{A})$ be a formal system given by its symbols

$$\mathcal{S} = \{V, U, L, (\Phi_m)_{m\in\mathbb{N}}\,(P_m)_{m\in\mathbb{N}}\}$$

and by its axioms $\mathscr{A}$. We say that the formal system τ is *complete* if for every closed relation $R \in \mathscr{R}$ of τ, one of the relations

$$R, \neg R$$

is a theorem of τ.

Example 4.8.31. (a) The formal system of group theory (4.4.5) is not complete. Indeed, the relation R

$$\forall \xi \, \forall \eta \, (\xi\eta = \eta\xi)$$

does not satisfy in this system the condition 4.8.30, since there exist commutative as well as non-commutative groups.

(b) Neither the formal system of commutative groups is complete. Indeed, the relation

$$\forall \xi \, (\xi + \xi = 0)$$

does not satisfy in this system condition 4.8.30 since there exist groups with elements of finite order, as well as free groups.

Example 4.8.32. *Gödel's incompleteness theorem.* The formal system of arithmetics (4.7.2.d)) is not complete. This surprising result represents one of the most remarkable contributions to the study of formal systems.

PROPOSITION 4.8.33. *If the formal system τ is complete, then any two semantic models of this system are elementarily equivalent.*

Proof. Let $M_1 = (\{0, 1\}^{J_1^V}, \mu_1)$, $M_2 = (\{0, 1\}^{J_2^V}, \mu_2)$ be two semantic models of the formal system τ and let R be a closed relation of τ which is valid in the model M_1. It is sufficient to show that this relation is also valid in the model M_2 But τ being complete, one of the relations R, $\neg R$ is a theorem in τ. Since R is valid in M_1, it follows that R is a theorem, consequently R is valid also in M_2.

PROPOSITION 4.8.34 (Vaught R. L.). *Let τ be an equalitary non-contradictory formal system and let α be the cardinal of the set of its axioms. Assume that the following conditions are fulfilled:*

1. *Every semantic model of τ has an infinite underlying set.*

2. *There exists a cardinal $\alpha' \geqslant \alpha$ such that all semantic models of τ whose underlying sets have the cardinal α' are isomorphic* (4.8.24).

Under these assumptions, the formal system τ is complete.

The proof of this proposition makes use of two preliminary results which we state as lemmas:

LEMMA 4.8.35. *If a formal system τ has a semantic model with an infinite underlying set, then it has models with underlying sets of any cardinal.*

This lemma is a consequence of Proposition 4.8.22 (cf. P. J. Cohen, *Set Theory and the Continuum Hypothesis*, Chap. I, § 4).

LEMMA 4.8.36. *(The reductio ad absurdum method). Let τ be a formal system, let R be a relation of τ and consider the formal system τ' with the same symbols as τ and having as axioms the axioms of τ and the relation $\neg R$. If τ' is contradictory, then R is a theorem of τ.*

(See for the proof N. Bourbaki, *Théorie des ensembles*, Chap. I, § 3, no. 3).

Proof of Proposition 4.8.34. Let R be a relation of the formal system τ. By *reductio ad absurdum*, assume that neither R, nor $\neg R$ are theorems of the formal system τ. Consider the formal systems

$$\tau' = (\mathscr{S}, \mathscr{A} \cup \{R\}), \tau'' = (\mathscr{S}, \mathscr{A} \cup \{\neg R\}).$$

In view of Lemma 4.8.36, it follows that the systems τ', τ'' are not-contradictory. On the other hand, in view of conditions 1—2, Proposition 4.8.22 (Gödel's theorem) and Lemma 4.8.35 show that these systems have the models M_1, M_2 whose underlying sets have the cardinal α'. But from condition 2 it follows that these two models are isomorphic, hence in each of them the relations R and $\neg R$ are both valid, which is absurd.

PROPOSITION 4.8.37. *The formal system of algebraic fields of a given characteristic is closed.*

Proof. We show that the conditions of Propositions 4.8.34 are fulfilled for the formal system of algebraic closed fields.

First, it is easy to see that in the formal system of field theory suitable predicates can be chosen, such that the set of axioms of the formal system of algebraically closed fields be countable. Consequently, we have $\alpha = \aleph_0$. On the other hand, every algebraically closed field is infinite, hence condition 1 from Proposition 4.8.34 is fulfilled. In order to show that condition 2 is fulfilled too, we prove the following.

PROPOSITION 4.8.38. *Any two algebraically closed fields of the same characteristic and of the power of the continuum are isomorphic.*

Proof. Let K_1, K_2 be two such fields of characteristic p. Obviously, they are extensions of the prime field $\mathbb{F}_p$ ($\mathbb{F}_0 = \mathbb{Q}$). Let $(x'_i)_{i \in I}$ be a transcendence basis of the extension $K_i \supset \mathbb{F}_p$ and let $(x''_i)_{i \in I''}$ be a transcendence basis of the extension $K_2 \supset \mathbb{F}_p$. Since K_1 and K_2 have the power of the continuum, the sets I', I'' obviously have the same property. Let

$$\varphi : I' \to I''$$

be a bijection. Then there exists an injection

$$\psi : K_1 \to K_2$$

such that $\psi(x'_i) = x''_{\varphi(i)}$. Since K_2 is algebraically closed, ψ is an isomorphism.

PROPOSITION 4.8.39. (Lefschetz's principle). *If R is a closed relation in the formal system of field theory which is valid for the field $\mathbb{C}$ of complex numbers, then R is a theorem of the formal system of algebraically closed fields of zero characteristic, consequently R is valid for every algebraically closed field of zero characteristic.*

4.8.40 *m-complete formal systems.* Let $(\mathcal{S}, \mathcal{A})$ be an equalitary formal system and let $(\{0, 1\}^{J^V}, \mu)$ be a semantic model of this system. According to the general definition 4.8.12, for every relation $R \in \mathcal{R}$ of the formal system $(\mathcal{S}\ \mathcal{A})$, $\mu(R)$ is an element of the Boole algebra $(\{0, 1\}^{J^V}, \mu)$. In particular, if R contains the set of free variables $X = \{x_1, x_2, \ldots, x_n\}$, if $i : X \hookrightarrow V$ is the canonical inclusion and if $\rho : V \to J$ is an arbitrary map, then $\mu(R)$ (ρ) actually depends only on the maps

$$\rho^i : X \to J$$

$$x_i \to a_i, \quad i = 1, 2, \ldots, n.$$

The element $\mu(R)(\rho)$ of the lattice $\{0, 1\}$ will be denoted by $R(a_1, a_2, \ldots$ $\ldots, a_n)$. If $R(a_1, a_2, \ldots, a_n) = 1$, we say that R is valid in the system $a_1, a_2, \ldots, a_n$ with respect to the model $(\{0, 1\}^{J^V}, \mu)$. If $X = \emptyset$, then either $\mu(R) = 1$ or $\mu(R) = 0$, and in the first case we simply say that the closed relation R is valid in the given model.

With every semantic model $\mathfrak{M} = (\{0, 1\}^{J^V}, \mu)$ of an equalitary formal system $(\mathscr{S}, \mathscr{A})$ a new formal system $(\mathscr{S}_{\mathfrak{M}}, \mathscr{A}_{\mathfrak{M}})$ can be associated as follows.

Let

$$\mathscr{S} = (V, U, L, (\Phi_m)_{m \geqslant 0}, (P_m)_{m \geqslant 0})$$

be the symbols of the formal system $(\mathscr{S}, \mathscr{A})$. We define the symbols of the formal system $(\mathscr{S}_{\mathfrak{M}}, \mathscr{A}_{\mathfrak{M}})$ by putting

$$\mathscr{S}_{\mathfrak{M}} = (V, U, L, (\Phi'_m)_{m \geqslant 0}), (P_m)_{m \geqslant 0}),$$

where $\Phi'_0 = \Phi_0 + J$ (the disjoint union of the two sets).

The axioms $\mathscr{A}_{\mathfrak{M}}$ are defined as follows:

First of all, every relation of the formal system $(\mathscr{S}, \mathscr{A})$ is actually a relation of $(\mathscr{S}_{\mathfrak{M}}, \mathscr{A}_{\mathfrak{M}})$. Moreover, if $R(x_1, x_2, \ldots, x_m)$ is a relation of the formal system $(\mathscr{S}, \mathscr{A})$, and if $a_i \in J$ $i = 1, 2, \ldots, m,$ then $R(a_1, a_2, \ldots, a_m)$ is a relation of the formal system $(\mathscr{S}_{\mathfrak{M}}, \mathscr{A}_{\mathfrak{M}})$, since every element of Φ'_0 is an elementary term (cf. 4.5, 4.6, b)).

The family $\mathscr{A}_{\mathfrak{M}}$ will contain the family $\mathscr{A}$ as well as the family of relations $\mathscr{D}_{\mathfrak{M}}$ defined as follows:

For every relation $R(x_1, x_2, \ldots, x_m)$ of the formal system $(\mathscr{S}, \mathscr{A})$, $R(a_1, a_2, \ldots, a_m) \in \mathscr{D}_{\mathfrak{M}}$ if and only if R is valid in the system $a_1, a_2, \ldots$ $\ldots a_m$ with respect to the model $\mathfrak{M}$.

PROPOSITION 4.8.41. *If $\mathfrak{M} = (\{0, 1\}^{J^V}, \mu)$ is a semantic model of the equalitary formal system $(\mathscr{S}, \mathscr{A})$ and if $\mathfrak{M}' = (\{0, 1\}^{M^V}, \nu)$ is a semantic model of the formal system $(\mathscr{S}_{\mathfrak{M}}, \mathscr{A}_{\mathfrak{M}})$, then obviously M is the underlying set of a model for $(\mathscr{S}, \mathscr{A})$ and there exists an injection*

$$f: J \hookrightarrow M$$

which realizes a homomorphism of models for the formal system $(\mathscr{S}, \mathscr{A})$.

Proof. First of all, for every $a \in J$, the relation

$$\exists\, \xi (\xi = a)$$

is obviously a theorem in the formal system $(\mathscr{S}_{\mathfrak{M}}, \mathscr{A}_{\mathfrak{M}})$.

For every $a \in J$, consider the relation

$$R_a(x) = (x = a),$$

It follows that there exists a unique element $\alpha = f(a) \in M$, such that the relation $R_a(x)$ is valid in the system (α) with respect to the model $\mathfrak{M}'$. We get in this way a map

$$f: J \to M.$$

Now consider a relation $R(x_1, x_2, \ldots, x_n)$ of the system $(\mathscr{S}, \mathscr{A})$ and the elements $a_1, a_2, \ldots, a_n$ in J such that $R(x_1, x_2, \ldots, x_n)$ is valid in the system $a_1, a_2, \ldots, a_n$ with respect to the model $\mathfrak{M}$. It follows that $R(a_1, a_2, \ldots, a_n) \in \mathscr{A}_{\mathfrak{M}}$, in particular $R(a_1, a_2, \ldots, a_n)$ is a theorem in the formal system $(\mathscr{S}_{\mathfrak{M}}, \mathscr{A}_{\mathfrak{M}})$.

On the other hand, in an equality system we have (4.7.2, example c)

$$\forall\xi_1\forall\xi_2 \ldots \forall\xi_n(((\xi_1 = a_1) \wedge (\xi_2 = a_2) \wedge \ldots$$
$$\ldots \wedge (\xi_n = a_n) \Rightarrow (R(a_1, a_2, \ldots, a_n) \Leftrightarrow R(\xi_1, \ldots, \xi_n))).$$

From this, it follows immediately that R is valid in the system $\alpha_1, \alpha_2, \ldots, \sigma_n$ with respect to the model $\mathfrak{M}'$.

Indeed, the following relations are simultaneously valid:

$$(\alpha_1 = a_1) \wedge (\alpha_2 = a_2) \wedge \ldots \wedge (\alpha_n = a_n)$$
$$(\alpha_1 = a_1) \wedge (\alpha_2 = a_2) \wedge \ldots \wedge (\alpha_n = a_n) \Rightarrow$$
$$\Rightarrow (R(a_1, a_2, \ldots, a_n) \Leftrightarrow R(\alpha_1, \alpha_2, \ldots, \alpha_n)).$$

This implies the validity of the relation

$$R(a_1, a_2, \ldots, a_n) \Leftrightarrow R(\alpha_1, \alpha_2, \ldots, \alpha_n);$$

but $R(a_1, a_2, \ldots, a_n)$ is a theorem in the formal system $(\mathscr{S}_{\mathfrak{M}}, \mathscr{A}_{\mathfrak{M}})$, whence the validity of the relation $R(\alpha_1, \alpha_2, \ldots, \alpha_n)$ follows.

In this way, f is a homomorphism of models for the formal system $(\mathscr{S}, \mathscr{A})$. In order to prove that f is an injection it is sufficient to remark that from what has been already proved, the implication

$$\forall\, a, b((a, b \in J) \text{ and } a \neq b \Rightarrow f(a) \neq f(b))$$

follows.

DEFINITION 4.8.42. We say that the formal system $(\mathscr{S}, \mathscr{A})$ is *m-complete* if for every semantic model $\mathfrak{M}$ of this system, the formal system $(\mathscr{S}_{\mathfrak{M}}, \mathscr{A}_{\mathfrak{M}})$ is complete (Definition 4.8.30) (In what follows, only semantic models will be considered).

PROPOSITION 4.8.43. *In order that the formal system $(\mathscr{S}, \mathscr{A})$ be m-complete, the following condition is necessary and sufficient:*

If $\mathfrak{M}'$, $\mathfrak{M}$ are models of the formal system $(\mathscr{S}, \mathscr{A})$ and if $\mathfrak{M}'$ is an extension of $\mathfrak{M}$, then $\mathfrak{M}$ and $\mathfrak{M}'$ are elementarily equivalent (4.8.10).

Proof. Let $(\mathscr{S}, \mathscr{A})$ be an *m*-complete formal system, let $\mathfrak{M}, \mathfrak{M}'$ be two models of this system, such that $\mathfrak{M}'$ is an extension of $\mathfrak{M}$ and let R be a closed relation of $(\mathscr{S}, \mathscr{A})$ which is valid in $\mathfrak{M}$. We have to show that R is also valid in $\mathfrak{M}'$. But either R, or $\neg R$ is a theorem in the formal system $(\mathscr{S}_{\mathfrak{M}'}, \mathscr{A}_{\mathfrak{M}'})$. By *reductio ad absurdum*, assume that $\neg R$ is a theorem in the formal system $(\mathscr{S}_{\mathfrak{M}'}, \mathscr{A}_{\mathfrak{M}'})$. Then it would obviously follow that $\neg R$ is valid in the model $\mathfrak{M}'$, which is absurd. Hence R is a theorem in the formal system $(\mathscr{S}_{\mathfrak{M}}, \mathscr{A}_{\mathfrak{M}})$.

PROPOSITION 4.8.44. *The formal system of algebraically closed fields of a given characteristic is m-complete.*

The proof of this proposition can be found in A. Robinson, *Introduction to Model Theory and to the Metamathematics of Algebra*, North-Holland Publ. Co., Amsterdam 1963, Chap. IV, theorem 4.3.3.

4.8.48. *Elimination of quantifiers.* Let $(\mathscr{S}, \mathscr{A})$ be a formal system and consider a relation $R \in \mathscr{R}$ of this system. We say that the formal system $(\mathscr{S}, \mathscr{A})$ allows the elimination of quantifiers in the relation R if there exists a quantifiers-free relation R' of $(\mathscr{S}, \mathscr{A})$ such that $R \Leftrightarrow R'$ be a theorem of the formal system.

Examples (a) Let $(\mathscr{S}, \mathscr{A})$ be the formal system of partially ordered sets with a (unique!) last element 1, and let R be the relation defined by

$$R = \forall \xi (\xi \leqslant x).$$

This formal system allows the elimination of quantifiers in the relation R. Indeed, we have

$$\forall \xi (\xi \leqslant x) \Leftrightarrow (x = 1).$$

(b) Let $(\mathscr{S}, \mathscr{A})$ be the formal system of ordered fields and let R be the relation defined by

$$R = \exists\, \xi(\xi^2 = x).$$

This formal system allows the elimination of the quantifiers in the relation R. Indeed, we have

$$\exists\, \xi(\xi^2 = x) \Leftrightarrow (x \geqslant 0).$$

DEFINITION 4.8.49. We say that the formal system τ has the property of elimination of quantifiers if it allows the elimination of quantifiers in every relation of τ. In the following, it will be shown that the formal system of really closed fields has the property of elimination of quantifiers.

§ 5. *Primitive recursive functions*

5.1. *The recursive device of defining a function.* We use the notation from 1.1, i.e. $\mathbb{N}$ is the set of natural numbers, $0 \in N$ is the null element and $s: \mathbb{N} \to \mathbb{N}$ is the successor function.

PROPOSITION 5.1.1. *Let*

$$\chi: \mathbb{N}^{r+1} \to \mathbb{N}$$

$$\psi: \mathbb{N}^{r-1} \to \mathbb{N},\ r \geqslant 1$$

be two arbitrary functions. There exists a unique function

$$\varphi: \mathbb{N}^r \to \mathbb{N}$$

such that the following conditions be fulfilled:

$$i)\ \varphi(0, x_2, x_3, \ldots, x_r) = \psi(x_2, x_3, \ldots, x_r);$$

$$ii)\ \varphi(y + 1, x_2, x_3, \ldots, x_r) = \chi(y, \varphi(y, x_2, \ldots$$

$$\ldots, x_r), x_2, x_3, \ldots, x_r).$$

(In the following, the function φ will be sometimes denoted by $\chi \top \psi$).

Proof. Existence of the function φ. Consider the set E of all couples (P, f), where P is a subset of $\mathbb{N}$ and $f: P \times \mathbb{N}^{r-1} \to \mathbb{N}$ is a function, such that

(α) $0 \in P$;

(β) ($x \in P$ and $y \leqslant x$) $\Rightarrow y \in P$;

(γ) $f(0, x_2, \ldots, x_r) = \psi(x_2, x_3, \ldots, x_r)$;

(δ) if $y \in P$ and if $y + 1 \in P$, then

$$f(y+1, x_2, \ldots, x_r) = \chi(y, \varphi(y, x_2, \ldots, x_r), x_2, \ldots, x_r).$$

The set E is obviously non-empty and it has a natural structure of inductively ordered set. If (P_0, f_0) is a maximal element of this set, then obviously $P_0 = \mathbb{N}$ and f_0 satisfies the conditions stated for φ. The uniqueness of φ is straightforward.

5.2. *Elementary functions.* Consider the set

$$\mathscr{F} = \bigcup_{r \geqslant 0} (\mathbb{N}^{\mathbb{N}^r})$$

whose elements are functions of the form

$$f: \mathbb{N}^r \to \mathbb{N}.$$

We intend to point out some remarkable subsets of $\mathscr{F}$. To this end, we begin by considering the set $\mathscr{F}_e$ which will be called the set of elementary functions and which contains the following elements:

— the successor function $s: \mathbb{N} \to \mathbb{N}$;

— for every $r \geqslant 0$ and every $i \in \{1, 2, \ldots, r\}$, the function

$$\mathrm{pr}_i^r: \mathbb{N}^r \to \mathbb{N}$$

$$(x_1, x_2, \ldots, x_r) \to x_i;$$

in particular, $\mathrm{id}_{\mathbb{N}} = \mathrm{pr}_1^1 \in \mathscr{F}_e$;

— the constant functions $\mathbb{N}^r \to \mathbb{N}$.

PROPOSITION 5.2.1. *There exists a least subset of $\mathscr{F}$ which will be denoted by $\mathscr{F}_{pr}$, satisfying the following conditions:*

(I) $\mathscr{F}_e \subset \mathscr{F}_{pr}$;

(II) (a) *If* $\varphi_i: \mathbb{N}_i^r \to \mathbb{N}$, $i = 1, 2, \ldots, m$; $\varphi: \mathbb{N}^m \to \mathbb{N}$ *and if* $\varphi_1, \varphi_2, \ldots, \varphi_m, \varphi \in \mathscr{F}_{pr}$, *then the function*

$$\varphi \circ \left(\prod_{i=1}^{m} \varphi_i \right) : \prod_{i=1}^{m} \mathbb{N}^{r_i} \to \mathbb{N}$$

belongs to $\mathscr{F}_{pr}$;

(b) *If* $\varphi_1, \varphi_2, \ldots, \varphi_m: \mathbb{N}^r \to \mathbb{N}, \varphi: \mathbb{N}^m \to \mathbb{N}$, *and if* $\varphi_1, \varphi_2, \ldots, \varphi_m, \varphi \in \mathscr{F}_{pr}$, *then the function*

$$\mathbb{N}^r \to \mathbb{N}$$

$$(x_1, x_2, \ldots, x_r) \mapsto \varphi(\varphi_1(x_1, \ldots, x_r), \ldots, \varphi_m(x_1, \ldots, x_r))$$

belongs to F_{pr};

(III) *If* $\chi: \mathbb{N}^{r+1} \to \mathbb{N}$, $\psi: \mathbb{N}^{r-1} \to \mathbb{N}$ *and if* $\chi, \psi \in \mathscr{F}_{pr}$, *then the function* $\chi \top \psi \in \mathscr{F}_{pr}$ (cf. Proposition 5.1.1) (The elements of the set $\mathscr{F}_{pr}$ will be called *primitive recursive functions*.

Proof. Let

(5.2.1.1) $\quad f_1, f_2, \ldots, f_n$

be a sequence of elements in $\mathscr{F}$. We say that this sequence is a sequence of *primitive recursive construction* if for every $i \in \{1, 2, \ldots, n\}$, at least one of the following conditions is fulfilled:

(*i*) $f_i \in \mathscr{F}_e$, in other words, f_i is an elementary function;

(*ii*) there exists $i_0, i_1, \ldots, i_m < i$ such that either

$$f_{i_j}: \mathbb{N}^{r_j} \to \mathbb{N}, \; j = 1, 2, \ldots, m$$

$$f_{i_0}: \mathbb{N}^m \to \mathbb{N}$$

$$f_i = f_{i_0} \circ \left(\prod_{j=1}^{m} f_{i_j} \right)$$

or $r_1 = r_2 = \ldots = r_m = r$ and f_i is the function

$$\mathbb{N}^r \to \mathbb{N}$$
$$(x_1, \ldots, x_r) \mapsto f_{i_0}(f_{i_1}(x_1, \ldots, x_r), \ldots, f_{i_m}(x_1, \ldots, x_r));$$

(*iii*) there exist, $j, k < i$ such that $f_j \colon \mathbb{N}^{r+1} \to \mathbb{N}$, $f_k \colon \mathbb{N}^{r-1} \to \mathbb{N}$ and $f_i = f_j \top f_k$.

Let $\mathscr{G}$ be the subset of $\mathscr{F}$ consisting of all functions which can be inserted in a sequence of primitive recursive construction. Obviously, $\mathscr{G}$ satisfies conditions I, II, III. Moreover, if $\mathscr{F}'$ is a subset of $\mathscr{F}$ satisfying conditions I, II, III, then $\mathscr{F}' \supset \mathscr{G}$, consequently $\mathscr{G}$ is the least subset of $\mathscr{F}$ satisfying these conditions.

COROLLARY 5.2.2. *The necessary and sufficient condition for a function to be primitive recursive is that it may be inserted in a sequence of primitive recursive construction.*

COROLLARY 5.2.3. *The set* $\mathscr{F}_{\text{pr}}$ *of primitive recursive functions is countable.*

Proof. The set $\mathscr{S}_{\text{pr}}$ of all sequences of primitive recursive construction is obviously countable, since it can be embedded in the free monoid generated by a countable set, namely $\mathscr{F}_e \cup \{\top, \prod, \ldots\}$; on the other hand, we have a map $\mathscr{F}_{\text{pr}} \to \mathscr{S}_{\text{pr}}$ whose fibres are finite, assigning to every primitive recursive function a sequence of primitive recursive construction to which it belongs.

COROLLARY 5.2.4. *There exist functions* $f \colon \mathbb{N}^r \to \mathbb{N}$ *which are not primitive recursive.*

5.2.5. *Examples of primitive recursive functions:*

(a) The sum function

$$\sigma \colon \mathbb{N} \times \mathbb{N} \to \mathbb{N}$$

is primitive recursive.

Indeed, let $\chi \colon \mathbb{N}^3 \to \mathbb{N}$ be defined by

$$\chi(x_1, x_2, x_3) = s(x_2)$$

and let $\psi \colon \mathbb{N} \to \mathbb{N}$ be defined by

$$\psi = \mathrm{id}_{\mathbb{N}}.$$

We have $\chi = s \circ \mathrm{pr}_2^3$ and $\sigma = \chi \top \psi$. In other words, we get the following sequence of primitive recursive construction for σ:

$$s, \mathrm{pr}_2^3, s \circ \mathrm{pr}_2^3, \sigma.$$

(b) The product function:

$$\pi: \mathbb{N} \times \mathbb{N} \to \mathbb{N}$$

is primitive recursive. A sequence of primitive recursive construction for π can be obtained as follows:

$$\mathrm{pr}_2, \mathrm{pr}_3, \sigma, 0: \mathbb{N} \to \mathbb{N}.$$

Indeed, the function π is characterized by the relations:

$$\pi(0, x) = 0, \ \forall x \in \mathbb{N},$$

$$\pi(y + 1, x) = x(y, x) + x.$$

Consequently we can apply the recursive device of defining a function starting with the functions

$$\psi: \mathbb{N} \to \mathbb{N}$$

$$\chi: \mathbb{N} \times \mathbb{N} \times \mathbb{N} \to \mathbb{N}$$

given by

$$\psi = 0$$

$$\chi(x_1, x_2, x_3) = \sigma(x_2, x_3)$$

$$\chi(y + 1, \pi(y, x), x) = \sigma(\pi(y, x), x) = \pi(y, x) + x =$$

$$= \pi(y + 1, x).$$

5.3. *The category of primitive recursive functions*

DEFINITION 5.3.1. A function

$$u: \mathbb{N}^r \to \mathbb{N}^s$$

is said to be primitive recursive if for every $i \in \{1, 2, \ldots, s\}$ the function

$$\mathrm{pr}_i^s \circ u : \mathbb{N}^r \to \mathbb{N}$$

is primitive recursive.

PROPOSITION 5.3.2. *The composition of two primitive recursive functions is a primitive recursive function.*

Proof. Let $u : \mathbb{N}^r \to \mathbb{N}^s$, $v : \mathbb{N}^s \to \mathbb{N}^t$ be primitive recursive functions. We have to prove that for every $i \in \{1, 2, \ldots, t\}$ the function $\mathrm{pr}_i^1 \circ v \circ u$: $\mathbb{N}^r \to \mathbb{N}$ is primitive recursive. To this end, it is sufficient to use property *ii.*

PROPOSITION 5.3.3. *Let* $\mathcal{R}$ Ens *be the subcategory of* Ens *whose objects are the sets* $\mathbb{N}^r$, $r \geqslant 0$, *and whose morphisms* $\mathrm{Hom}_{\mathcal{R}\,\mathrm{Ens}}(\mathbb{N}^r, \mathbb{N}^s)$ *are primitive recursive functions defined on* $\mathbb{N}^r$ *with values in* $\mathbb{N}^s$. This category has finite fibre products.

The proof is a straightforward consequence of Proposition 5.3.2 and 5.2.1.

§ 6. *Recursive functions*

PROPOSITION 6.1. *(Cantor's diagonal method)*
Let

$$(6.1.1) \qquad f_1, f_2, \ldots, f_n, \ldots$$

be a sequence of functions

$$f_i : \mathbb{N} \to \mathbb{N}$$

defined on the set of natural numbers and taking values in the set of natural numbers. The function

$$\varphi : \mathbb{N} \to \mathbb{N}$$

defined by the formula

$$(6.1.2) \qquad \varphi(n) = f_n(n) + 1$$

does not belong to the sequence (6.1.2), *in other words*

$$\forall n, \varphi \neq f_n.$$

Proof. By *reductio ad absurdum*, assume the existence of an $n_0 \in \mathbb{N}$ such that

$$\varphi = f_{n_0}.$$

This implies

$$\varphi(n_0) = f_{n_0}(n_0) + 1 = f_{n_0}(n)$$

which is absurd.

COROLLARY 6.1.3. *Let*

$$f_1, f_2, \ldots, f_r, \ldots$$

be the sequence of primitive recursive functions defined on $\mathbb{N}$ *and taking values in* $\mathbb{N}$ (Corollary 5.2.3). *The function* $\varphi: \mathbb{N} \to \mathbb{N}$ *defined by*

$$\varphi(n) = f_n(n) + 1$$

is not primitive recursive.

6.2. Let the symbols

$$\mathscr{S} = (V, U, L, (\Phi_m)_{m \in \mathbb{N}} (P_m)_{m \in \mathbb{N}})$$

of a formal system be defined as follows:

$U = \emptyset$, $\Phi_0 = 0$, $\Phi_1 \ni'$, Φ_i *are given* for $i \geqslant 1$;

$P_2 = \{=\}$, $P_i = \emptyset$ for $i \neq 2$. (We assume that the sets $\Phi_1, \Phi_2, \ldots$ contain a sufficiently high number of elements).

According to the general definition (cf. 4.5), the set $\mathscr{T}(\mathscr{S})$ of the elementary terms of the formal system $\mathscr{S}$ is characterized by the conditions:

(*i*) $0 \in \mathscr{T}(\mathscr{S})$;
(*ii*) $x \in V \Rightarrow x \in \mathscr{T}(\mathscr{S})$;
(*iii*) if $T_1, T_2, \ldots, T_m \in \mathscr{T}(\mathscr{S})$ and $\varphi \in \Phi_m$, $m \geqslant 1$, then $\varphi(T_1, T_2, \ldots, T_m) \in \mathscr{T}(\mathscr{S})$.

In the following, we denote as usually by n the term defined inductively

$$1 = 0'$$

$$n + 1 = n'.$$

The terms obtained in this way will be called *numerical terms.*

Let $\mathscr{E}$ be the set of elementary relations associated to $\mathscr{S}$ (cf. 4.6) which contain no element of L. If $R \in \mathscr{E}$, then obviously R has the form

$$T = S$$

where T and S are terms associated with the system $\mathscr{S}$.

DEFINITION 6.3. Let $E = \{E_1, E_2, \ldots, E_r\} \subset \mathscr{E}$ be a finite subset of the set $\mathscr{E}$. We say that a sequence

$$R_1, R_2, \ldots, R_n$$

of elements in $\mathscr{E}$ is deduced from E, if for every $i \in \{1, 2, \ldots, n\}$ at least one of the following conditions is fulfilled:

(1) $R_i \in E$.

(2) There exists a $j < i$ such that R_i is obtained from R_j by substituting numerical terms for one or several free variables occurring in R_j.

(3) There exist $j, k < i$ such that $R_k = (h(n_1, n_2, \ldots, n_p) = n)$, R_j does not contain free variables and R_i is obtained from R_j by substituting n for one or several occurrences of $h(n_1, n_2, \ldots, n_p)$ in R_j.

We say that a relation R is deduced from the set E if it can be inserted in a sequence deduced from E.

Example 6.3.1. Let E be the following subset of $\mathscr{E}$:

$$E = \{f(x') = h(x, f(x)), f(0) = 4, \ h(y, z) = 7\},$$

where, obviously $f \in \Phi_1$, $h \in \Phi_2$, $x, y, z \in V$.

The sequence of relations

$$(f(x') = h(x, f(x)), \ f(1) = h(0, f(0)), f(0) = 4$$

$$f(1) = h(0, 4), h(0, 4) = 7, \ f(1) = 7)$$

is a sequence deduced from E.

Definition 6.4. We say that

$$\varphi: \mathbb{N}^r \to \mathbb{N}$$

is a *recursive function* if there exists a finite subset E of $\mathscr{E}$ and an operation $f \in \Phi_r$, such that the following conditions be fulfilled:

(1) if $(n_1, n_2, \ldots, n_r) \in \mathbb{N}^r$ is an arbitrary element of $\mathbb{N}^r$ and if $n = \varphi(n_1, n_2, \ldots, n_r)$, then the relation $f(n_1, n_2, \ldots, n_r) = n$ can be deduced from E;

(2) if $m \in \mathbb{N}$ is such that the relation $f(n_1, n_2, \ldots, n_r) = m$ can be deduced from E, then $n = m$.

Examples 6.4.1. (a) The successor function $s: \mathbb{N} \to \mathbb{N}$ is recursive. Indeed, it is sufficient to consider the subset $E \subset \mathscr{E}$ defined as follows:

$$E = \{f(x) = x'\},\ f \in \Phi_1,\ x \in V.$$

(b) For every $r \geqslant 1$ and every $1 \leqslant i \leqslant r$, the function

$$\mathrm{pr}_i^r: \mathbb{N}^r \to \mathbb{N}$$

$$(x_1, x_2, \ldots, x_r) \to x_i$$

is recursive.

Indeed, it is sufficient to consider the subset $E \subset \mathscr{E}$ defined by

$$E = \{g(x_1, x_2, \ldots, x_r) = x_i\},\ \text{where } x_i \in V \text{ and } g \in \Phi_r.$$

(c) Every constant function: $\mathbb{N}^r \to \mathbb{N}$ is recursive.

Indeed, it is sufficient to consider the subset $E \subset \mathscr{E}$ defined by

$$E = \{h(x_1, x_2, \ldots, x_r) = n_i\},\ \text{where } x_i \in V \text{ and } h \in \Phi_r.$$

The results stated in Examples 6.4.1, a, b, c can be summed up as

Proposition 6.5. *Every elementary function* (5.2) *is recursive.*

Proposition 6.6. *Every primitive recursive function is recursive.* The proof is left to the reader.

6.7. *The "least number" operator*. Consider a subset P of $\mathbb{N}^s$, ($s \geqslant 1$) and an integer $i \in \{1, 2, \ldots, s\}$ such that the following condition be fulfilled:

For every element $(n_1, n_2, \ldots, n_{n-1}, n_{i+1}, \ldots, n_s) \in \mathbb{N}^{s-1}$ there exists an element $n \in \mathbb{N}$, such that $(n_1, n_2, \ldots, n_{i-1}, n, n_{i+1}, \ldots, n_s) \in \mathbb{N}^s$.
We define a function

$$(\mu\xi_i)\, P : \mathbb{N}^{s-1} \to \mathbb{N}$$

as follows:

$$[(\mu\xi_i)P](n_1, n_2, \ldots, n_{s-1}) = \inf \{\xi_i \in \mathbb{N}/(n_1, n_2, \ldots$$

$$\ldots, n_{i-1}, \xi_i, n_i, \ldots, n_{s-1}) \in P\}.$$

(We leave to the reader to specify the sense of this definition in the case $s = 1$).

Example 6.7.1. Let $Q \subset \mathbb{N}^2$ be defined by

$$Q = \{(n_1, n_2) |\, n_1 - n_2 > 8\}.$$

We have

$$[(\mu\xi_1)Q](n) = n + 9.$$

In the following, we use the notation from § 5.

PROPOSITION 6.8. *There exists a least subset of $\mathscr{F}$ which will be denoted by $\mathscr{F}_r$, such that the following conditions be fulfilled:*

(I) $\mathscr{F}_{\mathrm{pr}} \subset \mathscr{F}_r$.

(II) (A statement similar to condition II of Proposition 5.2.1)

(III) If $P \subset \mathbb{N}^s$ *is a primitive recursive subset* (i.e. *a subset whose characteristic function is primitive recursive) and if* $i \in \{1, 2 \ldots, s\}$ *satisfies conditions* 6.7, *then the function* $(\mu\xi_i)P$ *belongs to* $\mathscr{F}_r$.

The proof is similar to that of Proposition 5.2.1.

PROPOSITION 6.8.1. *The necessary and sufficient condition for a function to be recursive is that it be an element of* $\mathscr{F}_r$.

For the proof, see S. Kleene, *Introduction to Metamathematics*, 1952, D. Van Nostrand Co., Inc., New York, Toronto, § 58.

DEFINITION 6.8.2. A sequence of functions in $\mathscr{F}$:

$$f_1, f_2, \ldots, f_n$$

is said to be a recursive construction (or recursive description) sequence if for every $i \in \{1, 2, \ldots, n\}$ one at least of the following conditions is fulfilled (cf. 5.2.1):

(*i*) $f_i \in \mathscr{F}_{\text{pr}}$ (f_i is primitive recursive);

(*ii*) there exist $i_0, i_1, \ldots, i_m < i$ such that either

$$f_{i_j}\colon \mathbb{N}^{r_j} \to \mathbb{N},\ j = 1, 2, \ldots, m, f_{i_0}\colon \mathbb{N}^m \to \mathbb{N},$$

$$f_i = f_{i_0} \circ \left(\prod_{j=1}^{m} f_{i_j} \right)$$

or $r_1 = r_2 = \ldots = r_m = r$ and f_i is the function

$$\mathbb{N}^r \to \mathbb{N}$$

$$(x_1, x_2, \ldots, x_r) \mapsto f_{i_0}(f_{i_1}(x_1, x_2, \ldots, x_r), \ldots$$

$$\ldots, f_{i_m}(x_1, \ldots, x_m));$$

(*iii*) there exists $j < i$ such that, if

$$f_j\colon \mathbb{N}^{m+1} \to \mathbb{N},$$

then the following two conditions hold:

(a) there exists $i \in \{1, 2, \ldots, m+1\}$ such that for every $(\xi_1, \xi_2, \ldots, \xi_m) \in \mathbb{N}^m$, there exists $\xi \in \mathbb{N}$ for which

$$f_j(\xi_1, \xi_2, \ldots, \xi_{i-1}, \xi, \xi_{i+1}, \ldots, \xi_m) = 0;$$

(b) $f_j(n_1, n_2, \ldots, n_m) = (\mu\xi_i)(f_i(n_1, \ldots, n_{i-1}, \xi_i, n_{i+1}, \ldots, n_m) = 0)$.

PROPOSITION 6.8.3. *A function $f \in \mathscr{F}$ is recursive if and only if it can be inserted in a recursive description sequence.*

6.9. Using these elements, the category of recursive functions can be defined (cf.5.3).

6.10. *The concept of algorithm.* A concept closely related to that of recursive function is the concept of algorithm, introduced and developed by the Soviet mathematician A. A. Markov *(Theory of Algorithms*, Proceedings of the Steklov Institute, 1954, XLII).

Let E, F be two finite sets, which in the framework of this theory will be called alphabets, and let $i_E: E \to \mathbb{N}$, $i_F: F \to \mathbb{N}$ be two injections. We denote by $\mathcal{M}(E)$ (or $\mathcal{M}(F)$) the free monoid generated by E (or F). Let

$$\gamma_E: \mathcal{M}(E) \to \mathbb{N}$$
$$\gamma_F: \mathcal{M}(F) \to \mathbb{N}$$

be the canonical injections associated with the injections i_E, i_F. We shall be interested in the maps

$$\alpha: \mathcal{M}(E) \to \quad (F)$$

such that the map

$$\beta: \mathbb{N} \to \mathbb{N}$$

which makes commutative the diagram

$$\begin{array}{ccc} \mathcal{M}(E) & \xrightarrow{\alpha} & \mathcal{M}(F) \\ \downarrow{\scriptstyle \gamma_E} & & \downarrow{\scriptstyle \gamma_F} \\ \mathbb{N} & \xrightarrow{\beta} & \mathbb{N} \end{array} \tag{6.10.1}$$

be recursive. Such maps are realized by means of algorithms.

For the sake of simplicity, assume that the set E is identical to the set F. In this case, roughly speaking, an algorithm is a successive transformation "process" (rule) which can be applied to every word of the free monoid $\mathcal{M}(E)$, the result of every transformation of a given word being a new word of this monoid.

After several transformations of a word, a new word may be obtained which is stable with respect to this process, in the sense that it is left unchanged by a further application of the process. In this case, the algorithm is said to be applicable to such a word. It follows that an algorithm generates a map from the subset of the words of $\mathcal{M}(E)$ to which the algorithm is applicable, into the set $\mathcal{M}(E)$. In fact, the interest in the exact mathematical description which will be sketched below for the concept of algorithm, more precisely in the "normal algorithm" in the sense of Markov, consists in the fact that the resulting functions from $\mathbb{N}$ into $\mathbb{N}$ are recursive and conversely.

In a more precise formulation, an algorithm is given by a finite number of substitution formulae of the type

(*) $A \to B$

(*) $A \to \cdot B$

where A, B are elements of the monoid $\mathcal{M}(E)$. These substitution formulae define an algorithm in the above sense as follows.
Let $C \in \mathcal{M}(E)$ be a word of $\mathcal{M}(E)$ and let

$$A \to B$$

be the first substitution formula. We look for all subwords of C which coincide with B. If such subwords exist, we substitute A for the first such subword and we proceed in the same way with the resulting word. If the word C contains no subword identical to B, we go over to the second substitution formula (the substitution formulae are supposed to be written in a precise order) and we proceed in the same way.

The word C will be obviously considered as stable in the case when no substitution formula is applicable to it. But a word is also considered as stable if a substitution formula of the type

$$A \to \cdot B.$$

is applicable to it.

By the equivalence between the theory of recursive functions and the theory of algorithms we mean that every function so obtained from $\mathbb{N}$ into $\mathbb{N}$ is recursive and conversely.

A remarkable result of the theory of algorithms is the fact that every function

$$\beta : \mathbb{N} \to \mathbb{N}$$

which makes commutative the diagram (6.10.1), where α is an algorithm in the above sense, is necessarily recursive.

§ 7. *Turing machines*

In this section we intend to explain the importance of recursive functions from the point of view of contemporary computer science.

Assume that a recursive function

$$f: \mathbb{N}^s \to \mathbb{N}$$

is given. One can show that there exists a computer, more precisely a Turing machine — a concept which will be seen to yield a satisfactory approximation of modern electronic computers — allowing the calculation of $f(n_1, n_2, \ldots, n_s)$ for a given $(n_1, n_2, \ldots, n_s)$. We start with some general concepts, definitions and considerations.

DEFINITION 7.1. *(The concept of machine).* We call a machine a couple (C, Φ) consisting of a set C and a map

$$\Phi: C \to C,$$

of the set C into itself. The elements of C are called the configurations of the machine (C, Φ), and the map Φ is called the program of the machine.

A configuration $x \in C$ is said to be stable if it is a fixed point for the map Φ:

$$\Phi(x) = x.$$

Example 7.1.1. The couple $(\mathbb{N}, s)$ is a machine without stable configurations. If the time is assumed to be divided in a sequence of successive moments:

$$t_0, t_1, t_2, \ldots, t_n, \ldots$$

then the configuration $\Phi(x)$ can be interpreted as the configuration of the machine at time t_{i+1} if x was the configuration of the machine at time t_i. Usually, x is defined (described) by means of some technical coordinates.

Example 7.1.2. Let X be the set defined as follows:

$$X = \{n \in \mathbb{N} \mid n > 0\} \cup \{(m, n) \mid m, n \in \mathbb{N}, m > 0, n > 0\} \cup$$
$$\cup \{(m, n, r, 1) \mid r \in \mathbb{N}, m \in \mathbb{N}, n \in \mathbb{N}, m > 0, n > 0\} \cup$$
$$\cup \{(m, n, r, 2 \mid r \in \mathbb{N}, n \in \mathbb{N}, m > 0, n > 0\} \cup$$
$$\cup \{(m, n, p, 3) \mid m, n, p \in \mathbb{N}, m > 0, n > 0, p > 0\}.$$

The map $\Phi: X \to X$ is defined by

$$\Phi(n) = n,$$

$$\Phi(m, n) = (m, n, 0, 1),$$

$$\Phi(m, n, r, 1) = (m, n, \text{ remainder of the division of } m \text{ by } n, 2)$$

$$\Phi(m, n, r, 2) = \begin{cases} n \text{ if } r = 0 \\ (m, n, r, 3) \text{ otherwise} \end{cases}$$

$$\Phi(m, n, p, 3) = (n, p, p, 1)$$

For every pair (m, n) of positive integers, a positive integer k obviously exists, such that $\Phi^k(m, n) = d$, where d is the greatest common divisor of the integer m, n.

In other words, the program Φ allows the computation of the greatest common divisor of two given integers.

Example 7.1.3. (The algorithm allowing the effective computation of the maximum of a map defined on a finite set of natural numbers and having natural numbers for values). Consider $C = \{(a_0, a_1, \ldots \ldots, a_n) | n \in \mathbb{N}, a_n \in \mathbb{N}\}$.

We define the map

$$\psi: C \to C$$

by putting

$$\psi(a_0, a_1, \ldots, a_n) = \begin{cases} (a_n) \text{ if } n = 0 \\ (\max(a_{n-1}, a_n)) \text{ if } n = 1 \\ (a_0, a_1, \ldots, a_{n-2}, \max(a_{n-1}, a_n)) \text{ if } n \geqslant 2. \end{cases}$$

It is immediately seen that

$$\psi^n(a_0, a_1, \ldots, a_n) = \max\{a_0, a_1, \ldots, a_n\}.$$

DEFINITION 7.2. *(Turing machines)*. A Turing machine $\mathcal{T}$ is a system

$$\mathcal{T} = (A, E, s_0, \alpha),$$

where:

— A is a finite set containing the element s_0, whose elements will be called symbols of the Turing machine $\mathcal{T}$.

— E is a finite set, whose elements are called interior states of the machine $\mathcal{T}$.

— α is a map

$$\alpha : A \times E \to A \times E \times \mathbb{Z}$$

called the functional scheme of the machine.

We denote by $\alpha_1, \alpha_2, \alpha_3$ the maps

$$\alpha_1 = \mathrm{pr}_A \circ \alpha : A \times E \to A,$$

$$\alpha_2 = \mathrm{pr}_E \circ \alpha : A \times E \to E,$$

$$\alpha_3 = \mathrm{pr}_{\mathbb{Z}} \circ \alpha : A \times E \to \mathbb{Z}.$$

We show that with every Turing machine one can associate a machine in the sense of Definition 7.1. Given the Turing machine $\mathcal{T} = (A, E, s_0, \alpha)$, we construct a machine (C, Φ) as follows:

— $C = F \times E \times \mathbb{Z}$, where F is the set of all maps

$$\xi : \mathbb{Z} \to A$$

taking the value s_0 for almost all elements of $\mathbb{Z}$. (Note in particular that the set C is countable).

— The map $\Phi : C \to C$ is defined by

$$\Phi((\xi, e, n)) = (\xi', e', n'),$$

where

$$\xi'(i) = \begin{cases} \alpha_1(\xi(n), e), \text{ if } i = n, \\ \xi(i) \text{ if } i \neq n, \end{cases}$$

$$e' = \alpha_2(\xi(n), e),$$

$$n' = n + \alpha_3(\xi(n), e).$$

(From the definition of ξ', it follows that ξ' coincides with ξ for every integer i distinct from n, consequently $\xi' \in F$, i.e. ξ' takes the value s_0 for almost every element of $\mathbb{Z}$).

A technical model for the Turing machine $\mathscr{T} = (A, E, s_0, \alpha)$ can be obtained by means of the following elements:

— A linear tape, infinite in both directions and divided into successive identical and numbered cells (Fig. 7.2.1) (by the condition of the tape to be infinite

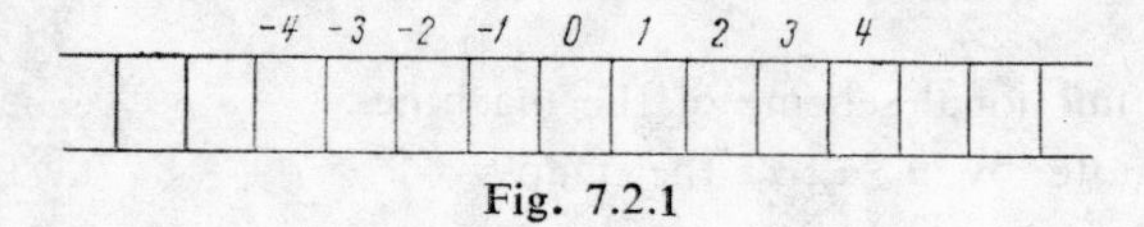

Fig. 7.2.1

we mean that it is sufficiently large in order for all subsequent operations to be performable). Every cell is capable of having printed upon it the material symbols $s_0, s_1, \ldots, s_n$ corresponding to the elements of the set A. The empty symbol is usually assumed to be represented by the absence of any material symbol. The cells of the tape are also called *memory cells* of the Turing machine $\mathscr{T}$. (From a technological point

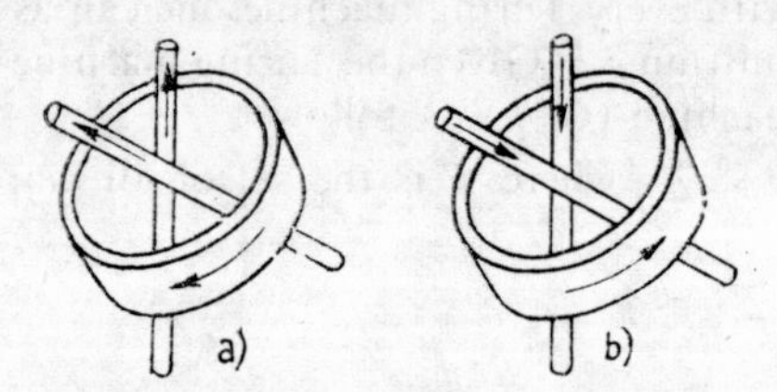

Fig. 7.2.2

of view, an elementary memory cell is realized by means of a magnetic torus crossed by some number of wires through which an electric current flows. According to the sense of the current, this torus receives a magnetization sense).

— A reading and recording device, movable with respect to the tape and which can be placed in front of any cell of the tape S.

— A control device K, able to find itself in a number of internal states $e_0, e_1, \ldots, e_m$ which are in a one-to-one correspondence with the elements of the set E.

A configuration of the machine (C, Φ) associated to the Turing machine $\mathscr{T} = (A, E, s_0, \alpha)$ corresponding to the triplet (ξ, e, n) is realized as follows:

1. In the cell of index i the material symbol s_i corresponding to the element $\xi(i)$ of the set A is printed. (The map ξ, described by the corresponding symbols printed on the tape of memory cells represents the data which the machine has to process and, in the usual practice, is read on perforated cards).

2. The reading and recording device is placed in front of the cell of rank n (fig. 7.2.3).

3. The internal state e of the control device K is realised.

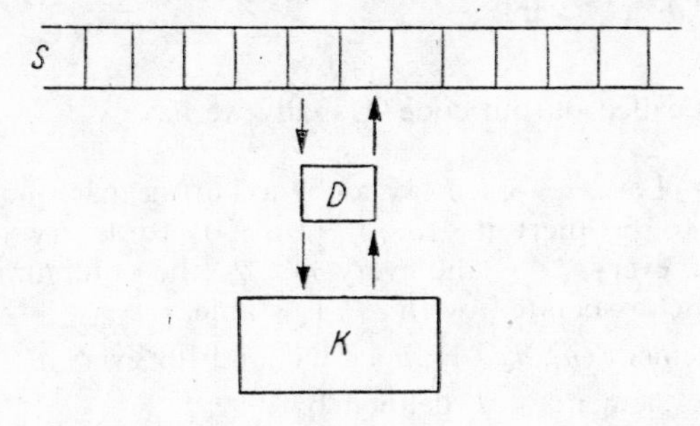

Fig. 7.2.3

The transition from the configuration of the machine corresponding to the triplet (ξ, e, n) to the state corresponding to the triplet $\Phi(\xi, e, n)$ is realized as follows:

The reading and recording device D sends to the control device K the information that the cell of rank n of the tape S contains the symbol corresponding to the element $\xi(n)$. The control device K which finds itself in the internal state e, sends to the device D the command to cancel the symbol $\xi(n)$ from the cell of rank n, to replace it by the symbol $\xi'(n) = \alpha_1(\xi(n), e)$, and to shift in front of the cell of rank $n' = n + \alpha_3(\xi(n), e)$. At the same time, the control device K goes over to the internal state $e' = \alpha_2(\xi(n), e)$. We can schematically represent the functioning of a Turing machine as shown in Fig. 7.2.3.

DEFINITION 7.3. *Machines with code.* Let $M = (C, \Phi)$ be any machine. A code on M is a triplet $(\theta_0, C_1, \theta_1)$, where:

— θ_0 is a map

$$\theta_0 : \mathbb{N}^r \to C.$$

Usually we have $r = 1$. The elements of the set $\theta_0(\mathbb{N}^r)$ are called initial configurations and the map θ_0 is called input code.

— C_1 is a subset of the set of stable configurations of C:

$$x \in C_1 \Rightarrow \Phi(x) = x, \ (C_1 \subset \{x \in C | \Phi(x) = x\}).$$

The elements of the set C_1 are called final configurations.

— θ_1 is a map of the set C^1 into $\mathbb{N}^s$

$$\theta_1 \colon C_1 \to \mathbb{N}^s$$

The map θ_1 is called output code. Usually we have $s = 1$.

Example 7.4. Let $\mathcal{T} = (A, E, s_0, \alpha)$ be a Turing machine. An element $e \in E$ is said to be inert if $\alpha(a, e) = (a, e, 0)$ for every $a \in A$. If e is inert, then for every $\xi \in F$ and every $n \in \mathbb{Z}$, the configuration (ξ, e, n) of the machine associated with $\mathcal{T}$ is stable.

For every element $(n_1, n_2, \ldots, n_k) \in \mathbb{N}^k$ and for every $a \in A$, we denote by $\xi^a_{n_1, \ldots, n_k}$ the element of F defined by

$$\xi^a_{n_1, n_2, \ldots, n_k}(i) = \begin{cases} s_0 \text{ if } i \leqslant 0 \text{ or if } i = n_1 + n_2 + \ldots \\ \qquad \ldots + n_e + l \quad l = 1, 2, \ldots, k \\ a \text{ in all other cases.} \end{cases}$$

Let (C, Φ) be the machine associated to a Turing machine $\mathcal{T} = (A, E, s_0, \alpha)$, let $a \in A$ be distinct from s_0, $e_0 \in E$ and let e_1 be an inert element of E. We define a code $(\theta_0, C_1, \theta_1)$ on the machine (C, Φ) as follows:

— the map $\theta_0 \colon \mathbb{N}^r \to C$ is defined by the formula:

$$\theta_0(n_1, n_2, \ldots, n_r) = (\xi^a_{n_1, n_2, \ldots, n_r}, e_0, 1);$$

— The set C_1 is defined by the formula

$$C_1 = \{(\xi^a_{n_1, n_2, \ldots, n_s}, e_1, n_1 + n_2 + \ldots + n_s + s) | (n_1, n_2, \ldots \\ \ldots, n_s) \in \mathbb{N}^s\};$$

— the map $\theta_1: C_1 \to \mathbb{N}^s$ is defined by the formula

$$\theta_1(\xi^a_{n_1,n_2,\ldots,n_s}, e_1, n_1 + n_2 + \ldots + n_s + s) = (n_1, n_2, \ldots, n_s).$$

Such a code is called a canonical code on the machine associated with the Turing machine $\mathscr{T}$. Usually we have $a = 1$.

DEFINITION 7.5. *Machine normal with respect to a code.* Let (C, Φ) be a machine and let $(\theta_0, C_1, \theta_1)$ be a code on the machine (C, Φ). We say that the machine (C, Φ) is normal with respect to the code $(\theta_0, C_1, \theta_1)$ if for every element $(n_1, n_2, \ldots, n_r) \in \mathbb{N}^r$ there exists a positive integer l such that $\Phi^l(\theta_0(n_1, n_2, \ldots, n_r)) \in C_1$.

With every normal machine one can associate a function

$$\varphi: \mathbb{N}^r \to \mathbb{N}^s$$

defined by

$$\varphi(n_1, n_2, \ldots, n_r) = (\theta_1 \circ \Phi^l \circ \theta_0)(n_1, n_2, \ldots, n_s),$$

where ℓ is a positive integer such that $\Phi^l(\theta_0(n_1, n_2, \ldots, n_s)) \in C_1$.

PROPOSITION 7.6. *For every recursive function*

$$\psi: \mathbb{N}^r \to \mathbb{N}^s$$

there exist a Turing machine $\mathscr{T}$ and a canonical code on the associated machine (C, Φ), such that the machine (C, Φ) is normal with respect to this code, and the associated function coincides with ψ.

Conversely, every function $\varphi: \mathbb{N}^r \to \mathbb{N}^s$, associated to a normal Turing machine (cf. Definition 7.5) is recursive.

The proof of this proposition will not be given here (see, for instance, S. C. Kleene, *Introduction to Metamathematics*, D. Van Nostrand Co, 1952, Chap. XIII, § 67). We shall merely explain the structure of Turing machines capable of computing some particular recursive functions.

Example 7.7. The Turing machine computing the successor function:

$$s: \mathbb{N} \to \mathbb{N}.$$

can be defined by taking

$$A = \{s_0, |\}$$

$$E = \{e_0, e_1\}$$

$$\alpha_1(s_0, e_0) = |,\ \alpha_1(|, e_0) = |,\ \alpha_1(s_0, e_1) = s_0,\ \alpha_1(|, e_1) = |$$

$$\alpha_2(s_0, e_0) = e_1,\ \alpha_2(|, e_2) = e_0,\ \alpha_2(s_0, e_1) = e_1,\ \alpha_2(|, e_1) = e_1$$

$$\alpha_3(s_0, e_0) = 1,\ \alpha_3(|, e_0) = |,\ \alpha_3(s_0, e_1) = 0,\ \alpha_3(|, e_1) = 0.$$

7.8. *Church's thesis.* If a recursive function

$$f: \mathbb{N}^r \to \mathbb{N}$$

is given, then obviously for every element $(n_1, n_2, \ldots, n_r) \in \mathbb{N}^r$, the value $f(n_1, n_2, \ldots, n_r)$ of the function f at this element is effectively calculable in the intuitive sense of the word (see for instance Definition 6.4 or Proposition 6.8). This conclusion is confirmed by the fact that every recursive function can be defined by means of an algorithm, e.g. in the sense of A. A. Markov (cf. 6.10), as well as by means of a Turing machine (cf. Proposition 7.6).

One can ask about the existence of some other reasonable classes of functions defined on a set of the form $\mathbb{N}^r$ and taking values in a set of the same type, satisfying the intuitive requirements of calculability.

Church's thesis states that every class of functions satisfying the intuitive requirements of calculability is a subclass of the class of recursive functions. Obviously, the statement of this thesis is a metamathematical one and therefore the question of its proof cannot be posed.

7.9. *Example of function which is not calculable in the intuitive sense.* (Following Hartley Rogers, Jr., *Theory of Recursive Functions and Effective Computability*, McGraw-Hill Co.).

Let $f: \mathbb{N} \to \mathbb{N}$ be the function defined by

$$f(n) = \begin{cases} 1 & \text{if in the decimal writing of } \pi \text{ there exist} \\ & n \text{ consecutive decimals, all equal to } 5 \\ 0 & \text{otherwise} \end{cases}$$

Remark 7.10. Concerning the computability of a function, we remark that the situation may occur in which a function is recursive but we have no means to effectively compute the values of this function. This can be due for instance to the fact that the function in question is chosen in a non-algorithmic (non-constructive) way from a set of recursive functions.

We shall place ourselves from the usual point of view of the fundamental mathematical concepts, where in general no constructivity requirements are imposed.

Example 7.10.1. The function f from Example 7.9 is primitive recursive. Indeed, either f is the constant function equal to 1, or there exist a fixed integer k, such that:

$$f(x) = \begin{cases} 1 \text{ for } x \leqslant k, \\ 0 \text{ for } x > k. \end{cases}$$

Unfortunately, it is a difficult question to decide which one among these two possibilities actually occurs. A simpler example of the same type is the following:

Let $g : \mathbb{N} \to \mathbb{N}$ be the function defined by

$$g(n) = \begin{cases} 1 \text{ if Goldbach's conjecture is true} \\ 0 \text{ if Goldbach's conjecture is false.} \end{cases}$$

Obviously, h is constant, hence recursive.

§ 8. *The concept of recursive (algorithmic) decidability*

8.1. Many mathematical problems can be reduced to the following general scheme:

Let X be a set and let A be a subset of X. An element $\xi \in X$ being given, decide whether ξ does belong to A.

Examples. (a) Let $\mathbb{R}$ be the set of real numbers and let $\mathbb{Q} \subset \mathbb{R}$ be the subset of rational numbers. A real number ξ being given (for instance by a convergent sequence of rational numbers) the problem of deciding whether ξ is rational is an extremely interesting mathematical problem. The following problem is still open:

Let

$$C = \lim_{n \to \infty} \left(1 + \frac{1}{2} + \ldots + \frac{1}{n} - \ln(n)\right),$$

be Euler's constant. Is C rational or not?

The following problem is open too: is the number $e + \pi$ rational or not? (Cf. S. Lang, *Algebra*, p. 117, Addison Wesley Publ. Co, 1971).

Nevertheless we mention the existence of rather strong criteria, allowing to decide whether a given real number is rational. We recall in particular the results obtained by Al. Froda, *Critères paramétriques d'irrationalité*, Mathematica Scandinavica, 12 (1963), pp. 199—208; *Eroare și paradox în matematică*, Editura Enciclopedică, București, 1971, Chap. 3, § D.

b) Let as above $\mathbb{R}$ be the set of real numbers and let $A \subset \mathbb{R}$ be the subset of all algebraic numbers (over $\mathbb{Q}$), i.e. the subset of real numbers ξ, for which there exists a polynomial in one variable with coefficients in $\mathbb{Q}$ such that ξ is a root of this polynomial. An element of $\mathbb{R}$ which does not belong to A is said to be transcendent. The problem of deciding whether a real number is algebraic is of major importance in mathematics. We recall some results in this line:

The real number $e = \lim_{n \to \infty} \left(1 + \frac{1}{n}\right)^n = \sum_{n=0}^{\infty} \frac{1}{n!}$ is transcendent (Ch. Hermite).

The number π, the quotient between the length of a circle and its radius, is transcendent (Lindemann).

If α is an algebraic number distinct from 0 and 1, and if β is an algebraic irrational number, then α^β is transcendent (A. O. Gelfond).

We mention that the last result, belonging to the Soviet mathematician Gelfond, has answered a problem posed by Hilbert (the VIIth problem) which had long waited for solution and which required among other things to decide whether $2^{\sqrt{2}}$ is algebraic.

Before exposing other examples, we remark in particular in connection with the last example that in order to decide whether some classes of real numbers are algebraic, specifical methods adapted to the case under consideration have been necessary. Therefore, it is scarcely probable that a uniform effective method could be indicated, which would be applicable to every real number and would efficiently allow to decide whether a real number is algebraic.

In the following, we try to elucidate the meaning of such terms as effective and uniform, which have been used in the above statement.

(c) Let M be the set of polynomials in one variable with integer coefficients of degree equal precisely to 1:

$$M = \{c_0 + c_1 X \mid c_0,\ c_1 \in \mathbb{Z},\ c_1 \neq 0\}$$

Also, let m be an integer $\neq 0, 1$ and let M_m be the subset of M consisting of all polynomials having at least a solution mod m:

$$c_0 + c_1 X \in M_m \leftrightarrow \exists \xi (\xi \in \mathbb{Z} \text{ and } c_0 + c_1 \xi \equiv 0 \bmod m).$$

In this case, a uniform method exists which is applicable to every such polynomial and which allows to decide in an "effective" way whether a polynomial belongs to M_m. This fact is a consequence of the following proposition:

PROPOSITION 8.1.1. *We have the equivalence:*

$$c_0 + c_1 X \in M_m \Leftrightarrow d = (c_1, m) / c_0.$$

The proof, which is very simple, is left to the reader.

In order to close the discussion of this section, we remark that given a set X, a subset A of X and an element $\xi \in X$ the problem of deciding whether ξ belongs to A is equivalent to the one of deciding whether the characteristic function

$$\chi_A : X \to \{0, 1\} \subset \mathbb{N}$$

takes at the point ξ the value 1 or 0.

DEFINITION 8.2. A subset A of $\mathbb{N}^r$ ($r \geqslant 1$) is said to be primitive recursive (recursive) if its characteristic function $\chi_A : \mathbb{N}^r \to \{0, 1\} \subset \mathbb{N}$ is primitive recursive (recursive) (Proposition 5.2.1). In this case, the mathematical problem defined by the couple $(\mathbb{N}^r, A)$ is said to be primitive recursively (recursively) decidable or algorithmically decidable.

Let A be a subset of a set X and assume that a bijection

$$\nu : X \to \mathbb{N}^r,\ r \geqslant 1$$

is given; A is said to be a primitive recursive (recursive) subset of X with respect to the bijection v if the subset $v(A)$ of $\mathbb{N}^r$ is primitive recursive (recursive). In this case, we say that the mathematical problem defined by the couple (X, A) is primitive recursively (recursively) decidable or algorithmically decidable with respect to the bijection v.

DEFINITION 8.3. Let

$$\mathscr{S} = (V, U, L, (\Phi_m)_{m \in \mathbb{N}}, (P_m)_{m \in \mathbb{N}})$$

be the symbols of a formal system, and denote by R its elementary relations, by R_0 its closed relations (4—6), by $\mathscr{A}$ its axioms, by $\mathscr{T}$ its theorems, and consider $\mathscr{T}_0 = \mathscr{T} \cap R_0$. If the sets V, U, P_R, Φ_m are countable and if some bijections

$$V \to \mathbb{N},\ U \to \mathbb{N},\ P_m \to \mathbb{N},\ \Phi_m \to \mathbb{N}$$

are given, then there exist the bijections

$$v : R \to \mathbb{N},\ v_0 : R_0 \to \mathbb{N}. \tag{8.3.1}$$

The formal system $(\mathscr{S}, \mathscr{A})$ is said to be recursively decidable or algorithmically decidable if the set $\mathscr{T}_0$ of R_0 is recursive with respect to the bijection v.

Example. The formal system of propositional logic (cf. 4.4.6, 4.7.2) *and the formal system of Aristotelian logic are algorithmically decidable**.

PROPOSITION 8.4. (Tarski) *The formal system of real-closed field theory is recursively decidable.*

We recall that a field K is said to be real-closed if it is totally ordered and if the following axiom is fulfilled: $P \in K[X], a, b \in K, a < b, P(a) \leqslant 0, P(b) \geqslant 0 \Rightarrow \exists \xi (a \leqslant \xi \leqslant b \text{ and } P(\xi) = 0)$.

The proof of this proposition will be given later.

Example 8.5. (a) *Hilbert's 10th problem.* Let $\mathbb{Z}[X_1, X_2, \ldots, X_n \ldots]$ be the set of polynomials in an arbitrary number of variables with integer coefficients. We have seen (Corollary 1.6) the existence of a

* The proof can be found in P. S. Novikov, *Elements of Mathematical Logic*, Chap. I, § 6, Chap. III, § 10—11

natural bijection

$$(8.5.1) \qquad \rho : \mathbb{Z}[X_1, \ldots, X_n, \ldots] \to \mathbb{N}.$$

Let A be the subset of $\mathbb{Z}[X_1, X_2, \ldots, X_n, \ldots]$ consisting of the polynomials having an integer solution.

Is the subset A recursive with respect to the bijection ρ?

This problem has recently be given a negative answer by the Soviet mathematician Yu. V. Matiyasevic. This result had been presented at the International Congress of Mathematicians held in 1970 at Nice (Actes du Congrès int. math., 1970, Tome I, pp. 235—238, *Diophantine Representation of Recursively Enumerably Predicates)* as well as at the Congress of Logic organized in 1971 in Bucharest.

In order to answer the previously stated problem of Hilbert, Yu. Matiyasevic starts with proving the following extremely interesting result:

PROPOSITION 8.5.1. *A function* $f : \mathbb{N}^r \to \mathbb{N}$ *is recursive if and only if it is diophantine.*

The proof can be found in Martin Davis, *Hilbert's Tenth Problem is Unsolvable*, The American Mathematical Monthly, 80, 3, pp. 233—269. We shall merely recall that a function $f: \mathbb{N}^r \to \mathbb{N}$ is diophantine if there exists a polynomial $P(X_1, X_2, \ldots, X_r, X_{r+1}, Y_1, Y_2, \ldots, Y_s) \in$ $\in \mathbb{Z}[X_1, \ldots, X_{r+1}, Y_1, \ldots, Y_s]$ such that

$$\{(X_1, \ldots, X_r, X_{r+1}) \in \mathbb{N}^{r+1} | \, X_{r+1} = f(X_1, \ldots, X_r)\} =$$

$$= \{(X_1, \ldots, X_r, X_{r+1}) \in \mathbb{N}^{r+1} | \, \exists (Y_1, \ldots, Y_s) \in$$

$$\in \mathbb{N}^s P(X_1, \ldots, X_{r+1}, Y_1, \ldots, Y_s) = 0\}.$$

(b) *The theorem of Ax.* Let B be the subset of $\mathbb{Z}[X_1, X_2, \ldots, X_n]$ consisting of the polynomials having a solution in $\mathbb{Z}_p^n$ for every prime number p.

The subset B of $\mathbb{Z}[X_1, X_2, \ldots, X_n]$ is primitive recursive with respect to the bijection $\rho_n: \mathbb{Z}[X_1, \ldots, X_n] \to \mathbb{N}$ (Corollary 1.6).

(c) The particular case of Hilbert's 10th problem, obtained by considering only polynomials in one variable with coefficients in $\mathbb{Z}$ obviously receives an affirmative answer. In other words, it is possible to indicate an algorithm which is applicable to every polynomial in one variable with integer coefficients, allowing to decide whether a given such polynomial has or not an integer solution.

This algorithm is an immediate consequence of the following well-known theorem of Algebra:

If the integer ξ is a root of the polynomial $P \in \mathbb{Z}[X]$, then ξ is a divisor of the free term of P.

The proof of this theorem is immediate. Indeed, put

$$P = a_n X^n + a_{n-1} X^{n-1} + \dots + a_0.$$

From the hypothesis it follows that

$$P = (X - \xi)(b_{n-1} X^{n-1} + b_{n-2} X^{n-2} + \dots + b_1 X + b_0)$$

where the coefficients $b_{n-1}, b_{n-2}, \dots, b_1, b_0$ can be determined from the formulae:

$$b_{n-1} = a_n$$

$$b_{n-2} = \xi b_{n-1} + a_{n-1}$$

$$\cdots\cdots$$

$$b_i = \xi b_{i+1} + a_{i+1}$$

$$\cdots\cdots$$

$$b_0 = \xi b_1 + a_1.$$

In other words, the coefficients $b_0, b_1, \dots, b_{n-1}$ are integers. Moreover, we have

$$(-\xi) b_0 = a_0$$

which yields precisely the required property.

The same algorithm allows also the treatment of the case of polynomials in one variable with rational coefficients and of rational roots. In order to decide algorithmically whether a polynomial in one variable with rational coefficients has a rational root it is sufficient to use the previous result together with the following two propositions.

PROPOSITION 8.5.2. *If the rational number* $q = \dfrac{b}{c} \in \mathbb{Q}$ *is a root of the polynomial*

$$P = X^n + a_{n-1}X^{n-1} + \dots + a_1X + a_0 (a_n = 1)$$

with integer coefficients, then q *is actually an integer.*

Proof. Assume that the integers b, c are mutually prime. From the hypothesis it follows that

$$\left(\frac{b}{c}\right)^n + a_{n-1}\left(\frac{b}{c}\right)^{n-1} + a_{n-2}\left(\frac{b}{c}\right)^{n-2} + \cdots$$

$$\dots + a_1\left(\frac{b}{c}\right) + a_0 = 0$$

whence

$$\frac{b^n}{c} = (a_{n-1}cb^{n-1} + a_{n-2}c^2b^{n-2} + \dots + a_1c^{n-1}b + a_0c^n).$$

Since the right-hand of the last relation is an integer, it follows that $c = \pm 1$.

PROPOSITION 8.5.3. *The polynomial* $P \in \mathbb{Z}[X]$,

$$P = a_nX^n + a_{n-1}X^{n-1} + \dots + a_1X + a_0$$

has a rational root if and only if the polynomial $S \in \mathbb{Z}[X]$:

$$S = X^n + a_{n-1}X^{n-1} + a_{n-2}a_nX^{n-2} + a_{n-3}a_nX^{n-3} + \dots$$

$$\dots + a_1a_n^{n-2}X + a_0a_n^{n-1}$$

has an integer root.

More precisely, if $\xi \in \mathbb{Z}$ *is an integer root of the polynomial* S, *then* $\dfrac{\xi}{a_n}$ *is a root of the polynomial* P *and conversely, if* $\eta \in \mathbb{Q}$ *is a rational* *root of the polynomial P, then $a_n^{n-1}\eta$ is an integer root of the polynomial S.*

Obviously, in order to solve completely the problem of algorithmically deciding whether a polynomial with rational coefficients has a rational root, it is sufficient to observe now that the roots of any such polynomial coincide with the roots of an associated polynomial with integer coefficients (by cancelling the denominators of the coefficients).

(d) An extremely interesting problem is that of algorithmically deciding whether a polynomial $P \in \mathbb{R}[X]$ with real coefficients has a solution in a given interval. This problem is answered by the famous theorem of Sturm.

PROPOSITION 8.5.4. *Let $P \in k[X]$ be a polynomial in one variable with coefficients in the field k. Assume that the characteristic of the field k is equal to zero. Then a polynomial $S \in k[X]$ exists such that the following conditions be fulfilled:*

(i) All roots of S are simple.

(ii) The set of the roots P (possibly in the algebraic closure of k) coincides with the set of the roots of S.

Proof. Let $D \in k[X]$ be the greatest common divisor of the polynomials P, P'. It is sufficient to take $S = P/D$.

DEFINITION 8.5.5. Consider $P \in \mathbb{R}[X]$. The finite sequence

$$P_0, P_1, \ldots, P_s \tag{8.5.5.1}$$

of polynomials in one variable with coefficients in $\mathbb{R}$ is said to be a Sturm sequence associated with the polynomial P if the following conditions are fulfilled:

(0) $P = P_0$.

(1) Any two consecutive polynomials of the sequence have no common roots.

(2) If $\alpha \in \mathbb{R}$ is a root of the polynomial P_k, $1 < k \leqslant s$, then $P_{k-1}(\alpha)$ and $P_{k+1}(\alpha)$ have opposite signs.

(3) The last polynomial P_s of the sequence has no real root.

(4) If α is a real root of the polynomial P, then the product PP_1 is increasing in the neighborhood of α.

PROPOSITION 8.5.6. *Let $P \in \mathbb{R}[X]$ be a polynomial without multiple roots. Then there exists at least a Sturm sequence*

$$P_0, P_1, \ldots, P_s$$

associated to the polynomial P.

Proof. First we put

$$P_0 = P, \; P_1 = P'.$$

Now assume that $i > 1$. We recurrently define the polynomial P_i by the condition that it coincides with $-R_i$, where R_i is the remainder obtained by dividing the polynomial P_{i-2} through the polynomial P_{i-1}:

$$P = P'Q_1 - P_2$$

$$P' = P_2Q_2 - P_3$$

$$P_2 = P_3Q_3 - P_4$$

$$\cdots\cdots$$

$$P_{i-2} = P_{i-1}Q_{i-1} - P_i$$

$$\cdots\cdots\cdots$$

Using exactly the same argument as in the case of Euclid's algorithm, it follows that there exists an integer s such that P_s is the greatest common divisor of P and P'. We show that the sequence

$$(8.5.6.1) \qquad P_0, P_1, P_2, \ldots, P_s$$

is a Sturm sequence associated with the polynomial P.

Condition 0 is obviously satisfied.

Condition 3 is also easily satisfied. From the definition of the polynomial P_s, it follows that it is a non-vanishing constant. Indeed, otherwise one of its roots (possibly in an extension of $\mathbb{R}$) would be a common root of the polynomials P, P', hence it would be a multiple root for P, contradicting the hypothesis.

Now we prove that for $i \in \{1, 2, \ldots, s\}$ the polynomials P_{i-1}, P_i have no common root. First of all, P and $P_1 = P'$ obviously have no common root since P is free of multiple roots. By *reductio ad absurdum*, let $i_0 > 0$ be the first number from the sequence $\{1, 2, \ldots, s\}$ for which P_{i_0-1}, P_{i_0} have a common root. Under this assumption, from the relation

$$P_{i_0-2} = P_{i_0-1}Q_{i_0-1} - P_{i_0}$$

it follows that the polynomials P_{i_0-2}, P_{i_0-1} have a common root, which is absurd.

Consequently, the sequence (8.5.6.1) satisfies condition 1 of Definition 8.5.5.

Now we check condition 2 of this definition. Let $\alpha \in \mathbb{R}$ be such that $P_k(\alpha) = 0$, $k = 1, 2, \ldots, s - 1$. From the relation

$$P_{k-1} = P_k Q_k - P_{k+1}$$

it follows that $P_{k-1}(\alpha) = -P_{k+1}(\alpha)$.

Since on the other hand condition 4 obviously follows from the relation $P_1 = P'$, Proposition 8.5.6 is proved.

Let $P \in \mathbb{R}[X]$ be free of multiple roots and let

$$(S) = (P_0, P_1, \ldots, P_s)$$

be a Sturm sequence associated to the polynomial P.

On the other hand, denote by $\sum_S$ the set of functions

$$f: \{0, 1, 2, \ldots, s\} \to \{+, -, 0\}.$$

A function

$$W: \textstyle\sum_S \to \mathbb{N}$$

can be defined as follows:

Assume $f \in \sum_S$. We begin by considering the sequence

$$(8.5.6.2) \qquad f(0), f(1), \ldots, f(i), f(i+1), \ldots, f(s).$$

in which we cancel all terms $f(i)$ such that $f(i) = 0$.

Let

(8.5.6.3) $\quad f(i_0), f(i_1,) \ldots, f(i_k), f(i_{k+1}), \ldots, f(i_r)$

be the sequence obtained after these cancelations. $W(f)$ is by definition the number of the elements k, $1 \leqslant k \leqslant r$, for which either

$$(f(i_{k-1}) = +, f(i_k) = -)$$

or

$$(f(i_{k-1}) = -, f(i_k) = +).$$

In other words, $W(f)$ represents the number of sign variations in the sequences (8.5.6.2) or (8.5.6.3).

Example:

$$W(0, +, +, -, -, 0, -, +, +) = 2,$$

$$W(+, +, -, -, +, +) = 2.$$

If (S) is a Sturm sequence associated to a polynomial P, a function

$$W_S \colon \mathbb{R} \to \mathbb{N}$$

can be defined as follows:

$$W_S(\xi) = W(\operatorname{sign}(P_0(\xi)), \operatorname{sign}(P_1(\xi)), \ldots, \operatorname{sign}(P_S(\xi)),$$

where, clearly, sign: $\mathbb{R} \to \{+, -, 0\}$ is defined by the relation

$$\operatorname{sign}(\alpha) = \begin{cases} 0 & \text{if } \alpha = 0, \\ + & \text{if } \alpha > 0, \\ - & \text{if } \alpha < 0. \end{cases}$$

PROPOSITION 8.5.7 (Sturm). *Consider a polynomial $P \in \mathbb{R}[X]$ an interval (a, b) of the real line and a Sturm sequence (S) associated to the polynomial P. Assume that $P(a) \neq 0 \neq P(b)$. Under these conditions, $W_S(a) \geqslant W_S(b)$ and the difference $W_S(a) \geqslant W_S(b)$ is equal to the number* *of (real) roots of the polynomial P which are contained in the open interval (a, b).*

Proof. Let M be the finite set of the points belonging to the interval (a, b) at which at least one polynomial P_i of the Sturm sequence associated to the polynomial P vanishes. The function W_S is obviously locally constant on the set $[a, b] - M$. This set is a union of intervals. Let ξ be an extremity of such an interval and let I'_ξ, I''_ξ be the two intervals whose intersection consists of the point ξ. We assume that for every $\alpha \in I'_\xi$ and every $\beta \in I''_\xi$ we have $\alpha < \beta$ (Fig. 8.5.7.1).

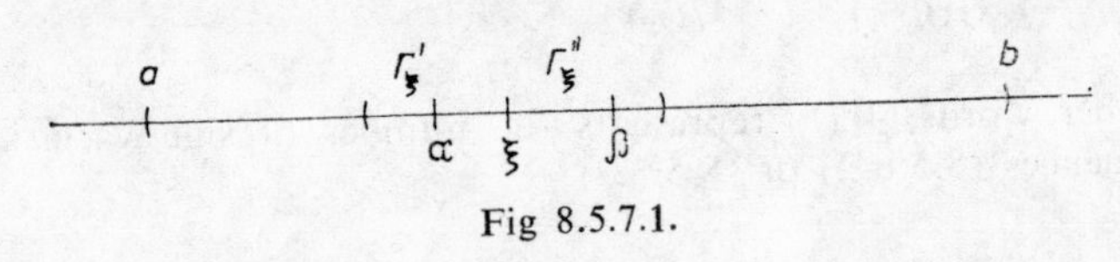

Fig 8.5.7.1.

In order to prove Proposition 8.5.7 it is sufficient to check the following two assertions:

(I) If ξ is not a root of the polynomial P, then $W_S(\alpha) = W_S(\beta)$ for every point $\alpha \in I'_\xi$ and every point $\beta \in I''_\xi$.

(II) If ξ is a root of the polynomial P then

$$W_S(\alpha) = W_S(\beta) + 1.$$

Concerning (I), it is sufficient to show that if ξ is a root of the polynomial P_k, $k \geqslant 1$, then the number of sign variations in the following two sequences

$$(8.5.7.1) \qquad P_{k-1}(\alpha)\, P_k(\alpha)\, P_{k+1}(\alpha),$$

$$(8.5.7.2) \qquad P_{k-1}(\beta)\, P_k(\beta)\, P_{k+1}(\beta).$$

is the same.

In view of condition 2 (Definition 8.5.5), the numbers $P_{k-1}(\xi)$, $P_{k+1}(\xi)$ have opposite signs. Obviously, we can assume that α and β are sufficiently close to ξ such that $P_{k-1}(\alpha)$, $P_{k-1}(\beta)$ (or $P_{k+1}(\alpha)$, $P_{k+1}(\beta)$) have the same sign, namely the sign of $P_{k-1}(\xi)$ (or $P_{k+1}(\xi)$).

It follows that as long as signs are concerned, we have the following possibilities:

$$+(-)\, P_k(\alpha)\, -(+),$$

$$+(-)\, P_k(\beta)\, -(+).$$

In both cases, one immediately sees that the number of sign variations in the two sequences is the same.

Concerning (II), we have to study the number of sign variations in the sequences

(8.5.7.3) $\quad P_0(\alpha), P_1(\alpha), \ldots$

(8.5.7.4) $\quad P_0(\beta), P_1(\beta), \ldots$

In view of condition 1 (Definition 8.5.5) it follows that $P_1(\xi) \neq 0$. Consequently we may assume that $P_1(\alpha)$ and $P_1(\beta)$ have the same sign. But in this case, condition 4 (Definition 8.5.5) shows that only the following sign variations are possible:

$$P_0(\alpha) < 0,\ P_1(\alpha) > 0\ (-\ +),$$

$$P_0(\beta) > 0,\ P_1(\beta) > 0\ (+\ +),$$

$$P_0(\alpha) > 0,\ P_1(\alpha) < 0\ (+\ -),$$

$$P_0(\beta) < 0,\ P_1(\beta) < 0\ (-\ -).$$

Taking also into account assertion (I), this leads to the relation $W_S(\alpha) = W_S(\beta) + 1$.

Using Proposition 8.5.7, it is possible to indicate an algorithm for deciding wheather a polynomial with real coefficients has a real root.

To this aim, it is sufficient to assign to every polynomial an interval such that no root of the polynomial lies outside this interval. Such an interval can be obtained using for instance the following proposition:

PROPOSITION 8.5.8. *Let* $P = a_nX^n + a_{n-1}X^{n-1} + \ldots + a_1X + a_0$ *be a polynomial with complex coefficients and let* ξ *be a root of* P. *Consider also* $A = \max\{|a_n|, |a_1|, \ldots, |a_{n-1}|\}$ *and* $M = \dfrac{A}{|a_n|} + 1$. *(Obviously, we suppose* $a_n \neq 0$*). Under these conditions we have*

$$|\xi| < M.$$

Proof. Assume by *reductio ad absurdum* that $|\xi| \geqslant M$. Hence we have

$$|a_0 + a_1\xi + \dots + a_{n-1}\xi^{n-1}| \leqslant |a_0| + |a_1|\,|\xi| + \dots$$

$$\dots + |a_{n-1}|\,|\xi|^{n-1} \leqslant A(1 + |\xi| + |\xi|^2 + \dots$$

$$\dots + |\xi|^{n-1}) = A\,\frac{|\xi|^n - 1}{|\xi| - 1}$$

The case $A = 0$ is trivial, hence we may assume that the previous relations yield

$$|a_0 + a_1\xi + \dots + a_{n-1}\xi^{n-1}| < A\,\frac{|\xi|^n}{|\xi| - 1} \leqslant$$

$$\leqslant A\,\frac{|a_n|\,|\xi|^n}{A} = |a_n|\,|\xi|^n$$

which is absurd.

Corollary 8.5.9. *If $P_1, P_2, \dots, P_n \in \mathbb{R}[X]$, then one can algorithmically decide whether the following relations are true (are theorems):*

$$\exists\xi((P_1(\xi) = 0) \vee (P_2(\xi) = 0) \vee \dots \vee (P_n(\xi) = 0)),$$

$$\exists\xi((P_1(\xi) = 0) \vee (P_2(\xi) = 0) \vee \dots \vee (P_n(\xi) = 0)).$$

Indeed, for the first relation it is sufficient to decide whether one of the polynomials $P_1, P_2, \dots, P_n$ has a real root. For the second one, it is sufficient to decide algorithmically whether the polynomial $P_1^2 + P_2^2 + \dots + P_n^2$ has a real root.

Example 8.6. *The homeomorphisms of polyhedra.* (a). Let $\mathscr{K}_f$ be the set of finite polyhedra. There exists a natural bijection (Proposition 1.9)

$$\gamma\colon \mathscr{K}_f \to \mathbb{N}$$

In the cartesian product $\mathscr{K}_f \otimes \mathscr{K}_f$, consider the subset A of all couples of the form (K, L) such that there exists a homeomorphism

$$K \xrightarrow{\sim} L.$$

The subset A is recursive with respect to the bijection

$$\gamma \times \gamma: \mathscr{K}_f \times \mathscr{K}_f \to \mathbb{N} \times \mathbb{N}.$$

(b) Let $\mathscr{V}_f$ be the set of finite polyhedra which are 2-dimensional topological varieties. Obviously, we have a natural bijection

$$\delta: \mathscr{V}_f - \mathbb{N}.$$

In the cartesian product $\mathscr{V}_f \times \mathscr{V}_f$, consider the subset B of all couples (K, L) such that there exists a homeomorphism

$$K \xrightarrow{\sim} L.$$

The subset B is recursive with respect to the bijection

$$\delta \times \delta: \mathscr{V}_f \times \mathscr{V}_f \to \mathbb{N} \times \mathbb{N}.$$

(For the proof, see H. Seifert, W. Threlfall, *Lehrbuch der Topologie*, Leipzig, Teubner, 1934).

Example 8.7. *The isomorphism of groups.* Let $\mathscr{T}$ be the set of finitely generated group presentations. Let C be the subset of $\mathscr{T} \times \mathscr{T}$ consisting of all couples of the form $(M, (r_1, r_2, \ldots, r_s))$, $(N, (s_1, \ldots, s_t))$ such that the groups $\mathscr{L}(M)/(r_1, \ldots, r_s)$, $\mathscr{L}(N)/(s_1, s_2, \ldots, s_t)$ are isomorphic. This subset is not recursive with respect to the canonical enumeration of $\mathscr{T} \times \mathscr{T}$.

Example 8.8. *The word problem in group theory.* Let A be a finite of (effectively) countable set and let $L(A)$ be the free group generated by A. The group $L(A)$ is effectively countable (cf. Proposition 1.11), and so is $L(A) \times L(A)$ (Lemma 1.4 and Proposition 1.2).

Now assume a subset R of $L(A)$ is given and let N_R be the least normal subgroup containing R. Denote by $W(R)$ the subset of $L(A) \times L(A)$ defined as follows: $(x, y) \in W(R) \overset{\text{Def}}{\Leftrightarrow} \pi(x) = \pi(y)$, where $\pi: L(A) \to \to L(A)/N_R$ is the canonical group homomorphism.

The word problem can now be stated in the following way: Is W_R a recursive subset of $L(A) \times L(A)$ (with respect to the standard enumerations)? P. S. Novikov has constructed subsets R for which $W(R)$ is not recursive.

Example 8.9. *The decision problem for a plane curve* (cf. A. Seidenberg' *New Decision Method for Elementary Algebra*, Ann. Math., 60 (1954) p. 366—369 . Now we give an important example of an algebraic problem which can be recursively (algorithmically) decided.

We shall not insist upon writing and analyzing the functions whose recursivity is to be shown, pointing out only the effectiveness (in an algorithmic sense) of the suggested method.

Problem 8.9.1. Let a plane curve C, rational over the field $\mathbb{R}$ of real numbers, be given. Decide *algorithmically* whether this curve has a rational point over $\mathbb{R}$.

In other words, let P be a polynomial in two variables with coefficients in the field $\mathbb{R}$. Decide algorithmically whether this polynomial has a root in the field of real numbers.

Using Euclid's algorithm, we may obviously assume that the polynomial P is a product of mutually distinct irreducible factors *. In order to show how this can be concretely realized, we first recall the important concept of primitive polynomial, which has been introduced by Gauss.

Consider a unique factorization domain D and let $F \in D[Y]$ be a polynomial in one variable with coefficients in D. The polynomial D is said to be primitive if its coefficients have no common divisor (distinct from the invertible elements of D).

Every polynomial $F \in D[Y]$ can obviously be written as

$$F = C(F)F_1,$$

where F_1 is a primitive polynomial and $C(F) \in D$ is an element of D, unique up to an invertible factor from D. In fact, $C(F)$ is the greatest common divisor of those coefficients of F which are distinct from 0, and is called the *content* of F.

In the case of interest, $D = k[X]$. (In fact, $k = \mathbb{R}$).

If $P \in k[X, Y]$, a new polynomial P_1, having only simple irreducible factors which are identical to the irreducible factors of P, can be determined as follows:

— First, we express the polynomial $P \in k[X, Y]$ as a polynomial in one variable Y with coefficients in $k[X]$, and, secondly, we write it as

$$P = c(P) = P_1,$$

* In a first version, the author had supposed that P is even irreducible. The reader is asked to explain why this assumption is not justified.

where $c(P) \in k[X]$ is the content of the polynomial $P \in k[X][Y]$ and P_1 is primitive.

Using Euclid's algorithm, we can determine a polynomial $d \in k[X]$ having simple irreducible factors coinciding with those of $c(P)$.

– We consider the polynomial P_1 as an element of the Euclidian ring $k(X)[Y]$.

Consequently, still using Euclid's algorithm, we can determine a polynomial P_2, having only simple irreducible factors which are identical to those of P_1.

– We express the polynomial P_2 as

$$P_2 = \frac{a(X)}{b(X)} S,$$

where $S \in k[X][Y]$ is a primitive polynomial.

– The polynomial we look for will be dS.

In order to solve the stated problem, we use the following direct consequence of the fact that the set $C(\mathbb{R})$ of the points of the curve C which are rational over $\mathbb{R}$ is closed.

Let $a = (\alpha, \beta) \in \mathbb{R}^2$ be an arbitrary point of the real plane $\mathbb{R}^2$. The curve C has a rational point over $\mathbb{R}(C(\mathbb{R}) \neq \Phi)$ if and only if C has a rational point π over $\mathbb{R}$, lying nearest the point a (Fig. 8.9.11).

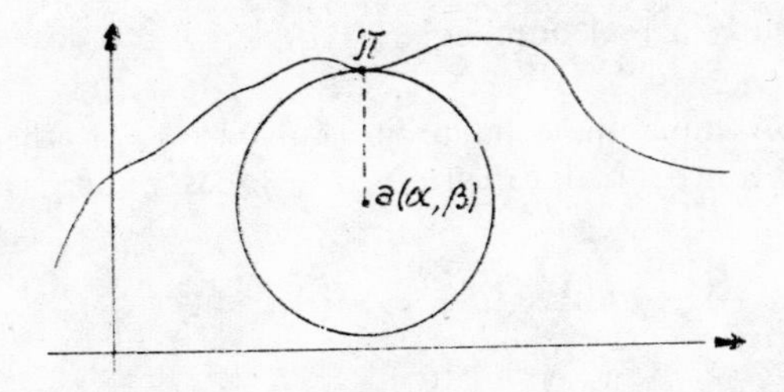

Fig. 8.9.11

It is also intuitively obvious that the tangent at π to the curve C coincides with the tangent at π to the circle with center a and radius $a\pi$.

We shall also make use of the following results, the first two of which are elementary facts of complex analytic geometry.

Let $\mathbb{C}^2$ be the complex plane and let $x = (\xi, \eta) \in \mathbb{C}^2$ be a point. By a real direction we mean a complex line passing through the origin of $\mathbb{C}^2$ and having a real slope. The equation of such a line has the form

$$Y = mx, \quad m \in \mathbb{R}.$$

LEMMA 8.9.2. *Let $u = (\mu, \nu) \in \mathbb{C}^2$ be a point of the complex plane $\mathbb{C}^2$. The necessary and sufficient condition for its coordinates μ, ν to be real is that the projections in two distinct real directions of the point u upon one coordinate axis be real.*

Proof. Necessity. Assume that μ, ν are real numbers and that

$$Y - \nu = m(x - \mu), \ m \in \mathbb{R}$$

is a real line, parallel to the real direction

$$Y = mX$$

and passing through the point $u = (\mu, \nu)$. Then the projection of the point $u = (\mu, \nu)$ on the x-axis satisfies the condition:

$$-\nu = mX - m\mu$$

We may obviously assume $m \neq 0$, since otherwise we are in presence of the projection upon the y-axis; consequently, we get for the projection of the point u upon the x-axis the coordinate

$$\frac{m\mu - \nu}{m}$$

which is obviously a real number.

Sufficiency. Assume that the projections X_1, X_2 of the point (μ, ν) corresponding to the real directions of slopes m_1, m_2 $(m_1 \neq m_2)$ are real. This yields

$$-\nu = m_1(X_1 - \mu),$$
$$-\nu = m_2(X_2 - \mu),$$
$$m_1, m_2, X_1, X_2 \in \mathbb{R},$$

consequently

$$\mu = \frac{m_1X_1 - m_2X_2}{m_1 \quad m_2},$$

$$\nu = \frac{m_1m_2(X_2 - X_1)}{(m_2 - m_1)}$$

hence the numbers μ, ν are real.

COROLLARY 8.9.3. *Let $a_1, a_2, \ldots, a_n$ be n points of the complex plane $\mathbb{C}^2$. Assume that projecting these points upon the x-axis by means of lines parallel to the real directions of mutually distinct slopes $m_1, \ldots, m_n, m_{n+1}$ we always obtain at least one real point.*

Then at least one of the points $a_1, a_2, \ldots, a_n$ has real coordinates.

Proof. Indeed, the hypothesis implies the existence of at least one point a_i and two directions $m_{i_1}, m_{i_2}, m_{i_1} \neq m_{i_2}$, such that the projections of a_i on the x-axis along the directions m_{i_1}, m_{i_2} are both real.

Going back to problem 8.9.1, we first remark that if α is an arbitrary real number, then the polynomial P has a real solution if and only if the system of equations

$$P(X, Y) = 0,$$

$$(8.9.4) \qquad (X - \alpha)\frac{\partial P}{\partial Y} - Y\frac{\partial P}{\partial X} = 0$$

has a real solution.

Indeed, if P has a real solution, then the curve C has a point σ lying nearest the point $(\alpha, 0)$. Assume first that (ξ, η) is a regular (non-singular) point of C (Fig. 8.9.3.1).

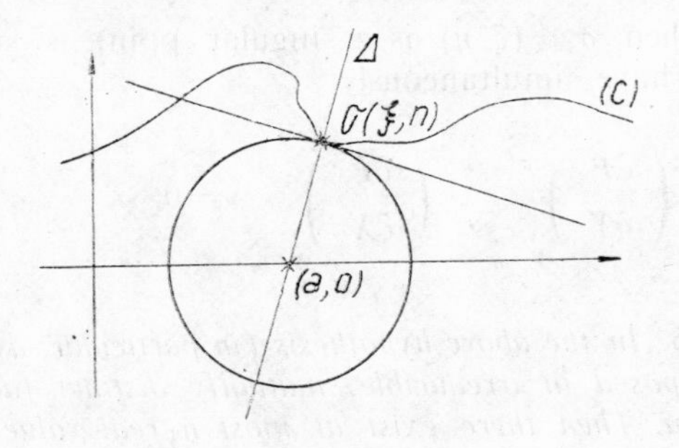

Fig. 8.9.3.1

The normal Δ to C at σ being also normal to the circle through σ, having $(\alpha, 0)$ as center, its equation is of the form

$$Y = m(X - \alpha).$$

On the other hand, the tangent at $\sigma(\xi, \eta)$ to the curve C has the equation

$$(X - \xi)\frac{\partial P}{\partial X} + (Y - \eta)\frac{\partial P}{\partial Y} = 0.$$

Consequently, from the obvious orthogonality condition, we get the relation

$$m\frac{\partial P}{\partial X} - \frac{\partial P}{\partial Y} = 0$$

i.e.

$$m = \frac{\partial P}{\partial Y} : \frac{\partial P}{\partial X},$$

hence the equation of Δ writes

$$(X - \alpha)\frac{\partial P}{\partial Y} - Y\frac{\partial P}{\partial X} = 0$$

Since the point σ belongs to both C and Δ, it follows that its coordinates satisfy system (8.9.4).

The case when $\sigma = (\xi, \eta)$ is a singular point is straightforward, since then we have simultaneously

$$\left(\frac{\partial P}{\partial Y}\right)_{\substack{X=\xi\\ Y=\eta}} = \left(\frac{\partial P}{\partial X}\right)_{\substack{X=\xi\\ Y=\eta}} = 0.$$

LEMMA 8.9.5. *In the above hypothesis (in particular, assuming that P can be decomposed in irreducible, mutually distinct factors), suppose that* $\deg P = n$. *Then there exist at most n real values $a_1, a_2, \ldots, a_n$ such that for every $i = 1, 2, \ldots, n$, the polynomials*

$$P, (X - a_i)\frac{\partial P}{\partial Y} - Y\frac{\partial P}{\partial X}$$

have a common factor.

Proof. Assuming the existence of $n+1$ real numbers $a_1, a_2, \ldots$ $\ldots, a_n, a_{n+1}$ with the above property, from the hypothesis $\deg P = n$ it follows the existence of at least two real numbers $a_i, a_j, a_i \neq a_j$ such that the polynomials

$$P,\ (X-a_i)\frac{\partial P}{\partial Y} - Y\frac{\partial P}{\partial X},\ (X-a_j)\frac{\partial P}{\partial Y} - Y\frac{\partial P}{\partial X}$$

have a common factor. Consequently, the polynomials

$$P,\ \frac{\partial P}{\partial Y},\ Y\frac{\partial P}{\partial X}$$

also have a common factor, which an elementary calculation shows to contradict the hypothesis that P is a product of irreducible, mutually disjoint factors.

COROLLARY 8.9.6. *Between the natural numbers* $0, 1, 2, \ldots, n$ *there exists at least one, say* a, *such that the resultant* $R\left(P,\ (X-a)\frac{\partial P}{\partial Y} - Y\frac{\partial P}{\partial X}\right)(X)$ *of the polynomials* $P, (X-a)\frac{\partial P}{\partial Y} - Y\frac{\partial P}{\partial X}$, *viewed as polynomials in one variable with coefficients in* $\mathbb{R}[X]$, *be distinct from* 0.

Solution of Problem 8.9.1. Now we are able to describe the "algorithm" which answers Problem 8.9.1. It consists in performing the following operations:

(I) One calculates the following resultants

$$R\left(P,\ X\frac{\partial P}{\partial Y} - Y\frac{\partial P}{\partial X}\right)(X),$$

$$R\left(P,\ (X-1)\frac{\partial P}{\partial Y} - Y\frac{\partial P}{\partial X}\right)(X), \ldots$$

$$\ldots, R\left(P,\ (X-n)\frac{\partial P}{\partial Y} - Y\frac{\partial P}{\partial X}\right)(X).$$

(II) One chooses from the numbers $0, 1, 2, \ldots, n$ one, say i, for which we have:

$$R\left(P, (X-i)\left(\frac{\partial P}{\partial Y} - Y\frac{\partial P}{\partial X}\right)(X) \neq 0.\right.$$

This is possible in view of Corollary 8.9.6.

In the following, we introduce the notation:

$$G = (X-i)\frac{\partial P}{\partial Y} - Y\frac{\partial P}{\partial X}.$$

(III) Let $m_1, m_2, \ldots, m_r$, $r = n^2 + 1$, be r real numbers which are mutually distinct * and distinct from the asymptotic directions of the curve C. For every $s \in [1, n^2 + 1]$, we perform the following change of variables:

$$X = x + \frac{1}{m_s} Y,$$

$$Y = y,$$

in the polynomials P, G. Let P_k, G_k be the resulting polynomials and let $R_k(P_k, G_k)(X)$ be the resultant of P_k, G_k viewed as polynomials in Y.

The roots of the polynomial $R_k(P_k, G_k)(X)$ are the projections on the x-axis of the common roots of the polynomials P, Q with respect to the real direction of slope m_k.

(IV) Using Sturm's theorem, one decides whether the polynomials $R_k(P_k, G_k)(X)$ have a real root.

(V) The polynomial P has a real root if and only if each polynomial $R_k(P_k, G_k)(X)$, $k = 1, 2, \ldots, s$, has a real root (straightforward consequence of Corollary 8.9.3).

Remark 8.9.6.1. One can prove that instead of $n^2 + 1$ projections (cf. step III) of the solution of Problem 8.9.1), a single projection can be used. To this aim, we first prove a lemma:

---

* The degree of the polynomial $R\left((P, (X-i)\frac{\partial P}{\partial Y} - Y\frac{\partial P}{\partial X}\right)(X)$ is at most n^2.

LEMMA 8.9.6.2. *If* $m \in \mathbb{R}$, $m \neq 0$, *is the slope of a real direction and if* $(\alpha, \beta) \in \mathbb{C}^2$ *satisfies the following conditions:*

(a) $\alpha \notin \mathbb{R}$;

(b) *the projection on the x-axis of the point* (α, β) *along the direction* m *is a real point, then* m *is given by the formula*

$$m = \frac{\bar{\beta} - \beta}{\bar{\alpha} - \alpha}.$$

Proof. Let $Y = mX + n$ be the line of slope m, passing through the point (α, β). We show that condition (b) implies $n \in \mathrm{R}$. Indeed first of all, we have

$$Y - \beta = m(X - \alpha),$$

whence it follows that the projection ξ of the point (α, β) on the x-axis along the direction m satisfies the relation

$$-\beta = m\xi - m\alpha$$

But we have

$$n = \beta - m\alpha = -m\xi \in \mathbb{R}$$

Consequently, together with the point (α, β), the line $Y = mX + n$ also contains the point $(\bar{\alpha}, \bar{\beta})$, hence

$$\bar{\beta} - \beta = m(\bar{\alpha} - \alpha)$$

from which the required formula follows.

In order to attain the objective of Remark 8.9.6.1, it will suffice to project along a real direction m which is distinct from every real number of the form $\dfrac{\bar{\beta} - \beta}{\bar{\alpha} - \alpha}$, $\alpha \neq \bar{\alpha}$, (α, β) solution of the system $P = 0, (X - i)\dfrac{\partial P}{\partial Y} - Y\dfrac{\partial P}{\partial X} = 0$, as well as from the asymptotic directions of the curve $P = 0$. Such a direction can be found as follows:

We first consider the resultant of the above system as a polynomial in X, let it be R_0, then as a polynomial in Y, let it be R_1. Afterwards.

we form the polynomial whose roots are $\frac{\bar{\beta}-\beta}{\bar{\alpha}-\alpha}$ (where β is a root of R_1 and α a root R_0) and the real asymptotic directions of the curve $P = 0$. This polynomial has real coefficients, which are simple functions of the coefficients of R_0 and R_1. We take for m any real number which is not a root of this polynomial (Proposition 8.5.7).

PROPOSITION 8.9.7. *Consider* $P \in \mathbb{R}[X, Y]$, $F \in \mathbb{R}[X]$, $F \neq 0$. *Starting with the polynomials, P, F, a polynomial $P_1 \in \mathbb{R}[X, Y]$ can be determined such as to have the equivalence*

$$\exists\xi\, \exists\eta((P(\xi, \eta) = 0) \wedge (F(\xi) \neq 0)) \Leftrightarrow$$

$$\Leftrightarrow \exists\xi\exists\eta(P_1(\xi, \eta) = 0).$$

Proof. First assume that the polynomials P, F have a common factor, let it be F_1:

$$P = F_1 P_2, \quad F = F_1 F_2.$$

It is sufficient to consider the case of the polynomials P_2, F. Indeed, we have

$$\exists\xi\, \exists\eta(((P(\xi, \eta) = 0) \wedge (F(\xi) \neq 0)) \Leftrightarrow$$

$$\Leftrightarrow \exists\xi\exists\eta(((P_2(\xi, \eta) = 0) \wedge (F(\xi) \neq 0)).$$

Consequently, we may suppose that the polynomials P, F have no common factors of positive degree. More precisely, the couple P, F can be replaced by a new couple P_2, F such that these last polynomials have no common factors of positive degree. This can be algorithmically realized by applying Euclid's algorithm for computing the greatest common divisor F_1 of the polynomials $F, a_0, a_1, \ldots, a_n \in \mathbb{R}[X]$, where $a_0, a_1, \ldots, a_n$ are the coefficients of the polynomial P viewed as a polynomial in Y:

$$P = a_0(X) + a_1(X)\, Y + \ldots + a_n(X) Y^n, a_n(X) \neq 0.$$

One can also take for P_2 the polynomial satisfying the relation $P = F_1P_2$. The two polynomials P_2, F have no common factors. So, assume P, F have no common factors.

Denote by $R_X(F, P) \in \mathbb{R}[Y]$ the resultant of the polynomials F, P, viewed as polynomials in one variable X with coefficients in $\mathbb{R}[Y]$. Since P and F have no common factors, it follows that $R_X(F, P)$ is distinct from 0. Let $c \in \mathbb{R}$ be a positive real number, which by Proposition 8.5.7 can be algorithmically determined such that the polynomial $R_X(F, P)$ has no real root outside the interval $(-c, c)$. Now consider the polynomials $F(X)$, $S(X, Y)$, where

$$S(X, Y) = P(X, Y + c).$$

In view of the definition of c, the polynomial F, S obviously have no common root on the x-axis.

Now consider $P_1(X, Y) = S(X, YF(X))$. We show that P_1 satisfies the conditions stated above.

To this aim, let $\xi, \eta \in \mathbb{R}$ be such that $P_1(\xi, \eta) = 0$. We show that putting $(\xi_1, \eta_1) = (\xi, \eta F(\xi) + c)$, we simultaneously have

$$F(\xi_1) \neq 0, \tag{8.9.7.1}$$

$$P(\xi_1, \eta_1) = 0. \tag{8.9.7.2}$$

Indeed, assume by *reductio ad absurdum* that $F(\xi_1) = 0$. Then we would have

$$0 = P_1(\xi, \eta) = P_1(\xi_1, \eta) = S(\xi_1, \eta F(\xi_1)) =$$

$$= S(\xi_1, 0) = P(\xi_1, c)$$

which is absurd.

Relation (8.9.7.2) follows from

$$P(\xi_1, \eta_1) = P(\xi, \eta F(\xi) + c) = S(\xi, \eta F(\xi)) =$$

$$= P_1(\xi, \eta) = 0.$$

Conversely, let $\xi, \eta \in \mathbb{R}$ be such that $P(\xi, \eta) = 0$ and $F(\xi) \neq 0$. We show that in these conditions we have $P_1\left(\xi, \frac{(\eta - c)}{F(\xi))}\right) = 0$. Indeed,

$$P_1\left(\xi, \frac{(\eta - c)}{F(\xi))}\right) = S(\xi, \eta - c) = P(\xi, \eta) = 0.$$

PROPOSITION 8.9.8. *Let $P \in \mathbb{R}[a_1, a_2, \ldots, a_m; X, Y]$ be a polynomial in $m + 2$ variables with coefficients in $\mathbb{R}$ and let $F \in \mathbb{R}[a_1, a_2, \ldots, a_m; X]$ be a polynomial in $m + 1$ variables with coefficients in $\mathbb{R}$.*

One can algorithmically determine the polynomials

$$G_i \in \mathbb{R}[a_1, a_2, \ldots, a_m], \qquad i = 1, 2, \ldots, s$$

$$g_i \in \mathbb{R}[a_1, a_2, \ldots, a_m; X], \quad i = 1, 2, \ldots, s.$$

such that for every $\alpha = (\alpha_1, \alpha_2, \ldots, \alpha_m) \in \mathrm{R}^m$, the following two assertions be equivalent:

(A) There exists a couple (ξ, η) of real numbers satisfying the conditions:

$$P(\alpha_1, \alpha_2, \ldots, \alpha_m; \xi, \eta) = 0,$$

$$F(\alpha_1, \alpha_2, \ldots, \alpha_m; \xi) \neq 0.$$

(B) For at least one $i \in \{1, 2, \ldots, s\}$, there exists $\xi \in \mathbb{R}$ such that the following two conditions are fulfilled:

$$G_i(\alpha_1, \alpha_2, \ldots, \alpha_m) \neq 0,$$
$$g_i(\alpha_1, \alpha_2, \ldots, \alpha_m; \xi) = 0.$$

The number s depends on the degrees of the polynomials P, F. In other words, we have the following equivalence:

$$\forall \alpha_1, \ldots, \forall \alpha_m \, \exists \xi \, \exists \eta(((P(\alpha_1, \alpha_2, \ldots, \alpha_m; \xi, \eta) = 0) \wedge$$
$$\wedge \, F(\alpha_1, \alpha_2, \ldots, \alpha_m) \neq 0) \Leftrightarrow \forall \alpha_1, \ldots, \forall \alpha_m \, \exists \xi((G_1(\alpha_1, \ldots$$
$$\ldots, \alpha_m) \neq 0) \wedge (g_1(\alpha_1, \ldots, \alpha_m, \xi) = 0) \vee \forall \alpha_1 \ldots$$

$$\ldots \forall\alpha_m \exists\xi((G_2(\alpha_1, \ldots, \alpha_m) \neq 0) \wedge (g_2(\alpha_1, \ldots$$

$$\ldots, \alpha_m, \xi) = 0)) \vee \ldots \forall\alpha_1 \ldots \forall\alpha_m \exists\xi((G_s(\alpha_1, \ldots$$

$$\ldots, \alpha_m) \neq 0)$$

$$g_s(\alpha_1, \ldots, \alpha_m, \xi) = 0).$$

Proof. Let $\alpha = (\alpha_1, \alpha_2, \ldots, \alpha_m) \in \mathbb{R}^m$ be an arbitrary element of $\mathbb{R}^m$. In view of Proposition 8.9.7 and of the solution of Problem 8.9.1, we can algorithmically decide whether two real numbers ξ, η exist such that

$$P(\alpha_1, \alpha_2, \ldots, \alpha_m; \xi, \eta) = 0$$

$$F(\alpha_1, \alpha_2, \ldots, \alpha_m; \xi) \neq 0$$

as follows:

– Consider the polynomials

$$P(\alpha_1, \alpha_2, \ldots, \alpha_m; X, Y) \in \mathbb{R}[X, Y],$$

$$F(\alpha_1, \alpha_2, \ldots, \alpha_m, X) \in \mathbb{R}[X].$$

Assume that we have simultaneously

$$P(\alpha_1, \alpha_2, \ldots, \alpha_m; X, Y) \neq 0,$$
$$F(\alpha_1, \alpha_2, \ldots, \alpha_m; X) \neq 0.$$

In this case, our problem is trivial.

– Proposition 8.9.7 is applicable to the polynomials

$$P(\alpha_1, \alpha_2, \ldots, \alpha_m; X, Y) \in \mathbb{R}[X, Y], F(\alpha_1, \alpha_2, \ldots$$
$$\ldots \alpha_m; X) \in \mathbb{R}[X]$$

In this case, we have to determine the polynomial P_1 satisfying the conditions of this proposition. As a first step, we have to determine a polynomial $P_2(X, Y) \in \mathbb{R}(X, Y)$, such that $P_2(X, Y)$, $F(\alpha_1, \ldots, \alpha_m; X)$ be free of common factors. This can be realized by a well-determined number of operations applied to the polynomial $F(\alpha_1, \ldots, \alpha_m; X)$ and to the coefficients in X of the polynomial $P(\alpha_1, \ldots, \alpha_m; X)(Y)$: the determination of the greatest common divisor F_1 of $F(\alpha_1, \ldots, \alpha_m; X)$ and of the coefficients of the polynomial $P(\alpha_1, \ldots, \alpha_m, X)(Y)$.

The specific form of the polynomial F_1 depends on the coefficients of $P(\alpha_1, \ldots, a_m; X)(Y)$ which are, after the substitution of the real numbers $\alpha_1, \alpha_2, \ldots, \alpha_m$ for the variables $a_1, a_2, \ldots, a_m$, non-zero as polynomials in one variable X with coefficients in $\mathbb{R}$.

This situation is accordingly characterized by a sequence of relations of the form

$$Q_i(\alpha_1, \alpha_2, \ldots, \alpha_m; X) \neq 0, \qquad i = 1, 2, \ldots, u$$

$$Q_i \in \mathbb{R}[a_1, a_2, \ldots, a_m; X], \qquad i = 1, 2, \ldots, u$$

or equivalently, writing each time the condition that the sum of the squares of the coefficients is non-vanishing:

$$S_i(\alpha_1, \alpha_2, \ldots, \alpha_m) \neq 0, \qquad i = 1, 2, \ldots, u,$$

$$S_i \in \mathbb{R}[a_1, a_2, \ldots, a_m], \qquad i = 1, 2, \ldots, u.$$

In fact, these relations can be written as a single one, by performing the product of the polynomials S_i:

$$S(\alpha_1, \alpha_2, \ldots, \alpha_m) \neq 0,$$

$$S \in \mathbb{R}[a_1, a_2, \ldots, a_m].$$

— Taking into account Proposition 8.9.7, we are led to a problem of the type considered in 8.9.1, for a suitable polynomial $P(X, Y)$.

In order to eliminate multiple irreducible factors from $P(X, Y)$, we again have to study a finite number of situations connected with the vanishing of some polynomials, hence described by some relations of the above type.

Indeed, the division algorithm of $U(a_1, \ldots, a_m; X, Y)$ through $V(a_1, \ldots, a_m; X, Y)$, both being considered as polynomials in Y, is influenced in an essential way by the leading coefficient of $U(a_1, \ldots$ $\ldots, a_m; \mathrm{T}, X)$.

— Finally, the solution of problem 8.9.1 leads each time to an ultimate condition of the type

$$\exists\xi(T(\alpha_1, \alpha_2, \ldots, \alpha_m; \xi) = 0),$$

where $T \in \mathbb{R}[a_1, a_2, \ldots, a_m; X]$.

Distinct elements $\alpha = (\alpha_1, \ldots, \alpha_m) \in \mathbb{R}^m$ naturally lead to distinct polynomials S, T, but it is clear that for all $\alpha \in \mathbb{R}^m$ only a finite number of conditions $S_1, S_2, \ldots, S_s$; $T_1, T_2, \ldots, T_s$ can result.

PROPOSITION 8.9.9. *Consider* $P \in \mathbb{R}^m[a_1, a_2, \ldots, a_m; X_1, X_2, \ldots, X_m]$. *The polynomials*

$$G_i(a_1, a_2, \ldots, a_m) \in \mathbb{R}[a_1, a_2, \ldots, a_m], \quad i = 1, 2, \ldots, s$$

$$g_i(a_1, a_2, \ldots, a_m; X) \in \mathbb{R}[a_1, a_2, \ldots$$

$$\ldots, a_m; X], \qquad i = 1, 2, \ldots, s$$

can be algorithmically determined such that, for every $\alpha = (\alpha_1, \alpha_2, \ldots$ $\ldots, \alpha_m) \in \mathbb{R}^m$ *the following two assertions be equivalent:*

(A) *There exists a system of real numbers* $(\xi_1, \xi_2, \ldots, \xi_n)$ *such that*

$$P(\alpha_1, \alpha_2, \ldots, \alpha_m; \xi_1, \xi_2, \ldots, \xi_n) = 0.$$

(B) *For at least one* $i \in \{1, 2, \ldots, s\}$ *there exists a* $\xi \in \mathbb{R}$ *such that*

$$G_i(\alpha_1, \alpha_2, \ldots, \alpha_m) \neq 0,$$

$$g_i(\alpha_1, \alpha_2, \ldots, \alpha_m; \xi) = 0.$$

Proof. We argue by induction on n. The proposition is trivial for $n = 1$ and is a particular case of Proposition 8.9.8 for $n = 2$. So, assume $n > 2$ and the proposition already proved for $n - 1$.

Consider the polynomial $P(a_1, a_2, \ldots, a_m; X_1, X_2, \ldots, X_n)$. By the induction hypothesis, we can algorithmically determine the polynomials

$$H_j \in \mathbb{R}[a_1, a_2, \ldots, a_m; X], \quad j = 1, 2, \ldots, t$$

$$U_j \in \mathbb{R}[a_1, a_2, \ldots, a_m; X, Y], \; j = 1, 2, \ldots, t$$

such that, for every system of real numbers $(\alpha_1, \alpha_2, \ldots, \alpha_m, \xi_m) \in \mathbb{R}^{m+1}$, the following two propositions be equivalent:

– There exists at least one system $(\xi_2, \xi_3, \ldots, \xi_n) \in \mathbb{R}^{n-1}$ of real numbers satisfying the condition

$$P(\alpha_1, \alpha_2, \ldots, \alpha_m, \xi, \xi_2, \ldots, \xi_n) = 0.$$

— For at least one $j \in \{1, 2, \ldots, t\}$ there exists $\eta \in \mathbb{R}$ such that the following two conditions be fulfilled:

$$H_j(\alpha_1, \ldots, \alpha_m, \xi) \neq 0,$$

$$U_j(\alpha_1, \ldots, \alpha_m, \xi, \eta) = 0.$$

For any $\alpha = (\alpha_1, \alpha_2, \ldots, \alpha_m) \in \mathbb{R}^m$, this implies the equivalence between the assertion (A) and the following one:

(C) For at least one $j \in \{1, 2, \ldots, t\}$ there exists a couple of real numbers $(\xi, \eta) \in \mathbb{R}^2$, satisfying the conditions

$$\begin{aligned} &H_j(\alpha_1, \ldots, \alpha_m, \xi) \neq 0, \\ &U_j(\alpha_1, \ldots, \alpha_m, \xi, \eta) = 0. \end{aligned} \tag{8.9.9.1}$$

The implication (A) $\Rightarrow$ (C) is straightforward.

In order to prove that (C) $\Rightarrow$ (A), let $j \in \{1, 2, \ldots, t\}$ and $(\xi, \eta) \in \mathbb{R}^2$ be such that the conditions 8.9.9.1 be fulfilled. Then, obviously, there exist $\xi_1, \xi_2, \ldots, \xi_n$ such that

$$P(\alpha_1, \alpha_2, \ldots, \alpha_m, \xi_1, \xi_2, \ldots, \xi_n) = 0.$$

Using Proposition 8.9.8, for every $j \in \{1, 2, \ldots, t\}$, we can determine the polynomials

$$\left.\begin{aligned} &G_{ji} \in \mathbb{R}[a_1, a_2, \ldots, a_m], \\ &g_{ji} \in \mathbb{R}[a_1, a_2, \ldots, a_m, X] \end{aligned}\right\} \quad i = 1, 2, \ldots, s(j)$$

such that for every $\alpha = (\alpha_1, \alpha_2. \ldots, \alpha_m) \in \mathbb{R}^m$ and for every $j \in \{1, 2, \ldots, t\}$, the following two assertions be equivalent:

— There exists a couple (ξ, η) of real numbers satisfying the conditions:

$$H_j(\alpha_1, \alpha_2, \ldots, \alpha_m; \xi) \neq 0,$$

$$U_j(\alpha_1, \alpha_2, \ldots, \alpha_m; \xi, \eta) = 0$$

— For at least one $i \in \{1, 2, \ldots, s(j)\}$, there exists a $\xi \in \mathbb{R}$ such that

$$G_{ji}(\alpha_1, \alpha_2, \ldots, \alpha_m) \neq 0,$$

$$g_{ji}(\alpha_1, \alpha_2, \ldots, \alpha_m, \xi) = 0$$

Obviously, G_{ji}, g_{ji} are the required polynomials.

PROPOSITION 8.9.10. *Let a finite family of polynomials*

$$P_i \in \mathbb{R}[\alpha_1, a_2, \ldots, a_m; X_1, X_2, \ldots, X_n], \; i = 1, 2, \ldots, \mu.$$

be given. A finite number of sets $G_1, G_2, \ldots, G_s$ whose elements are polynomials can be algorithmically constructed such that

$$G_j = \{F_1, \ldots, F_{u(j)}, \; S_1, \ldots, S_{v(j)} | F_i, S_i \in \mathbb{R}[a_1, a_2, \ldots \ldots, a_m]\}$$

and for every system $\alpha = (\alpha_1, \alpha_2, \ldots, \alpha_m) \in \mathbb{R}^m$ of real numbers, the following propositions are equivalents:

(A) *There exists a system $(\xi_1, \xi_2, \ldots, \xi_n) \in \mathbb{R}^n$ of real numbers satisfying the conditions:*

$$P_i(\alpha_1, \alpha_2, \ldots, \alpha_m; \; \xi_1, \xi_2, \ldots, \xi_n) = 0, \; i = 1, 2, \ldots, \mu.$$

(B) *For at least one $j \in \{1, 2, \ldots, s\}$ we have*

$$\begin{aligned} &F_i(\alpha_1, \alpha_2, \ldots, \alpha_m) > 0, \qquad i = 1, 2, \ldots, u(j), \\ &S_i(\alpha_1, \alpha_2, \ldots, \alpha_m) = 0, \qquad i = 1, 2, \ldots, v(j). \end{aligned} \tag{8.9.10.1}$$

Proof. In view of Proposition 8.9.9, it is sufficient to consider the case $n = \mu = 1$. But, given a polynomial $P \in \mathbb{R}[a_1, a_2, \ldots, a_m; X]$ and a system $(\alpha_1, \alpha_2, \ldots, \alpha_m) \in \mathbb{R}^m$ in order to decide whether there exists a real number ξ such that

$$P(\alpha_1, \alpha_2, \ldots, \alpha_m; \xi) = 0$$

we can use Sturm's theorem (Proposition 8.5.6). For every system $(\alpha_1, \alpha_2, \ldots, \alpha_m) \in \mathbb{R}^m$, the condition of this theorem can be expressed by a finite number of conditions of the type (8.9.10.1).

8.9.10.2. *Semialgebraic sets* (Cf. H. I. Levine, *Singularities of Differentiable Mappings*, Proceedings of Liverpool Singularities Symposium I, p. 1–90, Lecture Notes, Springer 192).

A subset M of $\mathbb{R}^p$ is said to be *semialgebraic* if there exists a finite number of polynomials $F_1, F_2, \ldots, F_u, G_1, G_2, \ldots, G_v \in \mathbb{R}[X_1, \ldots, X_p]$ such that

$$(*) \qquad M = \{(\alpha_1, \alpha_2, \ldots, \alpha_p) \in \mathbb{R}^p | F_i(\alpha_1, \ldots, \alpha_p) > 0,$$
$$i = 1, 2, \ldots, u.$$
$$G_j(\alpha_1, \ldots, \alpha_p) = 0, \; j = 1, 2, \ldots, v\}.$$

In the absence of the polynomials $F_1, \ldots, F_u$, we say that M is an algebraic set.

The existence of semialgebraic sets which are not algebraic is easily seen. An example of such a set is

$$\{\alpha \in \mathbb{R} | \; \alpha > 0\}.$$

A correspondence Γ between the spaces $\mathbb{R}^n$, $\mathbb{R}^m$ is said to be semialgebraic (or algebraic) if its graph is a semialgebraic (or algebraic) subset of $\mathbb{R}^{n+m} = \mathbb{R}^n \times \mathbb{R}^m$.

PROPOSITION 8.9.10.3. *Let K be a semialgebraic subset of $\mathbb{R}^n$ and let Γ be a semialgebraic correspondence between $\mathbb{R}^n$ and $\mathbb{R}^m$. The image of K through Γ is a union of semialgebraic sets of $\mathbb{R}^m$.*

Proof (In the particular case when the set K and the correspondence Γ are algebraic). The correspondence Γ and the set K are defined by some polynomials

$$P_i \in \mathbb{R}[Y_1, \ldots, Y_m; X_1, \ldots, X_n], \; i = 1, 2, \ldots, \mu.$$

(The polynomials which define the set K depend only on the variables $X_1, \ldots, X_n$).

Using Proposition 8.9.10 and the relations of this proposition, we have

$$\Gamma(K) = \bigcup_{j=1}^{s} \{(\alpha_1, \ldots, \alpha_m) \in \mathbb{R}^m | F_i(\alpha_1, \ldots, \alpha_m) > 0,$$
$$i = 1, 2, \ldots, u(j),$$
$$S_i(\alpha_1, \ldots, \alpha_m) = 0, \; i = 1, 2, \ldots, v(j)\},$$

which proves our proposition in this particular case. In the general case, the proof can be obtained using the theorem of "elimination of the quantifiers" for the formal system of ordered real-closed field theory (cf. the proof of Proposition 8.9.12).

Remark 8.9.10.4. The image of an algebraic set by an algebraic correspondence may not be algebraic.

For instance, let Γ be the correspondence between $\mathbb{R}$ and $\mathbb{R}$, defined by the graph

$$\{(x, x^2) | x \in \mathbb{R}\}$$

The image of $\mathbb{R}$ by Γ is obviously the subset of $\mathbb{R}$ consisting of all positive real numbers, which is obviously not algebraic.

Remark 8.9.11. The results obtained in the framework of Problem 8.9.1, in particular Proposition 8.9.10, hold for an arbitrary real closed field (cf. 8.4). This is due to the fact that the essential tool for the proof of these facts is Sturm's theorem (Proposition 8.5.6) which holds in any real closed field (cf. S. Lang, *Algebra*, chap. XI, § 2, p. 273–279).

This remark leads to the following fundamental proposition of Tarski.

PROPOSITION 8.9.12. (Tarski). *The formal system of real closed fields is recursively decidable.*

Proof. Let R be a closed relation. By induction on the number of bounded variables occurring in R, we show that one can algorithmically decide whether R is a theorem in the formal system of real closed fields.

In fact, we show by induction that every closed relation is equivalent to a closed relation without bounded variables. Such a relation is a relation between concrete rational numbers *.

Since the formal system of real-closed fields is logical and quantified, it follows that R is equivalent to a prenex form (Proposition 4.7.5). Consequently, R can be assumed to be already written in prenex form, i.e.

$$R = Q_1\xi_1 Q_2\xi_2 \ldots Q_n\xi_n(T(\xi_1, \xi_2, \ldots, \xi_n)),$$

where $Q_1, Q_2, \ldots, Q_n$ is one of the quantifiers $\exists$, $\forall$ while the relation $T(x_1, x_2, \ldots, x_n)$ involves no free variable distinct from $x_1, x_2, \ldots, x_n$

---

* Every closed relation without bounded variables is obviously algorithmically decidable in the formal system of real closed fields.

and no bounded variable. We may also assume the quantifier $\exists$ as occurring at least once between the quantifiers $Q_1, Q_2, \ldots, Q_n$. Indeed, otherwise, we replace the relation T by $\neg T$, i.e. we replace R by $\neg R$.

Now assume that $n = 1$. In this case, we have to decide upon the relation

$$R = \exists\xi(S(\xi)),$$

where $S(x)$ is an open relation, involving only the free variable x, and no bounded variable. Writing $S(x)$ under the normal disjunctive form, and taking into account Example 4.7.2.f, we may assume that $S(x)$ has the form

$$(0 = a_0 + a_1x + \ldots + a_px^p) \wedge (b_0 + b_1x + \ldots$$

$$\ldots + b_mx^m > 0),\ a_i,\ b_i \in \mathbb{Z}.$$

But the relation $b_0 + b_1x + \ldots + b_mx^m > 0$ can be written as

$$\exists\eta(\eta^2(b_0 + b_1x + \ldots + b_mx^m) = 1)$$

such that R can be assumed as having the form

$$(8.9.12.1) \qquad R = \exists\xi\ \exists\eta(P(\xi, \eta) = 0),\ \ P \in \mathbb{Z}[X, Y].$$

In this case, the decision upon R being or not a theorem is an immediate consequence of the solution of Problem 8.9.1.

More precisely, from the proposition already mentioned it follows that R is equivalent to a closed relation without bounded variables.

Now assume that the problem is already solved in the case of $n - 1$ variables and let $R = Q_1\xi_1\, Q_2\xi_2 - Q_n\xi_n(T(\xi_1, \xi_2, \ldots, \xi_n))$ be written in the prenex form.

As we have already remarked, we may assume $Q_n = \exists$. Using the same argument as in the case $n = 1$, the relation

$$Q_n\xi_n(T(x_1, \ldots, x_{n-1}, \xi_n))$$

can be assumed as having the form

$$\exists\xi\ \exists\eta(P(x_1, x_2, \ldots, x_{n-1}, \xi_1\, \eta) = 0),$$

where $P \in \mathbb{Z}[x_1, x_2, \ldots, x_{n-1}, \xi, \eta]$.

By Proposition 8.9.10, we can "eliminate quantifiers" in this relation, in such a way that (8.9.12.1) is equivalent to a relation with $n - 1$ free variables, hence R is equivalent to a relation with $n - 1$ bounded variables, which, by the induction hypothesis, concludes the proof of our proposition.

COROLLARY 8.9.13. *The formal system of real closed field theory is complete* (Definition 4.8.30).

Indeed, the above proof shows that every closed relation R of this formal system is equivalent to a closed relation without bounded variables. But if R is a closed relation without bounded variables in the formal system of real closed fields, then obviously either R or $\neg R$ is a theorem. It is sufficient to consider for instance a canonical disjunctive (conjunctive) form for R.

Proposition 8.9.10 can be used in order to obtain a simple proof of a theorem of Artin and Schreier, which answers a famous problem posed by Hilbert: the 17th problem of Hilbert (cf. Arch. Math. Phys., III, I (1901), 44–63, 213–237 or D. Hilbert, *Gesammelte Abhandlungen*, III, 1935, 290–329).

Having in view the proof of this theorem, we make some remarks connected with Proposition 8.9.10 and we recall some results from ordered field theory and from real closed field theory.

Remark 8.9.14. Let K_1, K_2 be two real closed fields such that $K_1 \subset K_2$ and consider $P \in K_1[X_1, X_2, \ldots, X_n]$. Then the polynomial P has a solution in K_1 if and only if it has a solution in K_2.

Indeed, the necessary and sufficient condition for P to have solution in K_2 can be expressed (cf. Proposition 8.9.10) by a finite number of polynomial conditions with elements in K_1.

PROPOSITION 8.9.15. *Every ordered field can be imbedded in a real closed field.*

(See for the proof S. Lang, *Algebra*, p. 273–279). By using Proposition 8.9.15 together with Remark 8.9.14 we obtain a stronger form of Remark 8.9.14.

Remark 8.9.16. Let K_1 be a real closed field, let K_2 be an ordered field containing K_1 and consider $P \in K_1[X_1, \ldots, X_n]$. The polynomial P has a solution in K_2 if and only if it has a solution in K_1.

DEFINITION. A field K is said to be *real* if the element -1 is not a sum of squares in K.

PROPOSITION 8.9.17. (a) *Every real closed field K has a unique ordered field structure: the positive elements of K are sums of squares.*

(b) *If the field K is real and $a \in K$ is an element of K which is not a sum of squares, then K has an ordered field structure such that a is negative.*

(c) *Every real field can be imbedded in a real closed field.*

(For the proof in S. Lang, *Algebra*, p. 273–274).

Example 8.9.18. If K is an ordered field, then $K(T_1, T_2, \ldots, T_n)$ is also an ordered field. If K is real closed, then $K(T_1, T_2, \ldots, T_n)$ is real.

Proof. It is sufficient to consider the case $n = 1$. If $R(T) \in K(T)$, we put by definition $R(T) \geqslant 0$ if and only if $\lim_{T \to \infty} R(T) = \infty$.

PROPOSITION 8.9.19. (Artin-Schreier). *Let L be a real closed field and let $R \in K(T_1, T_2, \ldots, T_n)$ be a rational fraction in n variables with coefficients in K. Assume that $R = \dfrac{P_1}{P_2}$, $P_1, P_2 \in K(T_1, \ldots, T_n]$, and that the following condition is fulfilled:*

$$\forall \xi_1\, \forall \xi_2, \ldots, \forall \xi_n \left(\left(((\xi_1, \xi_2, \ldots, \xi_n) \in K^n) \wedge \right.\right.$$

$$\left.\left. \wedge (P_2(\xi_1, \ldots, \xi_n) \neq 0)) \Rightarrow \frac{P_1(\xi_1, \xi_2, \ldots, \xi_n)}{P_2(\xi_1, \xi_2, \ldots, \xi_n)} \geqslant 0 \right) \right),$$

in other words, R takes only positive values.

Then R is a sum of squares (of elements from $K(T_1, T_2, \ldots, T_n)$).

Proof. Consider the proposition

$$(*) \qquad \exists \xi_1 \exists \xi_2 \ldots \exists \xi_n \left(\frac{(P_1(\xi_1, \xi_2, \ldots, \xi_n)}{(P_2(\xi_1, \xi_2, \ldots, \xi_n)} < \right.$$

$$\left. < 0 \vee (P_2(\xi_1, \ldots, \xi_n) = 0) \right)$$

By hypothesis, this proposition is obviously not valid (not true) in the field K. According to Remark 8.9.16 (transforming as usual the inequalities in equalities by introducing a new bounded variable, see for instance the proof of Proposition 8.9.12) it follows that this proposition is not valid in every ordered field containing the field K.

In particular, the proposition $(*)$ is not valid in the field $K(T_1, T_2, \ldots, T_n)$.

Now assume that the element $R = \dfrac{P_1}{P_2} \in K(T_1, \ldots, T_n)$ is not a sum of squares. Since $K(T_1, \ldots, T_n)$ is a real field (Example 8.9.18), from Proposition 8.9.17 a) it follows that there exists an order relation on $K(T_1, T_2, \ldots, T_n)$, with respect to which $R < 0$. In other words, we have

$$\frac{P_1(T_1, T_2, \ldots, T_n)}{P_2(T_1, T_2, \ldots, T_n)} < 0.$$

On the other hand, every order relation on $K(T_1, T_2, \ldots, T_n)$ induces the initial relation on K (Proposition 8.9.17 a)).

In this way, the proposition $(*)$ being valid in $K(T_1, T_2, \ldots, T_n)$, the Remark 8.9.16 shows that it is also valid in K, which is absurd.

§ 9. *On the formalist and the constructive point of view in mathematics*

We have seen that the development of "naive" set theory gave rise to some paradoxes. Of course the set theory as a whole was difficult to renounce at, since this theory had brought many elements which were precious for every branch of mathematics. As in other similar situations, a solution should be found, allowing simultaneously the revaluation and exploitation of the main results and concepts from naive set theory, as well as the avoiding of paradoxes.

One of these solutions has been the introduction of the concept of formal system. That is to say, the proposed solution was to present set theory as the study of theorems of a formal system: the formal system of set theory.

The basic idea in the framework of this system is the avoidance of every intuitive element in defining the concept of a set.

So, in this acception, a set is actually a term of a suitably defined formal system. Of course, some intuitive elements will nevertheless appear when one will have to formulate the axioms of this formal system.

In the following we intend to present, as an example, such a formal system for set theory.

Since we intend only to emphasize the main ideas of this system, we shall content ourselves to present a general formal system, with a very restrained number of axioms.

9.0. *The formal system of set theory.* The symbols

$$\mathscr{S} = (V, U, L, (\Phi_m)_{m \geqslant 0}, (P_m)_{m \geqslant 0})$$

of the formal system of set theory are defined by the conditions:

– V, U are finite sets;
– $\Phi_m = \emptyset$ for every $m \geqslant 0$;
– $P_2 = (=, \in)$;
– $P_m = \emptyset$ for $m \neq 2$.

In order to formulate the axioms of the formal system of set theory, we introduce the following notation:

– The relation $\forall\zeta((\zeta \in x) \Rightarrow (\zeta \in y))$ will be abbreviated as $x \subset y$ or $y \supset x$.

– If R is a relation defined by the system of symbols $\mathscr{S}$ containing the free variable x, we denote by $\mathrm{Coll}_x R$ the relation

$$\exists\eta\ \forall\xi((\xi \in \eta) \Leftrightarrow R(\xi)).$$

The description of the axioms $\mathscr{A}$ of the system $\varepsilon = (\mathscr{S}, \mathscr{A})$ of set theory is the following:

First of all $(\mathscr{S}, \mathscr{A})$ is an equalitary and quantified formal system. Besides the axioms specific to equalitary and quantified systems, the formal system of set theory has the following axioms:

9.1. *The extensionality axiom:* $\forall\xi\ \forall\eta(((\xi \subset \eta) \wedge (\eta \subset \xi)) \Rightarrow (\xi = \eta))$. Intuitively speaking, this axiom expresses the fact that any two sets with the same elements are equal.

9.2. *The axiom of the two elements set:*

$$\forall\xi\ \forall\eta\ \exists\zeta\ \forall\mu((\mu \in \zeta) \Leftrightarrow (\mu = \xi) \vee (\mu = \eta)).$$

Intuitively speaking, this axiom expresses the fact that if x and y are objects, then there exists a set whose objects are x and y.

9.3. *The axiom of the set of all subsets of a set:*

$$\forall\xi\ \exists\eta\ \forall\zeta((\zeta\in\eta)\Leftrightarrow(\zeta\subset\xi)).$$

This axiom asserts that for every set ξ, there exists a set η whose elements are the subsets of ξ.

9.4. *The union axiom:* $\forall\mu\ \exists\xi\ \forall\eta((\eta\in\xi)\Leftrightarrow\exists\zeta((\eta\in\zeta)\ \wedge\ (\zeta\in\mu)))$.
This axiom asserts that for every set μ, there exists the union of all sets which are elements of μ.

9.5. *The infinity axiom.* In order to simplify the writing of this axiom, we introduce the following abbreviation: $\chi(x)$ will designate the relation:

$$\forall\xi_1((\xi_1\in x)\Rightarrow\exists\xi_2((\xi_2\in x)\ \wedge\ (\xi_1\subset\xi_2)\ \wedge\ \neg(\xi_1=\xi_2))).$$

The relation $\chi(x)$ expresses the fact that for every ξ_1 which is an element of x, there exists a set strictly containing ξ_1 which is still an element of x.

With this notation, the infinity axiom writes as follows:

$$\exists\xi(\exists\eta((\eta\in\xi)\ \wedge\ \chi(\xi))).$$

In other words, the infinity axiom asserts the existence of a set ξ containing at least an element and having in addition the property that for every set ξ_1 which is an element of ξ there exists a set strictly containing ξ_1, which is an element of ξ.

Naively speaking, a set which satisfies the infinity axiom is the following:

$$\{\{0\},\{0,1\},\{0,1,2\},\{0,1,2,3\},\ldots\{0,1,2,\ldots n\}\ldots\}.$$

9.6. *The axiom of choice.* In order to formulate this axiom, we use the following abbreviations:

– $\theta(x)$ will designate the relation:

$$\forall\xi_1((\xi_1\in x)\Rightarrow\exists\xi_2(\xi_2\in\xi_1))$$

asserting that x is a set of non-empty sets;

– $i_0(x,y)$ will designate the formula:

$$\exists\zeta((\zeta\in x)\ \wedge\ (\zeta\in y))$$

asserting that the sets x, y have a non-empty intersection;

— $i_1(x, y)$ will designate the formula:

$$\forall\xi_1\ \forall\xi_2(((\xi_1 \in x) \wedge (\xi_1 \in y) \wedge (\xi_2 \in x) \wedge (\xi_2 \in \eta)) \Rightarrow$$
$$\Rightarrow (\xi_1 = \xi_2))$$

asserting that the intersection of the sets x, y contains at most an element;

— $\delta(x)$ will designate the formula

$$\forall\xi_1\ \forall\xi_2(((\xi_1 \in x) \wedge (\xi_2 \in x) \Rightarrow \neg i_0(\xi_1, \xi_2))$$

asserting that every two elements of x are disjoint sets. With these notation, the axiom of choice writes as follows:

$$\forall\zeta((\theta(\zeta) \wedge \delta(\zeta)) \Rightarrow \exists\xi\ \forall\eta((\eta \in \zeta) \Rightarrow (i_0(\xi, \eta) \wedge i_1(\xi, \eta))).$$

In the case of the formal system of set theory, the paradoxes described in § 4.2, § 4.3 appear no longer. Indeed, the first paradox (Russel's paradox) is deduced from the existence of a set a with the following property:

$$(x \in a) \Leftrightarrow (x \notin x)$$

For this paradox to appear in the formal system of the theory, the following relation of this system must be a theorem:

$$\exists\xi\ \forall\eta((\eta \in \xi) \Leftrightarrow (\eta \notin \eta)).$$

In other words, the relation

$$R = (x \notin x)$$

must be a collectivizing relation, a fact which is not at all obvious.

As for the second paradox (cf. 4.3), it cannot even be formulated in the framework of the formal system of set theory. Indeed, it is defined in terms which cannot be interpreted in the formal system of set theory.

Now we intend to show using some examples how the main results of set theory can be revaluated in the formal system of set theory. We have to introduce one further axiom.

9.7. *The substitution axiom.* Let $R(x, y)$ be a relation of the formal system of set theory, containing two variables x, y. We designate by R^0 the relation:

$$R^0 = \exists\xi\ \forall\eta_1\ \forall\eta_2((R(\xi, \eta_1) \wedge R(\xi, \eta_2)) \Rightarrow (\eta_1 = \eta_2)).$$

With this notation, the substitution axiom writes as follows:

$$\forall v(R^0 \Rightarrow \exists\zeta\ \forall\eta((\eta \in \zeta) \Leftrightarrow \exists\mu((\mu \in v) \wedge R(\mu, \eta))$$

which means that if $R(x, y)$ is a function-determining relation, then for every set v there exists the image of v through the function determined by $R(x, y)$.

9.8. An important consequence of the substitution axiom is the existence of a subset described by means of a property.

More precisely, let S be a relation in the formal system of set theory containing the free variable x. Under these conditions, the following relation is a theorem:

$$\forall\xi\ \exists\eta\ \forall\zeta((\zeta \in \eta) \Leftrightarrow (\zeta \in \xi) \wedge S(\zeta)), \tag{9.8.1}$$

in other words, for every set ξ, there exists a subset consisting of all elements of ξ satisfying the relation S.

Indeed, the relation R given by

$$R = ((x = y) \wedge S(x))$$

defines a function (cf. 9.7) i.e. R^0 is a theorem. Applying the substitution axiom, this yields the following relation:

$$\forall\xi\ \exists\eta\ \forall\zeta((\zeta \in \eta) \Leftrightarrow (\exists\mu((\mu \in \xi) \wedge R(\xi, \zeta))),$$

which is obviously equivalent to (9.8.1).

Theorem 9.8.1 is called the comprehension theorem.

9.9. The formal system of set theory which has been expounded above obviously allows the formulation and the proof of some theorems.

The further development of the theory in the direction of recovering the main results of naive set theory is carried through by introducing some supplementary symbols which have to designate, for instance, particular sets of functions, unions, intersections or products of sets, etc. In other words, we actually have to work in a richer formal system.

In the following, we give some details about this point of view in a somewhat more general context.

§ 10. *Derived terms in the theory of formal systems*

10.1. Let $(\mathcal{S}, \mathcal{A})$ be a formal system given by its symbols

$$\mathcal{S} = \{V, U, L, (\Phi_m)_{m\in\mathbb{N}}, (P_m)_{m\in\mathbb{N}}\}$$

and by its axioms $\mathcal{A}$.

In defining the terms and the relations of this formal system, we have used only the elements of the free monoid $\mathcal{M}(F)$ whose basis is the set F defined by

$$F = V \cup U \cup L \cup (\bigcup_{m\in\mathbb{N}} \Phi_m) \cup (\bigcup_{m\in\mathbb{N}} P_m)$$

The exclusive use of the elements of the monoid $\mathcal{M}(F)$ in developing the theory of a formal system leads to great difficulties of a typographical nature and would require unusual mental efforts. We have therefore to introduce some abbreviating symbols. Technically speaking, the introducing of these new symbols constituting the actual object of mathematical definitions is equivalent to a modification of the initial formal system $(\mathcal{S}, \mathcal{A})$. In this way, one actually studies a new formal system, which is of course closely related to the initial one and which allows the creation of new mathematical concepts. For the sake of exemplification, we content ourselves to describe the following three ways of introducing definitions or notations in the framework of a formal system.

(I) Let $R(x_1, x_2, \ldots, x_n)$ be a relation of the formal system $(\mathcal{S}, \mathcal{A})$, containing the free variables $x_1, x_2, \ldots, x_n$ and let ρ_R be an element which does not belong to the set $\mathcal{M}(F)$.

We define a new formal system $(\mathcal{S}', \mathcal{A}')$ as follows:

$$\mathcal{S}' = (V, U, L, (\Phi_m)_{m\in\mathbb{N}}, (P'_m)_{m\in\mathbb{N}}),$$

$$P'_m = P_m \text{ if } m \neq n,$$

$$P'_n = P_n \cup \{\rho_R\},$$

$$\mathscr{A}' = \mathscr{A} \cup \{R(x_1, x_2, \ldots, x_n) \Leftrightarrow \rho_R(x_1, x_2, \ldots, x_n)\}.$$

Example 10.2. Let $(\mathscr{S}, \mathscr{A})$ be the formal system of arithmetics and let $R(x, y)$ be the relation defined as follows:

$$R(x, y) = \exists\zeta(x + \zeta = y).$$

This relation is usually abbreviated as

$$R(x, y) \Leftrightarrow (x \leqslant y).$$

Example 10.3. In the framework of the formal system of set theory, one currently uses the abbreviation

$$(x \subset y) \Rightarrow \forall\zeta((\zeta \in x) \Rightarrow (\zeta \in y)).$$

(II) Let $\tau(x_1, x_2, \ldots, x_n)$ be a term of the formal equalitary system $(\mathscr{S}, \mathscr{A})$ and let t_τ be an element which does not belong to $\mathscr{M}(F)$. We define a new formal system $(\mathscr{S}', \mathscr{A}')$ by putting

$$\mathscr{S}' = (V, U, L, (\Phi'_m)_{m \in \mathbb{N}}, (P_m)_{m \in \mathbb{N}}),$$

$$\Phi'_m = \Phi_m \text{ if } m \neq n,$$

$$\Phi'_n = \Phi_n \cup \{t_\tau\},$$

$$\mathscr{A}' = \mathscr{A} \cup \{\tau(x_1, x_2, \ldots, x_n) = t_\tau(x_1, x_2, \ldots, x_n)\}.$$

Example 10.4. In the formal system of arithmetics, one uses the notation:

$$0' = 1,\ 0'' = 2,\ 0''' = 3, \ldots$$

Example 10.5. In the formal system of group theory, one uses the notation

$$xx = x^2,\ xxx = x^3, \ldots$$

(III) Let $R(x_1, x_2, \ldots, x_n, y)$ be a relation of the formal system $(\mathscr{S}, \mathscr{A})$. Assume that the relation

(10.5.1) $\qquad \exists\eta\, \forall\xi_1\, \forall\xi_2 \ldots \forall\xi_n(R(\xi_1, \xi_2, \ldots, \xi_n, \eta))$

is a theorem of this formal system.

Let ε_R be an element which does not belong to $\mathcal{M}(F)$. We define a new formal system $(\mathcal{S}', \mathcal{A}')$ by putting

$$\mathcal{S}' = (V, U, L, (\Phi')_{m \in \mathbb{N}}, (P_m)_{m \in \mathbb{N}}),$$

$$\Phi'_m = \Phi_m \text{ if } m \neq n,$$

$$\Phi'_n = \Phi_n \cup \{\varepsilon_R\},$$

$$\mathcal{A}' = \mathcal{A} \cup \{\forall \xi_1 \forall \xi_2, \ldots, \forall \xi_n (R(\xi_1, \ldots, \xi_n, \varepsilon_R(\xi_1, \ldots \\ \ldots, \xi_n)))\}.$$

In the case of equalitary systems, besides relation (10.5.1), one usually supposes that the relation

$$\forall \xi_1 \forall \xi_2 \ldots \forall \xi_n \forall \eta_1 \forall \eta_2 ((R(\xi_1, \xi_2, \ldots, \xi_n, \eta_1) \wedge \\ \wedge R(\xi_1, \ldots, \xi_n, \eta_2)) \Rightarrow (\eta_1 = \eta_2))$$

is a theorem. Intuitively speaking, in this last case, $\varepsilon_R(\xi_1, \ldots, \xi_n)$ is the single object satisfying the condition

$$\forall \xi_1 \forall \xi_2 \ldots \forall \xi_n (R(\xi_1, \ldots, \xi_n), \varepsilon_R(\xi_1, \ldots, \xi_n)).$$

Example 10.6. In the formal system of group theory, the theorem

$$\exists \eta \forall \xi (\xi \eta = \eta \xi = 1)$$

leads to the introduction of an element $i \in \Phi_1$ which is in turn abbreviated as

$$i(x) = x^{-1}.$$

Example 10.7. In the formal system of set theory, consider the relation

$$R(x, y, z) = \forall \mu ((\mu \in Z) \Leftrightarrow (\mu = x) \vee (\mu = y)).$$

The axiom of the two elements set shows that the relation

$$\exists \zeta \forall \xi \forall \eta (R(\xi, \eta, \zeta))$$

is a theorem.

Consequently, for every pair of free variables x, y we can consider — in an extended formal system — the term $\varepsilon_R(x, y)$ which is usually denoted by $\{x, y\}$. The resulting formal system can be further extended by considering terms of the form $\{\{x\}, \{y, z\}\}$.

The term $\{\{x\}, \{x, y\}\}$ is usually called the ordered pair defined by x, y and is denoted by (x, y).

In this system the following proposition holds:

PROPOSITION 10.7.1. *The following relation is a theorem:*

$$\{\{x\}, \{x, y\}\} = \{\{x'\}, \{x', y'\}\} \Leftrightarrow ((x = x') \wedge (y = y')).$$

Proof. The implication $\Leftarrow$ is trivial. In order to prove the converse, we first remark that this implication is obvious in the case $x = y$. Hence we may assume $x \neq y$. Using the extensionality axiom, the following two cases are possible:

$$\{x\} = \{x', y'\}, \{x, y\} = \{x'\},$$
$$\{x\} = \{x'\}, \{y\} = \{y'\}.$$

But, still using the extensionality axiom, the first case cannot occur.

In a similar way, one can introduce ordered triplets $(x, (y, z)) = (x, y, z)$, ordered 4-truples $(x, (y, z, t)) = (x, y, z, t)$, etc.

Example 10.8. Still in the formal system of set theory, or rather in the formal system obtained by its extension (cf. Example 10.3), consider the relation

$$P(x, y) = \forall\zeta((\zeta \in y) \Leftrightarrow (\zeta \subset x)).$$

By the axiom of the set of subsets, the relation

$$\exists\eta\ \forall\xi(P(\xi, \eta))$$

is a theorem. Consequently, a new formal system can be defined, such that for every free variable x, the term $\mathscr{P}(x)$, called the set of subset of the set x, is defined.

Example 10.9. Still in the formal system of set theory, consider the relation

$$U(x, y) = \forall\zeta(((\zeta \in y) \Leftrightarrow \exists\mu((\mu \in x) \wedge (\zeta \in \mu))).$$

As a consequence of the union theorem, for every set x, one can define the term usually denoted by $\bigcup_{\xi \in x} \xi$, i.e. the union of all sets which are elements of the set x.

In a formal system obtained from the system of set theory in which terms of the form $P(x)$, (x, y), $x \cup y$ have been introduced, we obviously have the following theorem:

$$(x \in a) \wedge (y \in b) \Rightarrow (x, y) \in \mathscr{P}(\mathscr{P}(a \cup b)).$$

Example 10.10. Let R be a relation in the formal system of set theory. Consider the relation:

$$\forall \xi (\xi \in y \Leftrightarrow R).$$

Assuming that the relation

$$\exists \eta \, \forall \xi (\xi \in y \Leftrightarrow R)$$

is a theorem (the comprehension theorem, cf. 9.8.1), then in a suitable formal system one can consider the term usually denoted by

$$\{x/R\}$$

such that the relation

$$(x \in \{x/R\}) \Leftrightarrow R,$$

is a theorem.

In particular, one can introduce the term $x \times y$ by the axiom

$$x \times y = \{(\xi, \eta) \in \mathscr{P}(\mathscr{P}(x \cup y)) | (\xi \in x) \wedge (\eta \in y)\}.$$

§ 11. *The constructive approach to set theory*

Besides the formal system theory, including as a particular case the formal system of set theory (cf. § 10), some other solutions have also been suggested for avoiding the rise of paradoxes in set theory. Among these, we shall briefly consider the constructive approach to set theory. This approach is opposite to the formalist one, i.e. to the formal system theory.

For this reason, the constructive approach is also rather difficult to expose.

Let us remark that most today mathematicians do share, as a means of avoiding paradoxes in set theory, the point of view of formal system theory, particularly developed in the papers of E. Zermelo, A. Frenkel, D. Hilbert, P. Bernays, N. Bourbaki.

The constructive point of view has for a long while been considered rather as a philosophical curiosity.

Strangely enough, the constructive direction has ultimately gained in interest and even in significance as consequence of the recent development of algorithmical mathematics.

Moreover, recent papers in category theory (Lawvere, Thierney) seem to add a further interest to this theory.

In the following, we try to point out the essentials of the constructive approach to set theory.

One major difference between the formalist approach and the constructive one refers to the meaning of the proposition "there exists an element with the property...".

Consider first the position of the formalist approach. To this aim, assume that we study the formal system of arithmetics (cf. 4.4.1, 4.7.2, d)). If R is an arithmetical relation which involves the free variable x and does not involve the bound variable ξ, then the assertion "there exist ξ, such that $R(\xi)$ is true" has the precise meaning that the relation

$$\exists\xi(R(\xi))$$

is a theorem in the formal system of arithmetics.

In the constructive approach, the situation is completely different. From this point of view, the assertion "there exists ξ such that $R(\xi)$ is true" means that we are able to give a method (an algorithm) in order to find out a positive integer n, for which $R(n)$ is true.

We can gain a certain insight into the constructive point of view upon the subsets of $\mathbb{N}$, by saying that a subset of $\mathbb{N}$ makes sense as a constructive set if it is a recursive subset of $\mathbb{N}$ (cf. Definition 8.2).

The constructive proof of a theorem is usually more intricate than the formalist one but it should be acknowledged that such a proof often throws more light on the true nature of the object whose existence was to be proved: recall for instance the proof of the existence of the greatest common divisor of two integers using Euclid's algorithm.

On the other hand, there are simple mathematical theorems for which one cannot hope to obtain constructive proofs.

Such an example is yielded by the following elementary theorem of analysis (cf. E. Bishop, *Foundations of Constructive Analysis)*: For every bounded sequence of rational numbers, there exists a least upper bound. In order to obtain the constructive proof of this theorem, we obviously have to indicate a method allowing, for every bounded sequence $\{x_i\}_{i\in\mathbb{N}}$ of rational numbers, the effective construction of a sequence *

$$b_1, b_2, \ldots, b_k, \ldots$$

of rational numbers and that of a sequence

$$m_1, m_2, \ldots, m_k, \ldots$$

of integers such that the following conditions be fulfilled:

(i) for every j and for every k, we have $x_j < b_k + \dfrac{1}{k}$;

(ii) for every k, we have $x_{m_k} > b_k - \dfrac{1}{k}$.

Now assume that the sequence $\{x_i\}_{i\in\mathbb{N}}$ satisfies the condition

$$x_i = 0 \text{ or } 1.$$

Taking $k = 3$, it would follow that for every such sequence $\{x_i\}_{i\in\mathbb{N}}$, we are able to find a natural number $m_3 = M$ with the following properties:

$$x_j < b_3 + \frac{1}{3} \quad \text{for every } j \in \mathbb{N}, \tag{11.1}$$
$$x_M > b_3 - \frac{1}{3}.$$

Obviously, two cases can occur:
(a) $x_M = 0$;
(b) $x_M = 1$.

* We obviously suppose that the map

$$x: \mathbb{N} \to \mathbb{Q}$$

defining our sequence satisfies the elementary effectiveness (constructivity) requirements.

In the case (a), condition (i) implies for every j

$$x_j < b_3 + \frac{1}{3} < \frac{1}{3} + \frac{1}{3} = \frac{2}{3}$$

i.e. $x_j = 0$ for every j.

Summing up, the existence of a constructive proof for the theorem stated above implies in particular the possibility of indicating a method allowing for every sequence of rational numbers such that $x_i = 0$ or 1 to effectively decide whether this sequence is identically vanishing or there exists a non-vanishing term of this sequence.

But some of the most difficult unsolved problems of mathematics can be reduced to just such a decision problem.

In order to justify this last assertion, consider "Fermat's last theorem". Let

$$f: \{1, 2, 3, \ldots, n, \ldots\} \times \{3, 4, 5, \ldots, n, \ldots\} \to \{0, 1\}$$

be the function defined as follows:

$$f(n, k) = \begin{cases} 0 \text{ if the equation } X^k + Y^k = Z^k \text{ has no} \\ \quad \text{integer non-trivial solution } (n_1, n_2, n_3) \text{ such} \\ \quad \text{that } 0 < n_i < n,\ i = 1, 2, 3 \\ 1 \text{ otherwise} \end{cases}$$

With this function, one obviously can associate a sequence such that the problem of solving Fermat's theorem is equivalent to the problem of deciding whether the least upper bound of this sequence is zero.

A similar setting could be used for the "four colours problem" or for "Riemann's hypothesis".

The requirements which constructive mathematics imposes to the "there exists" concept determines it to abandon some basic axioms of quantified formal systems.

For instance, the axiom

$$\neg(\forall\xi(\neg R(\xi))) \Rightarrow \exists\xi(R(\xi))$$

(see 4.7.2b), (μ)), cannot be accepted in the constructive approach, since in many cases this axiom would ensure the proof of the relation $\exists\xi(R(\xi))$ without any effective method being available for the construction of an object verifying the relation ξ.

The constructive approach differs from the formalist one in interpreting the logical implication $R \Rightarrow S$. From the constructive point of view, the relation

$$R \Rightarrow S$$

is considered as true if there exist a method for deducing a proof of S from a proof R.

(It is to be admitted that in the constructive approach the word "proof" has no precise meaning. For a better understanding of our formulation, we shall refer to examples of the above type).

Concerning the operation of disjoining relations, in the constructive approach, the relation $R \vee S$ is true if:

(a) one of the relations R, S is true;

(b) there exists an effective method allowing to indicate which one of the relations R, S is true.

From this point of view, it follows that the relation

$$R \vee (\neg R)$$

is not generally true for constructive mathematics, in other words the principle of the *excluded middle* does not hold in this approach.

For concluding, we point out that constructive mathematics accepts the implication

$$R \Rightarrow \neg(\neg R)$$

but rejects

$$\neg(\neg R) \Rightarrow R.$$

In order to give some idea of the constructive style in mathematics, we sketch in the following one of the constructive ways of introducing real numbers. We shall suitably adapt the usual construction of real numbers following Cantor.

Let $\rho: \mathbb{Q} \to \mathbb{N}$ be a bijection which will be kept fixed in all subsequent constructions and definitions. The recursivity concepts will be considered with respect to this bijection (cf. Definition 8.2).

DEFINITION 11.1. We say that a (constructive) real number is given if we give a couple (f, h) of recursive functions

$$f: \mathbb{N} \to \mathbb{Q}$$
$$h: \mathbb{Q} \to \mathbb{N}$$

such that the following conditions are fulfilled:

(a) $\forall k(h(k+1) \geqslant h(k) \geqslant k)$;

(b) $\forall k\, \forall n\, \forall m\left((m \geqslant n \geqslant h(k)) \Rightarrow |f(m)-f(n)| \leqslant \dfrac{1}{10^k}\right)$.

DEFINITION 11.2. Let (f, h), (f_1, h_1) be two real numbers. We say that they are recursively (constructively) equal if a recursive function

$$g: \mathbb{Q} \to \mathbb{N}$$

exists, such that the following conditions be fulfilled:

— $\forall k(g(k+1) \geqslant g(k) \geqslant k)$,

— $\forall k\, \forall n\left((n \geqslant k) \Rightarrow |f(n)-f_1(n)| \leqslant \dfrac{1}{10^k}\right)$.

DEFINITION 11.3. Let (f_1, h_1), (f_2, h_2) be two real numbers. We say that they are constructively unequal if a rational number $k > 0$, an integer M and a recursive function

$$\lambda: \mathbb{N} \to \mathbb{N}$$

exist, such that

$$\forall n((n \geqslant M) \Rightarrow |(f_1(\lambda(n)) - f_2(\lambda(n))| \geqslant k).$$

The arithmetics of constructive real numbers. The sum and the product of two constructive real numbers can be defined in an obvious way. Eventually, we can say that we get an ordered commutative ring with unit, which will be denoted by $\mathcal{K}_{\mathbb{R}}$. This ring is not a field; nevertheless, the following propositions hold:

PROPOSITION 11.4. *If (f, h) is a constructive real number which is constructively distinct from zero, then it is invertible in the ring $\mathcal{K}_{\mathbb{R}}$. The image of $\mathcal{K}_{\mathbb{R}}$ through the obvious ring homomorphism*

$$\beta: \mathcal{K}_{\mathbb{R}} \to \mathbb{R}$$

is a subfield of $\mathbb{R}$

For the proof, one makes use of the following result:

PROPOSITION 11.5. *If $\xi \in \beta(\mathcal{K}_{\mathbb{R}})$ and $\xi \neq 0$, then there exists $\eta \in \mathcal{K}_{\mathbb{R}}$ such that η is constructively distinct from zero and $\beta(\eta) = \xi$.*

In fact, $\beta(\mathcal{K}_{\mathbb{R}})$ is a countable real closed ordered field.

Chapter III

Introduction to modern algebraic geometry

§ 1. *Generalities*

Let k be a commutative field; denote by $k[X_1, X_2, \ldots, X_n]$ the polynomial ring in n variables with coefficients in k, by $\underline{a}$ an ideal of the ring $k[X_1, \ldots, X_n]$ and by K an extension of the field k.

The central problem of algebraic geometry is the study of the set

$$\mathscr{V}_{\underline{a}}(K) = \{(\xi_1, \ldots, \xi_n) \mid \xi_i \in K,\ i = 1, 2, \ldots n \text{ and}$$

$$P(\xi_1, \ldots, \xi_n) = 0,\ \forall\, P \in \underline{a}\}.$$

The ideal $\underline{a}$ being given, the set $\mathscr{V}_{\underline{a}}(K)$ depends on the field K. A classical theorem of algebra – the so-called Hilbert's Nullstellensatz – ensures that if the ideal $\underline{a}$ does not coincide with the whole ring $A = k[X_1, X_2, \ldots, X_n]$ and if the field K is algebraically closed, then $\mathscr{V}_{\underline{a}}(K)$ is a non-empty set. For the study of $\mathscr{V}_{\underline{a}}(K)$, various structures, especially of an algebraic or geometric nature, are traditionally used. As will be seen below, this set can be naturally endowed with a topology, the Zariski topology, and on the resulting topological space one can define a sheaf of rings. In other words, $\mathscr{V}_{\underline{a}}(K)$ is in particular the underlying set of a geometric structure: a ringed space.

If $L \supset K$ is an extension of K, we have a canonical injection

$$\mathscr{V}_{\underline{a}}(K) \to \mathscr{V}_{\underline{a}}(L).$$

This yields a remarkable functor

$$\mathscr{V}_{\underline{a}} : \mathscr{E}_k \to \text{Ens}$$

defined on the category $\mathscr{E}_k$ of the field extensions of k and taking value in the category of sets.

One can rightly say that the first important problem of algebraic geometry is the study of the functors of the type $\mathscr{V}_{\underline{a}}$.

For a given extension $k \subset K$, the set $\mathscr{V}_{\underline{a}}(K)$ depends on the ideal $\underline{a}$. In particular, we have the formulae:

(*i*) $\underline{a} \subset \underline{b} \Rightarrow \mathscr{V}_{\underline{a}}(K) \supseteq \mathscr{V}_{\underline{b}}(K)$,

(*ii*) $\mathscr{V}_{\bigvee_i a_i}(K) = \bigcap_i \mathscr{V}_{a_i}(K)$,

(*iii*) $\mathscr{V}_{\underline{a} \cap \underline{b}}(K) = \mathscr{V}_{\underline{a}}(K) \cap \mathscr{V}_{\underline{b}}(K)$.

In order to prove *iii*, consider an element $\xi = (\xi_1, \ldots, \xi_n) \in \mathscr{V}_{\underline{a} \cap \underline{b}}(K)$. We have to show that $\xi \in \mathscr{V}_{\underline{a}}(K) \cup \mathscr{V}_{\underline{b}}(K)$. By *reductio ad absurdum*, suppose that $\xi \notin \mathscr{V}_{\underline{a}}(K) \cup \mathscr{V}_{\underline{b}}(K)$. Then it would follow that there exists $P \in \underline{a}$ and $Q \in \underline{b}$ such that

$$P(\xi) \neq 0, \; Q(\xi) \neq 0.$$

But since $P \cdot Q \in \underline{a} \cap \underline{b}$, we have $P(\xi) \cdot Q(\xi) = (PQ)(\xi) = 0$, which is absurd.

We note that in general there exist distinct ideals $\underline{a}, \underline{b}$ such that, for every extension $k \subset K$, we have

$$\mathscr{V}_{\underline{a}}(K) = \mathscr{V}_{\underline{b}}(K),$$

i.e. such that the functors $\mathscr{V}_{\underline{a}}, \mathscr{V}_{\underline{b}}$ be equal.

Indeed, for every ideal $\underline{a}$ of the ring $k[X_1, \ldots, X_n]$, we have

$$\mathscr{V}_{\underline{a}}(K) = \mathscr{V}_{\sqrt{\underline{a}}}(K).$$

The formulae *ii*, *iii*, together with the obvious relations

$$\mathscr{V}_{(0)}(K) = K^n$$

$$\mathscr{V}_{k[X_1, \ldots, X_n]}(K) = 0$$

show that the sets of the form $\mathscr{V}_{\underline{a}}(K)$ are the closed sets of a topology on K^n, called the Zariski k-topology.

This topology induces a topology on the set $\mathscr{V}_{\underline{a}}(K)$, called the Zariski k-topology of $\mathscr{V}_{\underline{a}}(K)$.

As already remarked, on the topological space $\mathscr{V}_{\underline{a}}(K)$ one can introduce a sheaf of rings. To this end we consider first the ring $R_{\underline{a}} = k[X_1, \ldots$ $\ldots, X_n]/\underline{a}$, which is actually a k-algebra.

As it is well-known, for every commutative ring A, one considers the set Spec(A) of all its prime ideals. With these notation, we have a natural map

$$\lambda_k : \mathscr{V}_{\underline{a}}(K) \to \operatorname{Spec}(R_{\underline{a}})$$

assigning to every element $\xi \in \mathscr{V}_{\underline{a}}(K)$ the ideal $\underline{p}_\xi$ which is the kernel of the ring morphism

$$\dot{\varepsilon}_\xi : R_{\underline{a}} \to K,$$

$$\dot{P} \to \dot{P}(\xi).$$

In other words, we have

$$\underline{p}_\xi = \{\dot{P} \in R_{\underline{a}} \mid \dot{P}(\xi) = 0\}.$$

The ideal $\underline{p}_\xi$ is sometimes called the *locality* of the point. It is readily seen that the Zariski topology on $\mathscr{V}_{\underline{a}}(K)$ is the inverse image through λ_k of the Zariski topology on Spec$(R_{\underline{a}})$. The morphism λ_k can be further specified as follows:

First of all, if A is any commutative ring and if $\underline{p}$ is a prime ideal of A, then the local ring $A_{\underline{p}}$ obtained from A by localization with respect to the multiplicative system $S_{\underline{p}} = A/\underline{p}$ has the field of the quotients of the integrity domain $A/\underline{p}$ as its residual field. In particular if A is a k-algebra, then the residual field of the local ring is an extension of the field k, which will be denoted by $k(\underline{p})$. Consequently, we have

$$k \subset k(\underline{p}).$$

Assuming that the ring A is a k-algebra, for any field extension $k \subset K$, we define a subset $\operatorname{Spec}_{K/k}(A)$ as follows:

The prime ideal $\underline{p}$ is by definition an element of $\text{Spec}_{K/k}(A)$ if the field extension $k \subset k(\underline{p})$ is a subextension of the extension $k \subset K$, i.e. we have a morphism

$$k(\underline{p}) \hookrightarrow K$$

such that the following diagram

$$\begin{array}{ccc} k & \hookrightarrow & k(\underline{p}) \\ & \searrow & \downarrow \\ & & K \end{array}$$

be commutative.

It is readily seen that λ_k maps the set $\mathscr{V}_{\underline{a}}(K)$ onto the set $\text{Spec}_{K/k}(R_{\underline{a}})$.

The only fact needing a proof is: every ideal $\underline{p} \in \text{Spec}_{K/k}(R)$ is the locality of a point $\xi \in \mathscr{V}_{\underline{a}}(K)$. To this end, we first consider the homomorphism

$$\eta_{\underline{p}}: R_{\underline{a}} \to k(\underline{p}) \hookrightarrow K$$

whose kernel is obviously $\underline{p}$. Let $\xi = (\xi_1, \ldots, \xi_n) \in K^n$ be defined by

$$\xi_i = \eta_{\underline{p}}(X_i \bmod a), \quad i = 1, 2, \ldots, n.$$

We have $\xi \in \mathscr{V}_{\underline{a}}(K)$ and $\eta_{\underline{p}} = \varepsilon_\xi$, $\lambda_k(\xi) = \underline{p}$, in other words $\underline{p}_\xi = \underline{p}$.

One shows that on the topological space $\mathscr{V}_{\underline{a}}(K)$ endowed with the Zariski topology, a sheaf of rings $\mathcal{O}_{\underline{a}}$ can be introduced in such a way that for every point $\xi \in \mathscr{V}_{\underline{a}}(K)$ the following identity holds:

$$\mathcal{O}_{\underline{a},\xi} = (R_{\underline{a}})_{\underline{p}_\xi} \ (\underline{p}_\xi = \lambda_k(\xi) = \text{the locality of the point } \xi).$$

The description of the sheaf $\mathcal{O}_{\underline{a}}$ is particularly simple in the special case when the ideal $\underline{a}$ is prime. In this case, denote by $Q_{\underline{a}}$ the field of the quotients of the ring $R_{\underline{a}}$. For every prime ideal $\underline{p}$ of the ring $R_{\underline{a}}$, the local ring $(R_{\underline{a}})_{\underline{p}}$ is obviously a subring of the field $Q_{\underline{a}}$. For every open set U of $\mathscr{V}_{\underline{a}}(K)$, we put

$$\mathcal{O}_{\underline{a}}(U) = \bigcap_{\xi \in U} (R_{\underline{a}})_{\underline{p}_\xi}.$$

If U, V are two open sets with $U \subset V$, then there exists an obvious canonical inclusion

$$\mathcal{O}_{\underline{a}}(U, V): \bigcap_{\xi \in V} (R_a)_{\underline{p}_\xi} \to \bigcap_{\xi \in U} (R_a)_{\underline{p}_\xi}.$$

Summing up, the above constructions show that starting from an ideal $\underline{a}$ of the ring $k[X_1, \ldots, X_n]$, one can define in a natural way some geometric structures associated with the set $\mathscr{V}_{\underline{a}}(K)$ for every extension $k \subset K$ of the field k.

Although these constructions are of a natural (functorial) character, they nevertheless have the main shortcoming of depending not only on the ideal $\underline{a}$ but also on the field extension $k \subset K$.

Consequently, the problem has been posed of finding out a geometric object, preferably a ringed space, which should dominate the situation so described, in the sense of having an invariant character, i.e. of being independent of the field extension $k \subset K$, and which should nonetheless possess the structures described above. This invariant has been found out in the sixties by A. Grothendieck.

In order to introduce this invariant which is currently called an affine scheme, first note that the functor $\mathscr{V}_{\underline{a}}$ can be extended in a natural way to a functor still denoted by $\mathscr{V}_{\underline{a}}$ defined upon the category Alg/k of commutative k-algebras:

$$\mathscr{V}_{\underline{a}}: \mathrm{Alg}/k \to \mathrm{Ens}$$

$$A \to \{(\alpha_1, \ldots, \alpha_n) \mid \alpha_i \in A,\ i = 1, \ldots, n,$$

$$P(\alpha_1, \ldots, \alpha_n) = 0 \quad \forall P \in \underline{a}\}.$$

A first ground for considering this new functor consists in the fact that, unlike the category $\mathscr{E}_k$ of field extensions of k, the category Alg/k enjoys some reasonable properties: it has direct products and sums, kernels and cokernels, inductive and projective limits, etc. Moreover, the following important proposition holds:

PROPOSITION 1.1. *The functor*

$$\mathscr{V}_{\underline{a}}: \mathrm{Alg}/k \to \mathrm{Ens}$$

is representable.

In fact, the couple $(R_{\underline{a}}, \xi)$, where $\xi = (\xi_1, \ldots, \xi_n)$ is defined by the formula

$$\xi_i = X_i \bmod \underline{a}, \quad i = 1, 2, \ldots, n,$$

is a representation couple for the functor $\mathscr{V}_{\underline{a}}$.

The canonical bijection

$$\mathscr{V}_{\underline{a}}(A) \to \mathrm{Hom}_{\mathrm{Alg}/k}(R_a, A)$$

$$\xi \to \alpha_\xi$$

is defined by

$$\alpha_\xi(P) = \dot{P}(\xi).$$

Proposition 1.1 suggests the following definition:

DEFINITION 1.2. An *affine scheme* over the field k is any representable functor defined on the category Alg/k of commutative k-algebras and taking values in the category of sets.

We denote by $\mathscr{S}ch/k$ the full subcategory of representable functors defined on Alg/k and taking values in the category of sets.

Example 1.2.1. The n-dimensional affine space $\mathbb{A}_k^n$ over the field k is defined by the formula

$$\mathbb{A}_k^n(A) = A^n \text{ for every } k\text{-algebra } A.$$

The resulting functor is representable by the representation couple

$$(k[X_1, X_2, \ldots, X_n], (X_1, X_2, \ldots, X_n)).$$

1.2.2. *The multiplicative k-group M_k* is defined by the formula: $M_k(A) =$ the set of invertible elements of the k-algebra A. As a representation couple of this functor one can take

$$\left(k\left[X, \frac{1}{X}\right], X\right).$$

Remark. 1.3. From Definition 1.3 it follows that $\mathscr{S}ch/k$ is the dual category of Alg/k. Consequently the category $\mathscr{S}ch/k$ has direct sums, kernels, cokernels, etc.

§ 2. *The spectrum of a commutative ring*

Most definitions introduced so far are purely algebraic in character. One may wish to establish some connections between these concepts and some geometric concepts, in view of transferring geometric methods to the algebraic problems formulated above.

The main source of geometric objects is presently known to be the category of topological spaces with structure sheaves, a particular role in this direction being played by the category of ringed spaces, i.e. the category of all couples of the form $(X, \mathcal{O}_X)$, where X is a topological space and $\mathcal{O}_X$ is a sheaf of commutative rings on X.

For the convenience of the reader, we recall that given two ringed spaces $(X, \mathcal{O}_X)$, $(Y, \mathcal{O}_Y)$, a morphism

$$(X, \mathcal{O}_X) \to (Y, \mathcal{O}_Y)$$

is given by a couple (f, φ), where f is a continuous map

$$f: X \to Y$$

and φ is a morphism of sheaves on Y:

$$\varphi: \mathcal{O}_Y \to f_*(\mathcal{O}_X).$$

Such a morphism of ringed spaces induces in particular for every point $x \in X$ a ring homomorphism

$$\varphi_x: \mathcal{O}_{Y, f(x)} \to \mathcal{O}_{X, x}.$$

Example. Let $(\{*\}, \bar{A})$ be a ringed space defined on the space $\{*\}$ whose underlying set contains a single point, namely the point $*$, and whose sheaf of rings is the constant sheaf defined by the ring A. This sheaf will be denoted by $\bar{A}$.

Let $(X, \mathcal{O}_X)$ be an arbitrary ringed space. Any morphism of ringed spaces

$$(f, \varphi): (\{*\}, \bar{A}) \to (X, \mathcal{O}_X)$$

defines a point $x \in X$:

$$x = f(*).$$

The sheaf $f_*(\bar{A})$ is obviously defined by

$$f_*(\bar{A})(U) = \begin{cases} A \text{ if } x \in U, \\ \text{a set with a single element if } x \notin U. \end{cases}$$

It follows that φ defines a map which assigns to every open set U containing x, a ring homomorphism

$$\varphi(U): \mathcal{O}_X(U) \to A$$

such that, if $V \subset U$, the following diagram be commutative

$$(**) \qquad \begin{array}{ccc} \mathcal{O}_X(U) & \xrightarrow{\varphi(U)} & \\ \downarrow & & A \\ \mathcal{O}_X(V) & \xrightarrow{\varphi(V)} & \end{array}$$

This system of morphisms induces a unique morphism

$$(***) \qquad \mathcal{O}_{X,x} \to A.$$

By the definition of inductive limits, it follows that the systems of morphisms (**) are in a one-to-one correspondence with the morphisms of the form (***).

A first result obtained by Grothendieck shows that the category $\mathcal{S}ch/k$ can be identified to a subcategory of $\boldsymbol{Rg}$, more precisely to a subcategory of the category of ringed spaces whose structure sheaf is a sheaf of k-algebras.

In this way we are led to a geometric definition of the concept of affine scheme. Above we denoted by Spec(A) the set of prime ideals of the ring A, calling it the spectrum (or prime spectrum) of A.

A topology, called the Zariski topology is introduced on Spec(A) in the following way:

The closed subsets in the Zariski topology of Spec(A) are the subsets of the form

$$V(\underline{a}) = \{\underline{p} \in \operatorname{Spec}(A) | \underline{p} \supset \underline{a}\},$$

where $\underline{a}$ is an arbitrary ideal of the ring A.

Indeed, it is readily seen that the following relations hold:

(*i*) $V(A) = \emptyset$,

(*ii*) $V((0)) = \mathrm{Spec}(A)$,

(*iii*) $V(\underline{a} \cap \underline{b}) = V(\underline{a}) \cup V(\underline{b})$,

(*iv*) $V(\bigvee_i a_i) = \bigcap_i (a_i)$.

Among them, only *iii* requires a proof. So, let $\underline{p}$ be an element of $V(\underline{a} \cap \underline{b})$. We have to show that $\underline{p} \in V(\underline{a})\, UV(\underline{b})$. By ***reductio ad absurdum***, assume that we have simultaneously

$$\underline{p} \notin V(\underline{a}), \quad \underline{p} \notin V(\underline{b}).$$

It would follow that there exist $f \in \underline{a}$ and $g \in \underline{b}$ such that

$$f \notin \underline{p},\ g \notin \underline{p}.$$

But from this it follows that $f \cdot g \in \underline{a} \cap \underline{b}$, and, using the hypothesis, that $f \cdot g \in \underline{p}$, in contradiction with the fact that $\underline{p}$ is prime. For every element $f \in A$, consider the subset $D(f)$ of $\mathrm{Spec}(A)$, defined by

$$D(f) = \{p \in \mathrm{Spec}(A) | f \notin \underline{p}\}.$$

In other words, $D(f)$ consists of those prime ideals $\underline{p}$ for which the canonical image of f in the residual field of the local ring $A_{\underline{p}}$ is non-vanishing.

The following relations obviously hold:

$$D(f) = \mathrm{Spec}(A) - V((f)),$$

$$\mathrm{Spec}(A) - V(\underline{a}) = \bigcup_{f \in \underline{a}} (\mathrm{Spec}(A) - V((f)),$$

showing that the sets of the form $D(f)$ are open in the Zarisky topology and yield a basis for the open sets of this topology.

Example 2.1. The closed subsets of the space $\mathrm{Spec}\,(Z) = \{(0), (2), (3), (5), \ldots, (p), \ldots\}$, ($p$ positive integer) have the form

$$V((a)) = \{(p_1), (p_2), \ldots, (p_r)\},$$

where $a = p_1^{n_1} p_2^{n_2} \ldots p_r^{n_r}$.

Exemple 2.1 shows that the topological space Spec$A$ is, generally speaking, not separated. Indeed, every open set of Spec($\mathbb{Z}$) contains the point (0). Moreover, the adherence of (0) is the whole space.

DEFINITION 2.2. A topological space X is said to be separated in the sense of Kolmogorov if for every pair of distinct points in X, there exists an open set containing one of them and not containing the other one.

PROPOSITION 2.3. *The topological space Spec*(A) *is separated in the sense of Kolmogorov.*

Proof. Consider $\underline{p}, \underline{q} \in \operatorname{Spec} A$, $\underline{p} \neq \underline{q}$. Hence there exists an element $f \in A$ such that

$$(f \in \underline{p} \text{ and } f \notin \underline{q}) \text{ or } (f \in \underline{q} \text{ and } f \notin \underline{p}).$$

Assume that the first possibility holds. Then we have

$$\underline{q} \in D(f) \text{ and } \underline{p} \notin D(f).$$

PROPOSITION 2.4. *The topological space Spec*(A) *is quasicompact, i.e. from any open covering of X one can extract a finite subcovering.*

Proof. It is sufficient to show that if

$$\operatorname{Spec}(A) = \bigcup_{\alpha} D(f_\alpha), \tag{2.4.1}$$

then we can find a finite number of indices $\alpha_1, \ldots, \alpha_n$ such that

$$\operatorname{Spec}(A) = \bigcup_{i=1}^{n} D(f_{\alpha_i}).$$

First note that in view of relation (2.4.1) the ideal generated by the family $(f_\alpha)_\alpha$ coincides with the ring A. Indeed, we have

$$\emptyset = \complement\Big(\bigcup_{\alpha} D(f_\alpha)\Big) = \bigcap_{\alpha} \complement D(f_\alpha) = \bigcap_{\alpha} V((f_\alpha)) = V\Big(\bigvee_{\alpha} (f_\alpha)\Big). \tag{2.4.2}$$

Since by hypothesis A has a unit element and since every proper ideal of A is included in a maximal ideal, (2.4.2) shows that the ideal generated by $(f_\alpha)_\alpha$ coincides with A. Consequently, for some indices $\alpha_1, \ldots, \alpha_n$ we have

$$\bigvee_i (f_{\alpha_i}) = A,$$

whence

$$\mathrm{Spec}(A) = \bigcup_{i=1}^{n} D(f_{\alpha_i})$$

obviously follows.

Application 2.4.2.1. *The spectrum of a Boole ring. Let A be a Boole ring. Concerning the topological space SpecA and its open subsets D(f), the following properties hold:*

(1) $D(f) = D(g) \Leftrightarrow f = g$.

(2) $D(f) \cup D(g) = D_{f \cup g}$.

(3) $D(f)$ *is simultaneously open and closed.*

(4) *Spec*(A) *is separated.*

(5) *Every subset of* Spec(A) *which is simultaneously open and closed has the form* $D(f)$.

(6) *Spec*(A) *is totally disconnected.*

Proof. (1) From $D(f) = D(g)$ it follows that we have simultaneously $f^n = ug$, $g = vf^n$, consequently $f \geqslant g$, $g \geqslant f$, whence $f = g$.

(2) First we prove the equivalence:

$$\underline{p} \in D(f) \cup D(g) \Leftrightarrow \underline{p} \ni (1-f)\cdot(1-g).$$

Indeed, if $\underline{p} \in D(f)$, then $f \notin \underline{p}$ and $1 - f \in \underline{p}$, for $f.(1-f) = 0 \in \underline{p}$ and $\underline{p}$ is prime, consequently, $(1-f)(1-g) \in \underline{p}$. Conversely, if $(1-f)(1-g) \in \underline{p}$, then either $1 - f \in \underline{p}$, or $1 - g \in \underline{p}$, and so on. But we have

$$f \cup g = f + g + f\cdot g,$$

consequently we get the equivalence

$$\underline{p} \ni (1-f)\cdot(1-g) \Leftrightarrow \underline{p} \ni 1 - f \cup g \Leftrightarrow \underline{p} \not\ni f \cup g \Rightarrow$$
$$\Rightarrow \underline{p} \in D(f \cup g)$$

(3) Taking into account property (2) which has been already proved, the obvious relation $D(fg) = D(f) \cup D(g)$ and the relations

$$f.(1-f) = 0,$$

$$f + (1-f) = 1$$

we get

$$D(f) \cap D(1-f) = \emptyset$$

$$D(f) \cup D(1-f) = \operatorname{Spec}(A)$$

consequently the complement of $D(f)$ is $D(1-f)$.

(4) Consider $\underline{p}, \underline{q} \in \operatorname{Spec}(A)$, $\underline{p} \neq \underline{q}$. Since $\operatorname{Spec}(A)$ is separated in the sense of Kolmogorov (Proposition 2.3), it follows that there exists $f \in A$ such that $\underline{p} \in D(f)$ and $\underline{q} \notin D(f)$. Consequently, $D(1-f) \ni \underline{q}$, and $D(f) \cap D(1-f) = \emptyset$.

(5) Let $G \subset \operatorname{Spec}(A)$ be an open set. Since the family $(D(f))$ is a basis for the topology of $\operatorname{Spec}(A)$, it follows that we have $G = \bigcup_{i \in I} D(f_i)$. But G being also closed and $\operatorname{Spec}(A)$ being quasicompact, it follows that G is quasi-compact, hence we have $G = \bigcup_{j=1}^{n} D(f_{ij}) = D\left(\bigcup_{j=1}^{n} f_{ij}\right)$.

(6) Let E be a subset of $\operatorname{Spec}(A)$ containing two elements $\underline{p}, \underline{q}$ and consider $f \in A$ such that $D(f) \ni \underline{p}$ and $D(f) \ni \underline{q}$.
We have $D(1-f) \ni \underline{q}$ and $D(f) \cap D(1-f) = \emptyset$, $D(f) \cup D(1-f) = \operatorname{Spec}(A)$. Consequently

$$E = (E \cap D(f)) \cup (E \cap D(1-f)),$$

showing that E is not connected.

STONE'S REPRESENTATION THEOREM. *Every Boole algebra is isomorphic to the Boole algebra of the open and closed subsets of a compact and totally disconnected space.*

Proof. Consider a Boole algebra L, its associated Boole ring A, $X = \operatorname{Spec}(A)$ and the isomorphism $L \ni f \mapsto D(f) \subset X$.

Obviously, for every topological space, the set of the subsets which are both open and closed has a natural structure of Boole algebra.

DEFINITION 2.4.2.2. Let $(A, \cap, \cup, \neg)$ be a Boole algebra and let $(A, +, \cdot)$ be its associated Boole ring. The subset $M \subset A$ is said to be an ideal of the Boole algebra A if M is an ideal of the associated Boole ring.

PROPOSITION 2.4.2.3. *The subset M of A is an ideal of the Boole algebra A if and only if the following conditions are fulfilled:*

(*i*) $x, y \in M \Rightarrow x \cup y \in M$,

(*ii*) $x \in M$ şi $y \leqslant x \Rightarrow y \in M$.

The proof is straightforward.

DEFINITION 2.4.2.4. A non-empty set $\mathscr{F}$ of A is said to be a *filter* if the following conditions are fulfilled:

(*i*) $x, y \in \mathscr{F} \Rightarrow x \cap y \in \mathscr{F}$.

(*ii*) $x \in \mathscr{F}$ and $y \geqslant x \Rightarrow y \in \mathscr{F}$.

Remark. If the filter $\mathscr{F}$ contains the element 0, then $\mathscr{F} = A$. This filter is called the improper filter.

PROPOSITION 2.4.2.5. *The subset $M \subset A$ is an ideal of A if and only if the subset $\neg(M)$ is a filter of the Boole algebra A.*

The proof is straightforward.

PROPOSITION 2.4.2.6. *Let $M \subset A$ be an ideal of A and let $\mathscr{F} = \neg(M)$ be the associated filter. The ideal M is prime if and only if the filter $\mathscr{F}$ satisfies the condition:*

$$x \cup y \in \mathscr{F} \Rightarrow x \in \mathscr{F} \text{ or } y \in \mathscr{F}.$$

DEFINITION 2.4.2.7. A filter $\mathscr{F}$ on A is said to be prime if its associated ideal is prime.

PROPOSITION 2.4.2.8. Every prime filter $\mathscr{F}$ is maximal (in the set of proper filters).

Proof. We show that in a Boole ring every prime ideal is maximal. Indeed, if $\underline{p} \subset A$ is a prime ideal of the Boole ring A, and if $x \in A$, then the relation $x^2 = x$ implies $x(1 - x) = 0 \bmod \underline{p}$, hence $x \in \underline{p}$ or $x = 1 \bmod \underline{p}$, consequently $A/\underline{p}$ is a field isomorphic to the field $\mathbb{F}_2$ of the mod 2 integers.

Example 2.4.2.9. Let M be an arbitrary set and let A be the ring of functions defined on M and taking values in $\mathbb{F}_2$. The ring A is obviously a Boole ring: it actually is the Boole ring associated with the Boole algebra of all subsets of M.

We have a canonical injection

$$M \to \mathrm{Spec}(A),$$

$$m \mapsto \{f \in A \mid f(m) = 0\}.$$

consequently $\mathrm{Spec}(A)$ is a compact space which contains M.

Example 2.4.2.10. Let A be a commutative ring and let $I(A) = \{x \in A \mid x^2 = x\}$ be the set of its idempotent elements. $I(A)$ has a structure of Boole algebra:

$$e \vee f = e + f - ef$$

$$e \wedge f = ef$$

$$\neg e = 1 - e$$

$I(A)$ has also a structure of Boole ring, defined by

$$e \oplus f = (e - f)^2,$$

$$e \cdot f = ef.$$

2.4.2.11. On the topological space $\mathrm{Spec}(A)$ one can introduce a sheaf of rings $\tilde{A}$ whose fibres are local rings. More precisely:

$$\tilde{A}_{\underline{p}} = A_{\underline{p}}, \ \forall \underline{p} \in \mathrm{Spec}(A).$$

In order to describe the sheaf $\tilde{A}$, we recall some elementary results:

– The set of all nilpotent elements of a ring coincides with the intersection of all its prime ideals.

Indeed, let $f \in A$ be such that f belongs to every prime ideal of A and assume by *reductio ad absurdum* that f is not nilpotent. In this case the ring A is not reduced to zero and has a unit element, consequently it has at least one prime ideal. Let $\underline{p}$ be the inverse image of this ideal by the canonical ring homomorphism

$$A \xrightarrow{i_f} A_f.$$

Obviously $\underline{p}$ is a prime ideal of A which does not contain f, which is absurd.

– Let $f, g \in A$ be two elements of A such that $D(f) \subset D(g)$. Then the image of g by the canonical homomorphism

$$A \xrightarrow{i_f} A_f$$

is invertible in A_f. Consequently there exists a unique ring homomorphism

$$A_g \xrightarrow{i_{g,f}} A_f$$

such that the following diagram

$$\begin{array}{ccc} A_g & \xrightarrow{i_{g,f}} & A_f \\ & {}_{i_g}\nwarrow \quad \nearrow_{i_f} & \\ & A & \end{array}$$

is commutative.

For the proof, it is sufficient to show the existence of an integer n such that f^n belongs to the ideal (g). In other words, we have to show that the image of f in the ring $A/(g)$ is a nilpotent element.

On behalf of the previous result, it suffices to show that f belongs to all prime ideals of A containing g. But we have

$$f \in \bigcap_{\underline{p} \in \complement D(f)} \underline{p} \subseteq \bigcap_{\underline{p} \in \complement D(g)} \underline{p}$$

since the hypothesis implies

$$\complement D(f) \supset \complement D(g).$$

— Note that the previous result shows that if $f, g \in A$ are such that $D(f) = D(g)$, then the ring morphism defined above yields a canonical isomorphism between the rings A_f, A_g.

— For every point $\underline{p} \in \operatorname{Spec} A$, the set of open subsets of the form $D(f)$ containing $\underline{p}$ form a decreasing filtering system which will be denoted by $V_D(\underline{p})$.

Indeed, we have the general relation:

$$D(fg) = D(f) \cap D(g).$$

— In the constructions which follow, for every open set U of the form $D(f)$, we once for ever choose an element f such that $U = D(f_U)$. The invariants which will be defined will of course depend on this choice, but each time it is readily seen that changing the elements f, the corresponding invariants are still canonically isomorphic.

— With the conventions stated above, consider for every point $\underline{p}$ of $\operatorname{Spec} A$ the functor

$$A_{D,\underline{p}}: V_D(\underline{p}) \to Rg$$

$$D(f) \mapsto A_f$$

$$[D(f) \subset D(g)] \mapsto i_{g,f}.$$

We have a canonical isomorphism

$$\varinjlim A_{D,\underline{p}} \xrightarrow{\sim} A_{\underline{p}}.$$

Indeed, for every $D(f) \in V_D(\underline{p})$, we have a canonical homomorphism

$$A_f \to A_{\underline{p}}$$

since the multiplicative system defined by f belongs to the set $A \setminus \underline{p}$. On behalf of an obvious commutativity property, this yields the ring morphism

$$\varinjlim A_{D,p} \to A_p$$

which is readily seen to be an isomorphism.

– Let X be a topological space and let $\mathscr{B}$ be a set of open subsets of X, satisfying the following conditions:

(*i*) $\mathscr{B}$ is a basis of the topology of X.

(*ii*) $\mathscr{B}$ is stable with respect to finite intersections.

Suppose we are given a contravariant functor of sets, defined on the category associated with $\mathscr{B}$

$$F: \mathscr{B}^0 \to \text{Ens}$$

and satisfying the usual conditions for sheaves, obviously formulated only for elements of the set $\mathscr{B}$.

Then there exists a unique sheaf of sets $\mathscr{F}$ on the space X, such that for every set U in $\mathscr{B}$ we have

$$F(U) = \mathscr{F}(U).$$

It is sufficient to put for every open set V of X

$$\mathscr{F}(V) = \varprojlim_{\substack{U \subset V \\ U \in \mathscr{B}}} F(U).$$

The details of the proof are left to the reader. Using these results, in order to define the sheaf A on Spec(A), it is sufficient to put

$$\tilde{A}(D(f)) = A_f,$$

$$\tilde{A}(D(f) \subset D(g)) = i_{g,f}.$$

Note that if the ring $\tilde{A}$ is a k-algebra, then the sheaf $\tilde{A}$ defined above is actually a sheaf of k-algebras.

In the following we intend to use these constructions for identifying the category of schemes to a category of ringed spaces.

To this end we first point out a subcategory Ann loc of the category Ann of ringed spaces.

The objects of the category Ann loc of *local ringed spaces* are ringed spaces with the property that for every point $x \in X$, the fibre $\mathcal{O}_{X,x}$ of the structure sheaf $\mathcal{O}_X$ at the point x is a local ring.

The morphisms of local ringed spaces are by definition the morphisms (f, φ) of ringed spaces, for which the morphisms

$$(2.4.3) \qquad \varphi_x: \mathcal{O}_{Y,f(x)} \to \mathcal{O}_{X,x}$$

induced on the fibres are morphisms of local rings.

It is useful to consider also the category Ann loc/k of local ringed spaces whose fibres are k-algebras. Obviously, to the morphisms of this category one imposes the extra condition that they also be morphisms of k-algebras.

Example. Every differentiable variety $(X, \mathscr{D}_X)$ is a local ringed space whose fibres are $\mathbb{R}$-algebras.

If $(X, \mathscr{D}_X), (Y, \mathscr{D}_Y)$ are differentiable varieties and if $f: X \to Y$ is a differentiable map, then f obviously induces a morphism $(f, \varphi): (X, \mathscr{D}_X) \to$ $\to (Y, \mathscr{D}_Y)$ of Ann loc/$\mathbb{R}$.

Conversely, if $(X, \mathscr{D}_X)$, $(Y, \mathscr{D}_Y)$ are differentiable varieties and if $(f, \varphi): (X, \mathscr{D}_X) \to (Y, \mathscr{D}_Y)$ is a morphism of Ann loc/$\mathbb{R}$, then the map $f: X \to Y$ is differentiable and φ is, obviously, uniquely determined by f.

Indeed, consider a point $x_0 \in X$, an open set V of Y containing $f(x_0)$, an $\mathbb{R}$-valued differentiable map u on V and the ring homomorphism

$$\varphi(V): \mathcal{O}_Y(V) \to (f_* \mathcal{O}_X)(V) = \mathcal{O}_X(f^{-1}(V))$$

induced by φ.

We show that

$$\varphi(V)(u) = u_0(f_{|f^{-1}(V)}).$$

Take an $x \in f^{-1}(V)$. We have to show that $[\varphi(V)(u)](x) = u(f(x))$. Let $u_{f(x)}$ be the germ of differentiable map at the point $f(x)$ defined by u. If u_0 is the value taken by u at $f(x)$, we have

$$u_{f(x)} = u_0 + \mu,$$

where μ belongs to the maximal ideal $\underline{m}_{f(x)}$ of the local ring $\mathcal{O}_{Y,f(x)}$. It is sufficient to prove the relation

$$[\varphi_x(u_{f(x)})](x) = u(f(x)),$$

where

$$\varphi_x: \mathcal{O}_{Y,f(x)} \to \mathcal{O}_{X,x}$$

is the morphism of local rings and of $\mathbb{R}$-algebras induced by φ. But we have

$$\varphi_x(u_{f(x)}) = \varphi_x(u_0 + \mu) = \varphi_x(u_0) + \varphi_x(\mu) = u_0 + \varphi_x(\mu)$$

whence the required relation follows since $\varphi_x(\mu) \in \underline{m}_x$ and consequently

$$\varphi_x(\mu)(x) = 0.$$

Let $(X, \mathcal{O}_X)$ be a local ringed space and let x be a point of the topological space X. We denote by $k(x)$ the residual field of the local ring $\mathcal{O}_{X,x}$ and by $\underline{m}_x$ the maximal ideal of the local ring $\mathcal{O}_{X,x}$. With any global section $s \in \mathcal{O}_X(X)$ of the sheaf $\mathcal{O}_X$ two important invariants are associated, namely:

— The germ of the section s at the point x, i.e. the canonical image of s in the fibre $\mathcal{O}_{X,x}$.

This germ is usually denoted by s_x.

— The "value" of the section s at the point x, i.e. the canonical image of the germ s_x in the residual field $k(x)$, which will be denoted by $s(x)$.

Note that two distinct sections $s, t \in \mathcal{O}_X(X)$ may have the same values at every point of the space X. This fact has given rise to some perplexity at early stages of the development of scheme theory, but subsequent results have shown that this phenomenon is quite normal. First of all, we remark that the "pathology" we are dealing with can arise if and only if nilpotent elements are present in some fibres of the structure sheaf $\mathcal{O}_X$. But very elementary problems of a geometric nature exist, which lead in a natural way to schemes having nilpotent elements in their fibres. We shall content ourselves with the following two examples:

Let $D(k) = k[X]/(X^2)$ be the algebra of dual numbers over k. If $(\mathrm{Spec}(A), \tilde{A})$ is a ringed space, where A is a k-algebra, and if $\underline{p}$ is an arbitrary element of $\mathrm{Spec}(A)$, then the set of the morphisms of ringed spaces

$$(f, \varphi): (\mathrm{Spec}(D(k)),\ \widetilde{D(k)}) \to (\mathrm{Spec}(A), \tilde{A}),$$

for which $f(u) = \underline{p}$ coincides with the tangent vector space at $\underline{p}$ to $\mathrm{Spec}(A)$:

$$\mathrm{Hom}_k(m_p/m_p^2, k).$$

For any $\underline{p} \in \mathrm{Spec}(A)$, we obviously have a monomorphism of ringed spaces $(\mathrm{Spec}\,(A_p/(_pA_p)^n, (A_p/_pA_p)) \to (\mathrm{Spec}(A), \tilde{A})$, defining the n-th order infinitesimal neighborhood of the point $\underline{p}$ in $\mathrm{Spec}A$.

If $(X, \mathcal{O}_X), (Y, \mathcal{O}_Y)$ are two local ringed spaces, if

$$(f, \varphi): (X, \mathcal{O}_X) \to (Y, \mathcal{O}_Y)$$

is a morphism of local ringed spaces and if $t \in \mathcal{O}_Y(Y)$, then by means of the morphism $\varphi(Y): \mathcal{O}_Y(Y) \to f_*(\mathcal{O}_X)(Y)$ we can associate with the section t a section $s = \varphi(Y)(t) \in \mathcal{O}_X(X)$.

On the other hand, for every point $x \in X$ we have the morphism of local rings

$$\varphi_x: \mathcal{O}_{Y, f(x)} \to \mathcal{O}_{X,x}$$

inducing the field extension

$$\varphi(x): k(f(x)) \hookrightarrow k(x).$$

The values $s(x), t(f(x))$ satisfy the relation

$$\varphi(x)(t(f(x))) = s(x).$$

PROPOSITION 2.5. *We have a contravariant functor*

$$\mathrm{Spec}: \mathrm{Rg\,com}^\circ \to \mathrm{Top}$$

assigning to every commutative ring A its prime spectrum $\mathrm{Spec}(A)$.

Proof. We have only to define for every ring homomorphism

$$u: A \to B$$

a continuous map

$$\operatorname{Spec}(u)\colon \operatorname{Spec}(B) \to \operatorname{Spec}(A).$$

The map $\operatorname{Spec}(u)$, which in the following will be abbreviated to ${}^a u$, is defined by

$${}^a u(\underline{q}) = u^{-1}(\underline{q}).$$

The continuity of ${}^a u$ follows from the equality

$$(2.5.1) \qquad {}^a u^{-1}(V(E)) = V(u(E)). \quad \forall E \subset A.$$

The proof of (2.5.1) is a consequence of the following simple implications:

$$\underline{q} \in V(u(E)) \Rightarrow \underline{q} \supseteq u(E) \Rightarrow u^{-1}(\underline{q}) \supseteq u^{-1}(u(E)) \supset$$

$$\supset E \Rightarrow {}^a u(\underline{q}) \supset E \Rightarrow {}^a u(\underline{q}) \in V(E) \Rightarrow q \in {}^a u^{-1}(V(E)).$$

$$\underline{q} \in {}^a u^{-1}(V(E)) \Rightarrow {}^a u(q) \in V(E) \Rightarrow {}^a u(q) \supseteq E \Rightarrow u^{-1}(q) \supseteq$$

$$\supseteq E \Rightarrow \underline{q} \supseteq u(u^{-1}(q)) \supseteq E.$$

Remark 2.5.1. The inverse image of a maximal ideal may not be a maximal ideal. Consider for instance the inclusion

$$\mathbb{Z} \hookrightarrow \mathbb{Q}.$$

The inverse image of the (0) ideal from $\mathbb{Q}$ is not maximal in $\mathbb{Z}$.

COROLLARY 2.5.2. *Let $\underline{a} \subset A$ be an ideal of the ring A and let*

$$\pi\colon A \to A/\underline{a}$$

be the canonical homomorphism.

The map ${}^a\pi$ yields a homeomorphism of the space $\operatorname{Spec}(A/a)$ onto the subspace $V(\underline{a})$ of $\operatorname{Spec}(A)$.

Proposition 2.6. *The functor Spec defined in Proposition* 2.5 *can be extended to a contravariant functor still denoted by Spec*

$$\text{Spec} : \text{Rg com}^\circ \to \text{Ann loc}$$

from the dual of the category of commutative rings to the category of local ringed spaces.

The functor Spec yields an equivalence between the dual of the category of commutative unitary rings and the full subcategory of Ann loc generated by the local ringed spaces of the form $(\text{Spec } A, A)$ *where A is any commutative unitary ring.*

Proof. We have already seen how to assign to every ring homomorphism $u: A \to B$ a continuous map ${}^a u: \text{Spec}(B) - \text{Spec}(A)$. Now we show that to u we can also assign a morphism of sheaves of rings

$${}^u\varphi : \tilde{A} \to ({}^a u)_*(\tilde{B}).$$

In order to define ${}^u\varphi$, we write the relation

$${}^a u^{-1}(D(f)) = D(u(f)),$$

allowing to define ${}^u\varphi$ as follows:

For every open set $D(f)$ of $\text{Spec}(A)$, the morphism ${}^u\varphi(D(f))$ is the unique ring morphism

$$\tilde{A}(D(f)) = A_f \xrightarrow{{}^u\varphi(D(f))} B_{u(f)} = \tilde{B}(D(u(f))) = \tilde{B}({}^a u^{-1}(D(f))) =$$

$$= ({}^a u)_*(\tilde{B})\,(D(f))$$

for which the following diagram commutes:

$$\begin{array}{ccc} A & \xrightarrow{\quad u \quad} & B \\ {\scriptstyle i_f}\downarrow & & \downarrow{\scriptstyle i_{u(f)}} \\ A_f & \xrightarrow{{}^u\varphi(D(f))} & B_{u(f)} \end{array}$$

In the following, we have to assign to every morphism of ringed spaces

$$(2.6.1) \qquad (f, \varphi): (\mathrm{Spec}(B), \tilde{B}) \to (\mathrm{Spec}(A), \tilde{A})$$

a ring morphism

$$u : A \to B.$$

To this end, we put by definition

$$u = \varphi(\mathrm{Spec}(A)) : \tilde{A}(\mathrm{Spec}(A)) = A \to B = f_*(\tilde{B})(\mathrm{Spec}(A)).$$

The key point in proving that the functor so defined yields an equivalence between the considered categories consists in showing that in any morphism of the type (2.6.1), the component φ uniquely determines the continuous map f. It is noteworthy that in the proof of this property the fact that only morphisms of local ringed spaces have been considered plays an essential role.

Moreover, the result holds for more general morphisms of local ringed spaces, of the form

$$(f, \varphi): (X, \mathcal{O}_X) \to (\mathrm{Spec}(A), \tilde{A}).$$

Indeed, if $x \in X$ is any point of X, then we have

$$f(x) = \{a \in A | [\varphi(\mathrm{Spec}(A))(a)](x) = 0\}.$$

COROLLARY 2.7. *The category* Sch/k *of k-schemes is equivalent to the full subcategory of the category Ann loc/k generated by objects of the form (Spec (A), $\tilde{A}$), where A is a k-algebra.*

§ 3. *Schemes. Chevalley schemes*

The following definition, inspired by analogies from the theory of differentiable or analytic manifold, of fibre bundles, etc., is a natural one:

DEFINITION 3.1. A ringed space $(X, \mathcal{O})$ is said to be a *scheme* if it is locally isomorphic with a ringed space of the form (Spec$A$, $\tilde{A}$), i.e. to an affine scheme. We say that $(X, \mathcal{O})$ is a *k-scheme* if it is locally isomorphic to an affine scheme over the field k.

3.2. In this section we indicate a method of Chevalley for defining schemes.

First recall the following very elementary facts:

Let E be a set and let $\mathscr{T}$ be a collection of subsets of E. Assume that for every element M of $\mathscr{T}$, a topology τ_M on the set M is given. A topology τ on E can be introduced as follows:

A subset U of E is by definition open in the topology τ if for every element M of $\mathscr{T}$, the set $M \cap U$ is open in the topology τ_M.

It is readily seen that a subset F of E is closed in the topology τ if and only if for every element M of $\mathscr{T}$, the subset $M \cap F$ is closed in the topology τ_M.

Indeed, we obviously have

$$(\complement_E F) \cap M = \complement_M (F \cap M).$$

Moreover, for every element M of $\mathscr{T}$, the topology induced by τ on M is seen to be less fine than τ_M. In other words, the identical map

$$M \xrightarrow{\mathrm{id}_M} M$$

is continuous from the topology τ_M to the topology $\tau|_M$.

These elementary considerations will be applied to the following case:

For any field K, denote by $\mathscr{L}\mathrm{oc}(K)$ the set of the local subrings of K. We obviously have

$$\mathscr{L}\,\mathrm{co}(K) \subset \mathscr{P}(K).$$

For every subring A of K, consider the map

$$\mu_A \colon \mathrm{Spec}(A) \to \mathscr{L}\,\mathrm{oc}(K),$$

$$\underline{p} \mapsto A_{\underline{p}}.$$

μ_A is readily seen to set-up a bijection between $\mathrm{Spec}(A)$ and the subset $L(A)$ of $\mathscr{L}\,\mathrm{oc}(K)$ consisting of the local subrings of K of the form $A_{\underline{p}}$. Moreover,

$$A = \bigcap_{\underline{p} \in \mathrm{Spec}(A)} A_{\underline{p}}.$$

In this way, the subsets of the form $L(A)$ yield a collection of subsets of $\mathscr{L}\mathrm{oc}(K)$.

On the other hand, using μ, we can introduce by structure transport a topology on the subset $L(A)$ of $\mathscr{L}\mathrm{oc}(K)$, corresponding to the Zariski topology of $\mathrm{Spec}(A)$. As above, this allows to define a topology on $\mathscr{L}\mathrm{oc}(K)$, with respect to which a subset is open if and only if its intersection with $L(A)$ is open. The resulting topology on $\mathscr{L}\mathrm{oc}(K)$ will be called the Zariski topology.

DEFINITION 3.3. Let M, N be elements of $\mathscr{L}\mathrm{oc}(K)$. We say that N is a specialization of M, if there exists a prime ideal $\underline{p}$ of N such that $M = N_{\underline{p}}$.

Obviously, the prime ideal $\underline{p}$ of N is uniquely determined by M. If A is a subring of K and if $L \in \mathscr{L}\mathrm{oc}(K)$ is the field of the quotients of A, then for every prime ideal $\underline{p}$ of A, the local ring $A_{\underline{p}}$ is a specialization of L.

More generally, if $\underline{p}, \underline{q}$ are prime ideals of A and if $\underline{p} \subset \underline{q}$, then $A_{\underline{q}}$ is a specialization of $A_{\underline{p}}$. Indeed, we have the identity

$$A_{\underline{p}} = (A_{\underline{q}})_{\underline{p}A_{\underline{q}}}.$$

More precisely, if $\underline{p}, \underline{q}$ are prime ideals of A, then $A_{\underline{q}}$ is a specialization of $A_{\underline{p}}$ if and only if $\underline{p} \subset \underline{q}$.

For every element M of $\mathscr{L}\mathrm{oc}(K)$, we denote by $S(M)$ the set of the elements N which are specializations of M.

PROPOSITION 3.4. *The intersection $S(M) \cap L(A)$ is non-empty if and only if there exists a prime ideal $\underline{p}$ of A such that $M = A_{\underline{p}}$. In this case, we have*

$$S(M) \cap L(A) = \{A_{\underline{q}} \mid \underline{q} \supset \underline{p}\}. \tag{3.4.1}$$

Proof. Consider $N \in S(M) \cap L(A)$. It follows that N is a local ring of the form $A_{\underline{q}}$ and in addition $M = (A_{\underline{q}})_{\underline{p}A_{\underline{q}}}$ for some prime ideal $\underline{p}$ of A. In other words, $M = A_{\underline{p}}$, and obviously $\underline{p} \subset \underline{q}$. Formula (3.4.1) is obvious.

PROPOSITION 3.5. *Let M be an arbitrary element of the set $\mathscr{L}\mathrm{oc}(K)$. The adherence of the subset M in the Zariski topology on $\mathscr{L}\mathrm{oc}(K)$*

coincides with the set $S(M)$:

$$\{\overline{M}\} = S(M).$$

Proof. Proposition 3.4 shows that the set $S(M)$ is closed. If F is a closed subset of $\mathscr{L}\text{oc}(K)$ and if A is a subring of K, then, assuming that $F \cap L(A) \supset A_{\underline{p}}$, we have the inclusion:

$$F \cap L(A) \supset \{A_{\underline{q}} | \underline{q} \supset \underline{p}\}.$$

It follows that $S(M)$ is the least closed set containing M.

PROPOSITION 3.6. *Let A be a Noetherian subring of the field K. The topology induced on $L(A)$ by the Zariski topology of $\mathscr{L}\text{oc}(K)$ coincides with the topology induced on $L(A)$ by the Zariski topology of Spec(A) by means of the bijection*

$$\mu_A : \text{Spec}(A) \to L(A).$$

Proof. It is sufficient to prove that every closed set of $L(A)$ in the topology induced by μ_A is closed in the topology induced by the Zariski topology of $\mathscr{L}\text{oc}(K)$.

Let F be such a closed set. From the definition of the Zariski topology of Spec(A), it follows that F is of the form

$$F = \{A_{\underline{p}} | \underline{p} \supset \underline{a},\ \underline{a} \text{ ideal of } A\}$$

Let $\underline{p}_1, \underline{p}_2, \ldots, \underline{p}_r$ be the prime ideals associated with the ideal $\underline{a}$ We obviously have

$$F = \bigcup_{i=1}^{r} F_i,$$

where we have put

$$F_i = \{A_{\underline{p}} | \underline{p} \supset \underline{p}_i\}.$$

Consequently, it is sufficient to show that for every $i \in \{1, \ldots, r\}$ the set F_i is closed in the topology induced by the Zariski topology of $L(A)$. This follows from the equality

$$F_i = S(A_{\underline{p}_i}) \cap L(A).$$

Example 3.7.1. *Let K be a field and let $\mathscr{W}_d$ be the set of the discrete valuation subrings of K. We obviously have*

$$K \in \mathscr{W}_d \subset \mathscr{L}\mathrm{oc}(K).$$

Consider on $\mathscr{W}_d$ the topology induced by the Zariski topology of $\mathscr{L}\mathrm{oc}(K)$. If $A \in \mathscr{W}_d$ and $A \neq K$, then the subset $\{A\}$ is closed in the topology of $\mathscr{W}_d$.

Proof. We have to show (cf. Proposition 3.5) that if B is a discrete valuation subring of K, specialization of A, then $A = B$.

But from Definition 3.3 it follows that there exists a prime ideal of B, say $\underline{p}$, such that $B_{\underline{p}} = A$. But B being a discrete valuation ring, it has a single proper prime ideal, say $\underline{q}$. It follows that either $\underline{p} = (0)$, or $\underline{p} = \underline{q}$. In the first case, we have $B_{\underline{p}} = K$, and this case must be excluded by the hypothesis $A \neq K$, and in the second case, we have $B_{\underline{p}} = B_{\underline{q}} = B = A$.

Corollary 3.7.1. *Let M be a finite subset of the space $\mathscr{W}_d$ which does not contain the element K. The set $\mathscr{W}_d - M$ is open.*

3.7.2. Let k be a field and consider $K = k(X)$. The topology of the space $\mathscr{W}_{d,k}(K)$ of the discrete valuation rings of K which contain k can be described as follows:

A subset U of $\mathscr{W}_{d,k}(K)$ is open if and only if there exists a finite subset M of $\mathscr{W}_{d,k}(K)$ which does not contain the element K, such that $U = \mathscr{W}_{d,k}(K) - M$.

Proof. It follows from Corollary 3.7.1.1 that every set of the described form is open. In order to prove that the converse is also true, we recall the relations

$$\mathscr{W}_{d,k}(K) = L(k[X]) \cup L\left(k\left[\frac{1}{X}\right]\right),$$

$$L(k[X]) \cap L\left(k\left[\frac{1}{X}\right]\right) = L\left(k\left[X, \frac{1}{X}\right]\right),$$

$$\mathscr{W}_{d,k}(K) - L(k[X]) = k\left[\frac{1}{X}\right]_{\left(\frac{1}{X}\right)}$$

which show that $L(k[X])$ is open in $\mathscr{W}_{d,k}(K)$.

On the other hand, every open subset of $L(k[X])$ is obviously the complement of a finite subset which does not contain $k(X)$. If U is an arbitrary open subset of $\mathscr{W}_{d,k}(K)$, then $U \cap L(k[X])$ is open, hence it is the complement of a finite subset. In order to obtain U, one eventually adds $k\left[\frac{1}{X}\right]_{\left(\frac{1}{X}\right)}$.

3.7.3. On the space $\mathscr{W}_{d,k}(k(X))$ a canonical sheaf of rings $\mathcal{O}_k$ is obviously defined by putting

$$\mathcal{O}(U) = \bigcap_{M \in U} M.$$

The resulting ringed space is nothing else than $P_1(k)$.

3.7.4. Let $K \supset k$ be a field extension of finite type, having the transcendence degree 1 over k. We intend to describe the topological space $\mathscr{W}_{d,k}(K)$.

The open sets of this space are readily seen to have the same simple structure as in the previous case: they are the complements of finite sets which do not contain K. Indeed, choosing an element $X \in K$ which is transcendent over k, we obtain the following subrings of K:

$$k[X],\ k\left[\frac{1}{X}\right],\ \widetilde{k[X]},\ \widetilde{k\left[\frac{1}{X}\right]},$$

where $\widetilde{k[X]}$, $\left(\text{or } \widetilde{k\left[\frac{1}{X}\right]}\right)$ is the integral closure of the ring $k[X]$ $\left(\text{or } k\left[\frac{1}{X}\right]\right)$ in K, for which the following relation holds:

$$\mathscr{W}_{d,k}(K) = L(\widetilde{k[X]}) \cup L\left(\widetilde{k\left[\frac{1}{X}\right]}\right).$$

3.7.5. As in the previous case, a sheaf of rings $\mathcal{O}_K$ can be defined on the space $\mathscr{W}_{d,K}(K)$.

The resulting scheme will be denoted by C_K.

We have a morphism of schemes

$$C_K \to P_1(k)$$

determined by the choice of a transcendent element X of k with respect to k.

3.7.6. *Separated Chevalley schemes*

DEFINITION 3.7.6.1. Two elements $M, N \in \mathscr{L}oc(K)$ are said to be *related* if there exists an element $V \in \mathscr{L}oc(K)$ dominating both M and N.

DEFINITION 3.7.6.2. A subset S of $\mathscr{L}oc(K)$ (or of $\mathscr{L}oc_k(K)$) is said to be separated if the following relation is fulfilled:

$$(M, N \in S \text{ and } M, N \text{ related}) \Rightarrow M = N.$$

Example 3.7.6.2.1. For any field K, the subset $\mathscr{W}(K)$ of $\mathscr{L}oc(K)$ consisting of all valuations subrings is a separated subset of $\mathscr{L}oc\,(K)$.

PROPOSITION 3.7.6.3. *For any subring $A \subset K$, the subset $L(A)$ is separated.*

Proof. Let $\underline{p}, \underline{q} \in \mathrm{Spec}(A)$ be such that $A_{\underline{p}}, A_{\underline{q}}$ are related; we show that $\underline{p} = \underline{q}$. Indeed, assume that there exists a local subring $(V, \underline{v})$ of K such that the following domination relations hold:

$$V \succ A_{\underline{p}}, \quad V \succ A_{\underline{q}}.$$

In particular we have then the relations

$$\underline{v} \cap A_{\underline{p}} = \underline{p}A_{\underline{p}},$$
$$\underline{v} \cap A_{\underline{q}} = \underline{q}A_{\underline{q}}.$$

whence

$$\underline{v} \cap A = \underline{v} \cap A_{\underline{p}} \cap A = \underline{p}$$
$$\underline{v} \cap A = \underline{v} \cap A_{\underline{q}} \cap A = \underline{q},$$

i.e. $\underline{p} = \underline{q}$.

PROPOSITION 3.7.6.4. *The subset $P_n(k)$ of*

$$\mathscr{L}\mathrm{oc}_k\left(\mathrm{Quot}\left(k\left[\frac{X_1}{X_0}, \ldots, \frac{X_n}{X_0}\right]\right)\right)$$

is separated.

Proof. Denote by K the field $k\left(\frac{X_1}{X_0}, \frac{X_2}{X_0}, \ldots, \frac{X_n}{X_0}\right)$. We obviously have

$$k\left(\frac{X_1}{X_0}, \ldots, \frac{X_n}{X_0}\right) \subset k(X_0, X_1, \ldots, X_n),$$

$$K = k\left(\frac{X_1}{X_0}, \ldots, \frac{X_n}{X_0}\right) =$$

$$= k\left(\frac{X_0}{X_1}, \frac{X_2}{X_1}, \ldots, \frac{X_n}{X_1}\right) = \ldots =$$

$$= k\left(\frac{X_0}{X_n}, \frac{X_1}{X_n}, \ldots, \frac{X_{n-1}}{X_n}\right).$$

By definition, $P_n(k)$ is the subset of $\mathscr{L}\mathrm{oc}_k(K)$ defined by the relation

$$P_n(k) = L\left(k\left[\frac{X_1}{X_0}, \ldots, \frac{X_n}{X_0}\right]\right) \cup$$

$$\cup L\left(k\left[\frac{X_0}{X_1}, \frac{X_2}{X_1}, \ldots, \frac{X_n}{X_1}\right]\right) \cup \ldots$$

$$\ldots \cup L\left(k\left[\frac{X_0}{X_n}, \frac{X_1}{X_n}, \ldots, \frac{X_{n-1}}{X_n}\right]\right).$$

Assume that M, N are related elements of $P_n(k)$; we show that $M = N$. In view of the previous proposition, we may assume, after eventually changing notation, that

$$M \in L\left(k\left[\frac{X_1}{X_0}, \frac{X_2}{X_0}, \ldots, \frac{X_n}{X_0}\right]\right),$$

$$N \in L\left(k\left[\frac{X_0}{X_1}, \frac{X_2}{X_1}, \ldots, \frac{X_n}{X_1}\right]\right).$$

In other words, we have:

$$M = \left(k\left[\frac{X_1}{X_0}, \frac{X_2}{X_0}, \ldots, \frac{X_n}{X_0}\right]\right)_{\underline{p}},$$

$$N = \left(k\left[\frac{X_0}{X_1}, \frac{X_2}{X_1}, \ldots, \frac{X_n}{X_1}\right]\right)_{\underline{q}}.$$

By hypothesis, there exists a local subring V of K such that the domination relations hold:

$$V \succ M, \; V \succ N.$$

First note that the element $\dfrac{X_1}{X_0}$ does not belong to the ideal $\underline{p}$, since it is invertible in $V\left(\dfrac{X_0}{X_1} \in N \subset V\right)$. Consequently, we have the inclusion

$$k\left[\frac{X_0}{X_1}, \frac{X_2}{X_1}, \ldots, \frac{X_n}{X_1}\right] \subset$$

$$\subset \left(k\left[\frac{X_1}{X_0}, \frac{X_2}{X_0}, \ldots, \frac{X_n}{X_0}\right]\right)_{\underline{p}},$$

since

$$\frac{X_0}{X_1} = \frac{1}{\left(\dfrac{X_1}{X_0}\right)}, \quad \frac{X_2}{X_1} = \frac{\left(\dfrac{X_2}{X_0}\right)}{\left(\dfrac{X_1}{X_0}\right)}, \ldots, \frac{X_n}{X_1} = \frac{\left(\dfrac{X_n}{X_0}\right)}{\left(\dfrac{X_1}{X_0}\right)}.$$

We also have the relation

$$k\left[\frac{X_0}{X_1}, \frac{X_2}{X_1}, \ldots, \frac{X_n}{X_1}\right] - \underline{q} \subset$$

$$\subset \left(k\left[\frac{X_1}{X_0}, \frac{X_2}{X_0}, \ldots, \frac{X_n}{X_0}\right]\right)_{\underline{p}} -$$

$$- \underline{p}\left(k\left[\frac{X_1}{X_0}, \frac{X_2}{X_0}, \ldots, \frac{X_n}{X_0}\right]\right)_{\underline{p}}.$$

Indeed, an element of $k\left[\frac{X_0}{X_1}, \frac{X_2}{X_1}, \ldots, \frac{X_n}{X_1}\right]$ cannot belong to $\underline{p}k\left[\frac{X_1}{X_0}, \frac{X_2}{X_0}, \ldots, \frac{X_n}{X_0}\right]$ without belonging to $\underline{q}$, since it would simultaneously belong to $\underline{v}$ and to $k\left[\frac{X_1}{X_0}, \frac{X_2}{X_0}, \ldots, \frac{X_n}{X_0}\right]_{\underline{p}}$, while from the domination relation $V \succ k\left[\frac{X_1}{X_0}, \frac{X_2}{X_0}, \ldots, \frac{X_n}{X_0}\right]_{\underline{q}}$, we have $\underline{v} \cap k\left[\frac{X_0}{X_1}, \frac{X_2}{X_1}, \ldots, \frac{X_n}{X_1}\right] = \underline{q}$.

Consequently we have proved the relation

$$N \subset M,$$

whose converse can also be proved in a similar way.

3.7.6.5. *Example.* There exist subsets of $\mathscr{L}oc_k(K)$ which are not separated. It is sufficient to consider the subset consisting of a pair $\{A, V\}$ where A is not a valuation ring and V is a valuation ring which dominates A. Such a V exists for every A.

A concrete example is obtained as follows:

First consider the ring $k[X, Y]/(Y^2 - X^3)$. We have an isomorphism

$$\mathrm{Quot}(k[X, Y]/Y^2 - X^3) \to k(T)$$

$$X \bmod (Y^2 - X^3) \to T^2$$

$$Y \bmod (Y^2 - X^3) \to T^3$$

$$k(T) \to \mathrm{Quot}\,(k[X, Y]/(Y^2 - X^3))$$

$$T \to \frac{Y \bmod (Y^2 - X^3)}{X \bmod (Y^2 - X^3)}$$

In $k[X, Y]/(Y^2 - X^3)$, consider the ideal generated by $X \bmod (Y^2 - X^3)$, $Y \bmod (Y^2 - X^3)$. This is a prime ideal and the corresponding localized ring is not of discrete valuation (This ring can also be defined as the localized of the subring $k[T^2, T^3]$ of $k(T)$ with respect to the prime ideal generated by T^2 and T^3).

3.7.7. *The concept of completeness in the sense of Chevalley.* A subset S of $\mathcal{L}\mathrm{oc}_k(K)$ is said to be complete if every valuation ring of $\mathcal{L}\mathrm{oc}_k(K)$ dominates at least an element of S.

Example 3.7.7.1. *If A is a subalgebra of K, then the subset $L(A)$ of $\mathcal{L}\mathrm{oc}_k(K)$ is complete if and only if A is a field.*

Proof. We shall use the following algebraic results:

1. With the notation and assumptions of the proposition, if A is integral over K, then A is a field.
2. If R is a subring of K, then the intersection of all valuation rings containing R coincides with the integral closure of R in K.

Now the proof of the proposition runs as follows.

If $L(A)$ is complete, then the intersection of the valuation rings of K which contain k is bound to contain the intersection of the rings of the form $A_{\underline{p}}$:

$$A = \bigcap_{\underline{p} \in \mathrm{Spec}(A)} A_{\underline{p}} \subset \bigcap_{V \supset k} V.$$

But 2 shows that A is integral over k and 1 shows that it is a field.

Proposition 3.7.7.3. *The set $P_n(k)$ is complete.*

Proof. Let A_v be a valuation ring of $K = k\left(\frac{X_1}{X_0}, \frac{X_2}{X_0}, \ldots, \frac{X_n}{X_0}\right)$. Two cases are possible:

Case 1. $v\left(\frac{X_i}{X_0}\right) \geqslant 0$ for every $i \geqslant 1$. In this case we obviously have $A_v \supset k\left[\frac{X_1}{X_0}, \frac{X_2}{X_0}, \ldots, \frac{X_n}{X_0}\right]$, consequently A_v dominates an element of $L\left(k\left[\frac{X_1}{X_0}, \frac{X_2}{X_0}, \ldots, \frac{X_n}{X_0}\right]\right)$.

Case 2. $v\left(\frac{X_i}{X_0}\right) < 0$ for at least one $i \geqslant 1$. Since the group of a valuation is a totally ordered one, in this case it follows that there exists an i, let it be $i_0 \geqslant 1$, such that $v\left(\frac{X_{i_0}}{X_0}\right)$ takes a minimum.

We show that

$$A_v \supset k\left[\frac{X_0}{X_0}, \frac{X_1}{X_{i_0}}, \ldots, \frac{X_{i_0-1}}{X_{i_0}}, \frac{X_{i_0+1}}{X_{i_0}}, \ldots, \frac{X_n}{X_{i_0}}\right].$$

Indeed, we have

$$v\left(\frac{X_0}{X_{i_0}}\right) = -v\left(\frac{X_{i_0}}{X_0}\right) > 0,$$

$$v\left(\frac{X_j}{X_{i_0}}\right) = v\left(\frac{\frac{X_j}{X_0}}{\frac{X_{i_0}}{X_0}}\right) = v\left(\frac{X_j}{X_0}\right) - v\left(\frac{X_{i_0}}{X_0}\right) \geqslant 0,$$

from which the required relation follows.

PROPOSITION 3.7.7.4. *Any complete subset T of a separated subset S coincides with S.*

Proof. We have to show the inclusion

$$S \subset T.$$

To this end, consider an element M of S and a valuation ring V dominating M. The subset T being complete, it follows that there exists $N \in T$ which is dominated by V. Consequently M and N are related, and S being separated, it follows that $M = N$.

§ 4. *Elements of projective geometry*

Consider a commutative field k, a field extension $K \supset k$, an element ξ of the n-dimensional projective space $P_n(K)$ associated with the field K and a polynomial $P \in K[X_0, \ldots, X_n]$.

DEFINITION 4.1. We say that the polynomial P vanishes at the point ξ if every $(\xi_0, \xi_1, \ldots, \xi_n)$ K^{n+1} such that $\xi = (\xi_0 : \xi_1 : \ldots : \xi_n)$ satisfies the condition

$$P(\xi_0, \ldots, \xi_n) = 0.$$

If this condition is fulfilled, we write $P(\xi) = 0$.

PROPOSITION 4.2. Let $P = \sum_i P_i$, where P_i is the homogeneous component of degree i of P.

If the field K is infinite, then the polynomial P vanishes at the point ξ if and only if every homogeneous component P_i vanishes at ξ.

The proof is straightforward.

DEFINITION 4.3. An ideal $\underline{a} \subset k[X_0, \ldots, X_n]$ is called *homogeneous* if together with any element P, $\underline{a}$ contains every homogeneous component of P.

More generally, if S is a graded ring, an ideal $\underline{a}$ of S is called homogeneous if together with every element P, $\underline{a}$ contains every homogeneous component of P.

PROPOSITION 4.4. *An ideal $\underline{a}$ is homogeneous if and only if it has a system of generators consisting of homogeneous elements.*

The proof is straightforward.

PROPOSITION 4.5. *If $\underline{a}$ is a homogeneous ideal of* $K[X_0, \ldots, X_n]$, then the quotient ring $K[X_0, \ldots, X_n]/\underline{a}$ is a graded ring.

The proof is straightforward.

4.6. With every ideal $\underline{a} \subset k[X_0, \ldots, X_n]$ one can associate two homogeneous ideals of $k[X_0, \ldots, X_n]$:

— The ideal generated by the homogeneous elements of $\underline{a}$, which coincides with the greatest homogeneous ideal included in $\underline{a}$. This ideal will be denoted by $\underline{a}^*$. If $\underline{a}$ is primary (prime), then $\underline{a}^*$ is primary (prime).

— The ideal generated by the homogeneous components of the elements of $\underline{a}$, which coincides with the smallest homogeneous ideal including $\underline{a}$.

PROPOSITION 4.7. *If the ideal $\underline{q}$ of the graded ring S is primary and homogeneous, then its associated prime ideal is homogeneous.*

The proof is straightforward.

PROPOSITION 4.8. If the homogeneous ideal $\underline{a}$ of the graded ring S has the primary decomposition

$$\underline{a} = \bigcap_i q_i$$

then it also has the primary decomposition

$$\underline{a} = \bigcap_i q_i^*$$

COROLLARY 4.9. *The associated prime ideals of a homogeneous ideal are homogeneous.*

4.10. Let $\underline{a}$ be a homogeneous ideal of the ring $k[X_0, \ldots, X_n]$. By analogy with the case of affine algebraic varieties, for every field extension $K \supset k$ we consider the subset $\mathscr{P}_{\underline{a}}(K) \subset P_n(K)$ defined by

$$\mathscr{P}_{\underline{a}}(K) = \{\xi \in P_n(K) | P(\xi) = 0,\ \forall\, P \in \underline{a}\}.$$

In this way we get a functor

$$\mathscr{P}_{\underline{a}}: \mathscr{E}_k \to \text{Ens}.$$

4.11. With the homogeneous ideal $\underline{a} \subset k\,[X_0, \ldots, X_n]$ one can associate the graded ring

$$S_{\underline{a}} = k[X_0, \ldots, X_n]/\underline{a}.$$

If $\underline{a}$ is prime, the ring S is an integrity domain. Its field of quotients contains a subfield $k_{\underline{a}}$, which is an extension of k, consisting of all quotients of the form

$$\left\{\frac{f}{g}, f, g \in S_{\underline{a}}, g \neq 0, \text{ grad } f = \text{grad } g\right\}.$$

The field $k_{\underline{a}}$ will be called the field of rational fractions of the projective variety defined by the ideal $\underline{a}$. This terminology will be justified later.

Let $\xi_i = X_i \bmod \underline{a}$ and assume that $\xi_{i_0} \neq 0$. Then we have

$$k_{\underline{a}} = k\left(\frac{\xi_0}{\xi_{i_0}}, \ldots, \frac{\xi_n}{\xi_{i_0}}\right).$$

4.12. A map

$$\lambda_K : \mathscr{P}_{\underline{a}}(K) \to \mathscr{L}\text{oc}_k(k_{\underline{a}})$$

exists, assigning to every element $\mu \in \mathscr{P}_{\underline{a}}(K)$ a locality of the field extension $k_{\underline{a}} \supset k$, whose residual field is included in K.

The locality M_μ is defined as follows:

First consider the set U_μ of all homogeneous elements g of the ring S which do not vanish at the point μ. To this end we remark that if $P \in k[X_0, \ldots, X_n]$ is a homogeneous polynomial of degree m, if $(\mu_0, \ldots, \mu_n) \in K^{n+1}$, $\lambda \in K$, $\lambda \neq 0$ and $P(\mu_0, \ldots, \mu_n) \neq 0$, then $P(\lambda\mu_0, \ldots, \lambda\mu_n) = \lambda^m P(\mu_0, \ldots, \mu_n) \neq 0$.

Moreover, the set U_μ is multiplicatively closed. The subset of the field $k_{\underline{a}}$ consisting of the elements of the form $\frac{f}{g}$, where g is an element of U_μ, forms a ring M_μ containing the field k. The ring M_μ is local, having as maximal ideal the set of the fractions of the form $\frac{f}{g}$ for which $f(\mu) = 0$.

We denote by $C_{\underline{a}}$ the subset of $\mathscr{L}\text{oc}_k(k_{\underline{a}})$ consisting of all elements M for which there exists a field $K \supset k$ and a point $\mu \in \mathscr{P}_{\underline{a}}(K)$ such that

$$\lambda_K(\mu) = M.$$

PROPOSITION 4.13. *If* $X_i \notin \underline{a}$, $i = 0, 1, \ldots, n$ *and if* $\xi_i = X_i \bmod \underline{a}$, *then*

$$C_a = L\left(k\left[\frac{\xi_1}{\xi_0}, \ldots, \frac{\xi_n}{\xi_0}\right]\right) \cup L\left(k\left[\frac{\xi_0}{\xi_1}, \ldots, \frac{\xi_n}{\xi_1}\right]\right) \cup \ldots$$
$$\ldots \cup L\left(k\left[\frac{\xi_0}{\xi_n}, \ldots, \frac{\xi_{n-1}}{\xi_n}\right]\right).$$

Proof. First we prove the inclusion "$\subset$".

Consider an element $\mu \in \mathscr{P}_{\underline{a}}(K)$. Assume that $\mu = (\mu_0, \ldots, \mu_n)$ and $\mu_{i_0} \neq 0$. In these conditions, we show that

$$\lambda_K(\mu) \in L\left(k\left[\frac{\xi_0}{\xi_{i_0}}, \ldots, \frac{\xi_n}{\xi_{i_0}}\right]\right).$$

It is sufficient to prove that $\lambda_K(\mu) \supset k\left[\frac{\xi_0}{\xi_{i_0}}, \ldots, \frac{\xi_n}{\xi_{i_0}}\right]$, i.e. $\frac{\xi_0}{\xi_{i_0}}, \ldots$ $\ldots, \frac{\xi_n}{\xi_{i_0}} \in \lambda_K(\mu)$, which follows from the condition $\mu_{i_0} \neq 0$.

Conversely, we show that for every $i_0 \in \{0, 1, \ldots, n\}$ we have

$$L\left(k\left[\frac{\xi_0}{\xi_{i_0}}, \ldots, \frac{\xi_n}{\xi_{i_0}}\right]\right) \subset C_{\underline{a}}.$$

To this end, let $\underline{p}$ be a prime ideal of the ring $k\left[\frac{\xi_0}{\xi_{i_0}}, \ldots, \frac{\xi_n}{\xi_{i_0}}\right]$, let K be the residual field of the local ring $k\left[\frac{\xi_0}{\xi_{i_0}}, \ldots, \frac{\xi_n}{\xi_{i_0}}\right]_{\underline{p}}$ and let $\mu_0, \ldots, \mu_n$ be the elements of K defined by

$$\mu_i = \frac{\xi_i}{\xi_{i_0}} \bmod \underline{p}, \ i = 0, 1, \ldots, i_0 - 1, i_0 + 1, \ldots, n.$$

We show that $\mu = (\mu_0, \ldots, \mu_{i_0-1}, 1, \mu_{i_0+1}, \ldots, \mu_n) \in \mathscr{P}_{\underline{a}}(K)$. But for any homogeneous polynomial $P \in k[X_0, X_1, \ldots, X_n]$ of degree n, we have

$$P(\mu_0, \ldots, \mu_{i_0-1}, 1, \mu_{i_0+1}, \ldots, \mu_n) =$$

$$= P\left(\frac{\xi_0}{\xi_{i_0}}, \ldots, \frac{\xi_n}{\xi_{i_0}}\right) \bmod \underline{p} = \tag{4.13.1}$$

$$= \left(\frac{P(\xi_0, \ldots, \xi_n)}{\xi_{i_0}^n}\right) \bmod \underline{p}.$$

In other words, if $P \in \underline{a}$, then $P(\mu_0, \ldots, \mu_{i_0-1}, 1, \mu_{i_0+1}, \ldots, \mu_n) = 0$.

Finally, we show that $\lambda_K(\mu) = k\left[\frac{\xi_0}{\xi_{i_0}}, \ldots, \frac{\xi_n}{\xi_{i_0}}\right]_{\underline{p}}$.

This follows from the remarks:

– Considering the algebra morphisms

$$\varepsilon_{\underline{a}}: k[X_0, X_1, \ldots, X_n] \to k\left[\frac{\xi_0}{\xi_{i_0}}, \ldots, \frac{\xi_n}{\xi_{i_0}}\right]$$

$$X_i \mapsto \frac{\xi_i}{\xi_{i_0}}, \qquad i \in \{0, 1, \ldots, i_0 - 1, i_0 + 1, \ldots, n\}$$

$$X_{i_0} \mapsto 1.$$

$$\varepsilon_\mu: k[X_0, \ldots, X_n] \to k\left[\frac{\xi_0}{\xi_{i_0}}, \ldots, \frac{\xi_n}{\xi_{i_0}}\right] \to$$

$$\to k\left[\frac{\xi_0}{\xi_{i_0}}, \ldots, \frac{\xi_n}{\xi_{i_0}}\right]/\underline{p} \hookrightarrow K,$$

the image through $\varepsilon_{\underline{a}}$ of the multiplicatively closed system U_μ of $K[X_0, \ldots, X_n]$, consisting of all homogeneous polynomials $P \in k[X_0, \ldots$ $\ldots, X_n]$ for which $P(\mu) \neq 0$ obviously does not intersect the ideal $\underline{p}$:

$$\varepsilon_{\underline{a}}(U_\mu) \cap \underline{p} = \emptyset.$$

This yields, first, a homomorphism of the ring $(S_{\underline{a}})_{U_\mu}$ into the ring $k\left[\frac{\xi_0}{\xi_{i_0}}, \ldots, \frac{\xi_n}{\xi_{i_0}}\right]_{\underline{p}}$, then, taking the restriction, a homomorphism

$$\delta_\mu: \lambda_K(\mu) \to k\left[\frac{\xi_0}{\xi_{i_0}}, \ldots, \frac{\xi_n}{\xi_{i_0}}\right]_{\underline{p}}.$$

Our aim is to prove that δ_μ is an isomorphism.

– δ_μ *is injective*. Consider $P, Q \in k[X_0, \ldots, X_n]$, $Q(\mu_0, \ldots, \mu_{i_0-1}, 1, \mu_{i_0+1}, \ldots, \mu_n) \neq 0$, such that

$$\frac{P\left(\frac{\xi_0}{\xi_{i_0}}, \ldots, \frac{\xi_n}{\xi_{i_0}}\right)}{Q\left(\frac{\xi_0}{\xi_{i_0}}, \ldots, \frac{\xi_n}{\xi_{i_0}}\right)} = 0.$$

We show *that* $P \in \underline{a}$, i.e. $P \bmod \underline{a} = P(\xi_0, \ldots, \xi_n) = 0$. Indeed,

$$P\left(\frac{\xi_0}{\xi_{i_0}}, \ldots, \frac{\xi_n}{\xi_{i_0}}\right) = 0 \Rightarrow \xi_{i_0}^{\text{grad } P} P\left(\frac{\xi_0}{\xi_{i_0}}, \ldots, \frac{\xi_n}{\xi_{i_0}}\right) =$$

$$= 0 \Rightarrow P(\xi_0, \ldots, \xi_n) = 0.$$

—δ_μ *is surjective*. Let

$$\frac{P\left(\frac{\xi_0}{\xi_{i_0}}, \ldots, \frac{\xi_n}{\xi_{i_0}}\right)}{Q\left(\frac{\xi_0}{\xi_{i_0}}, \ldots, \frac{\xi_n}{\xi_{i_0}}\right)} \in k\left[\frac{\xi_0}{\xi_{i_0}}, \ldots, \frac{\xi_n}{\xi_{i_0}}\right]_{\underline{p}}$$

be given.

First consider the polynomials in $X_0, X_1, \ldots, X_n$:

$$X_{i_0}^{\text{grad } P + \text{grad } Q} P\left(\frac{X_0}{X_{i_0}}, \ldots, \frac{X_n}{X_{i_0}}\right)$$

$$X_{i_0}^{\text{grad } P + \text{grad } Q} Q\left(\frac{X_0}{X_{i_0}}, \ldots, \frac{X_n}{X_{i_0}}\right),$$

which are homogeneous and of the same degree.

It is readily seen that

$$X_{i_0}^{\text{grad } P + \text{grad } Q} Q\left(\frac{X_0}{X_{i_0}}, \ldots, \frac{X_n}{X_{i_0}}\right) \in U_\mu,$$

and the image through δ_μ of the element

$$\frac{X_{i_0}^{\text{grad } P + \text{grad } Q} P\left(\frac{X_0}{X_{i_0}}, \ldots, \frac{X_n}{X_{i_0}}\right)}{X_{i_0}^{\text{grad } P + \text{grad } Q} Q\left(\frac{X_0}{X_{i_0}}, \ldots, \frac{X_n}{X_{i_0}}\right)}$$

is just the given element of $k\left[\frac{\xi_0}{\xi_{i_0}}, \ldots, \frac{\xi_n}{\xi_{i_0}}\right]_{\underline{p}}$.

PROPOSITION 4.14. *Under the hypotheses of Proposition* 4.13 $(X_i \notin \underline{a})$, *the subset* $C_{\underline{a}}$ *of* $\mathscr{L}oc_k(k_{\underline{a}})$ *is separated.*

The proof is identical to that of Proposition 3.7.6.4 which is in fact a particular case of this proposition.

PROPOSITION 4.15. *Under the hypotheses of Proposition* 4.13, *the subset* $C_{\underline{a}}$ *of* $\mathscr{L}oc_k(k_{\underline{a}})$ *is complete.*

The proof is identical to that of Proposition 3.7.7.3, which is a particular case of this proposition.

PROPOSITION 4.16. *Let* $\underline{a} \subset k[X_0, \ldots, X_n]$, $\underline{b} \subset k[Y_0, \ldots, Y_m]$ *be prime homogeneous ideals, each of them fulfilling the condition of Proposition* 4.13.

Assume in addition that the following conditions are satisfied:

(i) $\deg \operatorname{trans}_k k(\underline{a}) = \deg \operatorname{trans}_k(k(\underline{b})) = 1$.

(ii) The elements of the sets $C_{\underline{a}}$, $C_{\underline{b}}$ *are regular, i.e. the projective curves defined by the ideals* $\underline{a}$, $\underline{b}$ *are non-singular.*

(iii) The field extensions $k_{\underline{a}} \supset k$, $k_{\underline{b}} \supset k$ *are isomorphic.*

In these conditions, the subsets $C_{\underline{a}}$, $C_{\underline{b}}$ *are canonically isomorphic.*

Proof. First of all, it is clear that the subsets $\mathscr{W}_{d,k}(k(\underline{a}))$, $\mathscr{W}_{d,k}(k(\underline{b}))$ are canonically isomorphic. On the other hand, using condition *ii*, we get the inclusions

$$C_{\underline{a}} \subset \mathscr{W}_{d,k}(k(\underline{a})),$$
$$C_{\underline{b}} \subset \mathscr{W}_{d,k}(k(\underline{b})).$$

But applying Propositions 4.15, 3.7.7.4, Example 3.7.7.1.1 and the hypothesis *i*, it follows that $C_{\underline{a}} = \mathscr{W}_{d,k}(k(\underline{a}))$, $C_{\underline{b}} = \mathscr{W}_{d,k}(k(\underline{b}))$.

§ 5. *The Chevalley scheme associated to an integral and irreducible scheme*

DEFINITION 5.1. A point ξ of a topological space X is said to be generic if $\overline{\{\xi\}} = X$. In other words, ξ is a generic point if it belongs to every non-empty open set of X.

Example 5.1.1. If A is an integral ring, then the point (0) is a generic point for the space Spec(A).

PROPOSITION 5.2. *A topological space which is separated in the sense of Kolmogorov has at most one generic point.*

Proof: straightforward.

DEFINITION 5.3. A topological space is said to be irreducible if the intersection of any two non-empty open subsets of X is non-empty.

PROPOSITION 5.4. *If the space X has a generic point, then it is irreducible.*

Proof: straightforward.

DEFINITION 5.5. A scheme $(X, \mathcal{O}_X)$ is said to be integral if every point x has a neighborhood U such that $(U, \mathcal{O}_{X|U})$ be an affine scheme of the type $(\mathrm{Spec}(A), \widetilde{A})$, where A is an integral ring.

PROPOSITION 5.7. *Let $(X, \mathcal{O}_X)$ be an integral and irreducible scheme. In these conditions, the topological space X has a unique generic point.*

Proof. Let $(\mathrm{Spec}(A), \widetilde{A})$ be an open subscheme of $(X, \mathcal{O}_X)$ such that A is an integral ring, and let $\xi_A \in \mathrm{Spec}(A)$ be the generic point of the space $\mathrm{Spec}(A)$. We show that ξ_A is the generic point of X. To this end, let V be an open subset of X; the scheme being irreducible, the intersection between V and $\mathrm{Spec}(A)$ is a non-empty open subset of $\mathrm{Spec}\, A$, which contains ξ_A since ξ_A is the generic point of $\mathrm{Spec}(A)$. The uniqueness follows from the fact that X is separated in the sense of Kolmogorov.

PROPOSITION 5.8. *If $(X, \mathcal{O}_X)$ is an integral and irreducible scheme and if ξ is a generic point of X, then $\mathcal{O}_\xi$ is a field and for every $x \in X$, $\mathcal{O}_x$ is a subring of $\mathcal{O}_\xi$, consequently $\mathcal{O}_\xi$ is the field of the quotients of $\mathcal{O}_x$.*

Proof. From the uniqueness of ξ, it follows that for any integral subscheme $(\mathrm{Spec}(A), \widetilde{A})$, the ring $\mathcal{O}_\xi$ is the field of the quotients of A.

DEFINITION 5.9. If $(X, \mathcal{O}_X)$ is an integral and irreducible scheme, then the subset of $\mathscr{L}\mathrm{oc}(\mathcal{O}_\xi)$ defined by the subrings $\mathcal{O}_x$ is called the Chevalley scheme associated with the scheme $(X, \mathcal{O}_X)$.

Chapter IV

Theory of Topoi

I. GENERAL THEORY OF TOPOI

§ 1. *Introduction*

Let $\mathscr{C}$ be a category. It is well known (cf. I. Bucur—A. Deleanu, *Introduction of the Theory of Categories and Functors)* that under rather general conditions on $\mathscr{C}$, one can define and study various algebraic structures upon the objects of $\mathscr{C}$.

In this way, one can define for instance the category $\mathrm{Ab}(\mathscr{C})$ of $\mathscr{C}$-Abelian groups, i.e. the category whose objects are objects of $\mathscr{C}$, endowed with a structure of Abelian groups, and whose morphism is morphism of $\mathscr{C}$-Abelian groups.

In the same way, one can define the category $\mathrm{Rg}(\mathscr{C})$ of the rings o the category $\mathscr{C}$, and for a given ring Λ of $\mathscr{C}$, the category $\mathrm{Mod}(\Lambda)$.

Unfortunately, compared with the corresponding categories in the case $\mathscr{C} = \mathrm{Ens}$ which has been systematically studied so far, these categories have in general pathological properties.

For instance, if $\mathscr{C} = \mathrm{Top}$ is the category of separated topological spaces and continuous maps, then Ab(Top) is not an Abelian category.

If $\mathscr{C}$ is an arbitrary category, a problem concerning the category $\mathrm{Ab}(\mathscr{C})$ arises in connection with the existence and the study of the cokernel of a morphism

$$A \xrightarrow{u} B$$

in $\mathrm{Ab}(\mathscr{C})$. This is due to the fact that while for the kernel $\mathrm{Ker}(u)$, the functor

$$h_{\mathrm{Ker}(u)} : \mathrm{Ab}(\mathscr{C})^0 \to \mathrm{Ab}$$

has a very simple description, namely

$$h_{\mathrm{Ker}(u)}(X) = \mathrm{Ker}\, h_X(u) = \mathrm{Ker}(h_X(A) \xrightarrow{h_X(u)} h_X(B)),$$

such a description is much more intricated for $h_{\mathrm{coker}(u)}(u)$.

Assuming that Coker(u) exists, one should not mean that it necessarily represents the functor

$$F: \mathrm{Ab}(\mathscr{C})^0 \to \mathrm{Ab}$$

$$X \rightsquigarrow \mathrm{Coker}(h_X(A) \xrightarrow{h_X(u)} h_\Lambda(B))$$

In the case when $\mathscr{C} = \mathrm{Ens}$, this functor is represented by Coker(u) if and only if Coker(u) is a free group.

If $\mathscr{C} = \mathrm{Ens}_f$ is the category of finite sets, then $\mathrm{Ab}(\mathrm{Ens}_f)$ is an Abelian category, for $\mathrm{Ab}(\mathrm{Ens}_f)$ actually coincides with the category of finite Abelian groups, but Ab(Ens) obviously does not have enough injective objects.

For the aims of homological algebra, it is extremely useful to find conditions on the category $\mathscr{C}$ in order for the category Mod(Λ) corresponding to any ring Λ of $\mathscr{C}$ to be Abelian and to have enough injective objects. This is always the case for a topos (see the exact definition below).

For any category $\mathscr{C}$ with a final element 1, one can work-out an arithmetics in the sense of the following definitions.

DEFINITION 1.1. An *arithmetical structure* on an object w of the category $\mathscr{C}$ is given if the morphism

$$1 \xrightarrow{0} w \qquad w \xrightarrow{s} w$$

are given such that the following conditions are fulfilled:

Fo any morphism

$$1 \xrightarrow{\varepsilon} X$$

and every endomorphism

$$X \xrightarrow{\xi} X$$

a unique morphism

$$w \xrightarrow{u} X$$

exists such that the following diagram commutes:

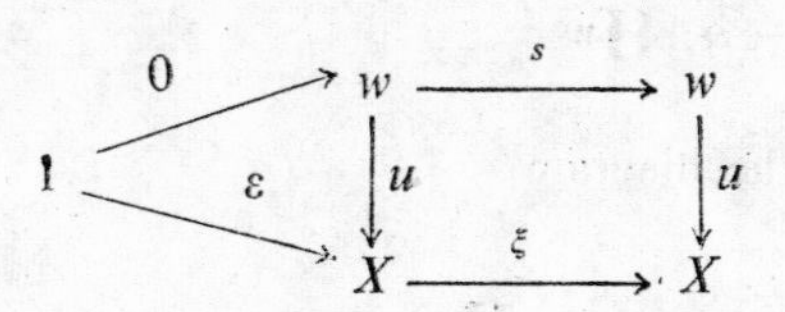

If $\mathscr{C} = \text{Ens}$, it is readily seen that the conditions of Definition 1.1 are equivalent to the classical axioms of Peano.

Indeed, first one finds that if $(1 \xrightarrow{0} w \xrightarrow{s} w)$, $(1 \xrightarrow{0'} w' \xrightarrow{s'} w')$ define two arithmetical structures, then there exists a canonical isomorphism $v: w \to w'$ such that the following diagram commutes:

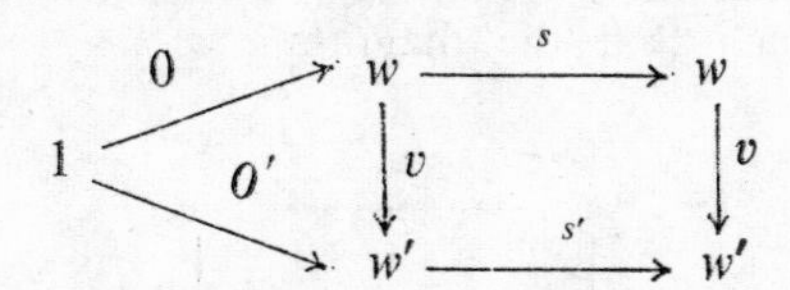

At the same time, one readily sees that if the system $(1 \xrightarrow{0} w \xrightarrow{w} w)$ satisfies in Ens the Peano axioms, it also satisfies the conditions of Definition 1.1.

Besides the univalence of arithmetics, Definition 1.1 allows the proof of several other elementary arithmetical properties. In the following, we give some examples:

PROPOSITION 1.2. *If* $(1 \xrightarrow{0} w \xrightarrow{s} w)$ *is an arithmetical structure and if the direct sum* $1 \coprod w$ *exists, then the morphism* $\delta: 1 \coprod w \to w$ *induced by the morphisms*

$$1 \xrightarrow{0} w$$

$$w \xrightarrow{s} w$$

is an isomorphism.

Proof. In order to define the inverse $\gamma: w \to 1 \coprod w$ of δ, consider first the morphism $\tau: 1 \coprod w \to 1 \coprod w$ defined by the morphisms

$$1 \xrightarrow{0} w \xrightarrow{i_w} 1 \coprod w$$

$$w \xrightarrow{s} w \xrightarrow{i_w} 1 \coprod w$$

On the one hand, the diagram

$$\begin{array}{ccc} 1 \coprod w & \xrightarrow{\tau} & 1 \coprod w \\ \downarrow \delta & & \downarrow \delta \\ w & \xrightarrow{s} & w \end{array}$$

obviously commutes.

On the other hand, using the condition from Definition 1.1, a morphism $\gamma: w \to 1 \coprod w$ exists, such that the diagram

$$\begin{array}{ccccc} & \overset{0}{\nearrow} & w & \xrightarrow{s} & w \\ 1 & & \downarrow \gamma & & \downarrow \gamma \\ & \underset{i_1}{\searrow} & 1 \coprod w & \xrightarrow{\tau} & 1 \coprod w \end{array}$$

commutes. This shows that γ is the inverse of δ.

PROPOSITION 1.3. *If* $(1 \xrightarrow{0} w \xrightarrow{s} w)$ *is an artthmetical structure, then the diagram*

$$w \underset{1_w}{\overset{s}{\rightrightarrows}} w \xrightarrow{p_w} 1$$

is exact.

Proof. Let $\xi: w \to X$ be such that $\xi s = \xi$. We have to prove that in this case a unique morphism $1 \xrightarrow{\xi'} X$ exists, such that the following diagram commutes:

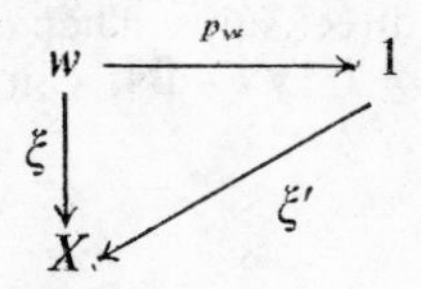

In order to prove the existence of ξ', it is sufficient to take $\xi' = \xi$. Indeed, the diagram

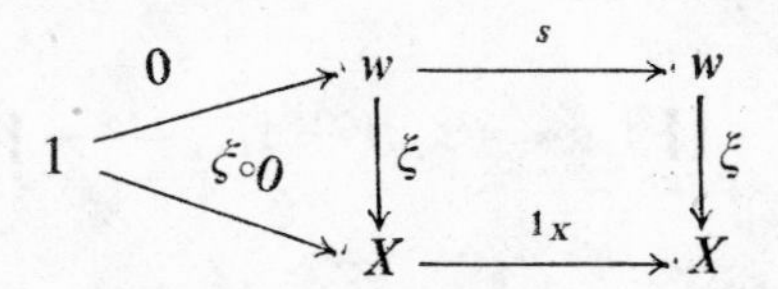

being obviously commutative, in view of the condition from Definition 1.1, we only have to verify the commutativity of the diagram

$$\begin{array}{ccc} w & \longrightarrow & w \\ {\scriptstyle \xi \circ p_w}\downarrow & & \downarrow{\scriptstyle \xi \circ p_w} \\ X & \xrightarrow{\ 1_X\ } & X \end{array}$$

which is obvious since $p_w s = p_w$. The uniqueness of ξ' follows from the evident fact that p_w is an epimorphism.

Starting from an object w endowed with an arithmetical structure $(1 \xrightarrow{0} w \xrightarrow{s} w)$, it would be important to develop a theory of integer numbers, of rational numbers, and, of course, of real numbers with respect to $\mathscr{C}$.

Unfortunately, for an arbitrary category, this theory presents abnormal features as compared with the particular case $\mathscr{C} =$ Ens which is systematically studied. In order to convince oneself, it is sufficient to consider the particular case of the category $\mathscr{Gr}$ of groups, in which an arbitrary arithmetical structure is of the form $(0 \xrightarrow{0} 0 \xrightarrow{s} 0)$.

The existence of an object w with an arithmetical structure in an arbitrary category is to be interpreted as an assumption similar to that of the existence of infinite sets in the case of set theory.

In order to illustrate a type of difficulty arising in the development of a theory of natural numbers in a category $\mathscr{C}$, assume, for instance, that $\mathscr{C}$ has a final object 1 and infinite direct sums. Then an arithmetical structure on the object $w = \coprod_{i \in \mathbb{N}} 1_i$, $1_i = 1$, $\forall i \in \mathbb{N}$, can be introduced as follows:

The morphisms

$$\sigma : w \to w, \; 0 : 1 \to w$$

are defined by the morphisms

$$1_j = 1 \xrightarrow{s_j} \coprod_{i \in \mathbb{N}} 1_i s_j = i_{j+i} : 1 = 1_{j+1} \xrightarrow{i_j+1} \coprod_{i \in \mathbb{N}} 1_i$$

respectively by the formula

$$0 = i_0 : 1 \to w$$

In order to define the sum

$$\sigma : w \times w \to w$$

we are of course tempted to use the existence of the usual sum on $\mathbb{N}$. Consequently, we have, starting from the sum in $\mathbb{N}$, to define a sum

$$\sigma : (\coprod_{i \in \mathbb{N}} 1_i) \times (\coprod_{i \in \mathbb{N}} 1_i) \to \coprod_{i \in \mathbb{N}} 1_i .$$

This would be possible if the objects

$$(\coprod_{i \in \mathbb{N}} 1_i) \times (\coprod_{i \in \mathbb{N}} 1_i), \quad \coprod_{(i, j) \in \mathbb{N} \times \mathbb{N}} (1_i \times 1_j)$$

could be canonically identified.

A canonical morphism

$$\alpha : \coprod_{(i, j) \in \mathbb{N} \times \mathbb{N}} (1_i \times 1_j) \to (\coprod_{i \in \mathbb{N}} 1_i) \times (\coprod_{i \in \mathbb{N}} 1_i)$$

actually exists between these objects, defined by the morphisms

$$1_{i_0} \times 1_{j_0} \xrightarrow{\alpha_{i_0 j_0}} (\coprod_{i \in \mathbb{N}} 1_i) \times (\coprod_{i \in \mathbb{N}} 1_i)$$

making commutative the diagrams

$$\begin{array}{ccc} 1_{i_0} \times 1_{j_0} & \xrightarrow{\alpha_{i_0 j_0}} & (\coprod_{i \in \mathbb{N}} 1_i) \times (\coprod_{i \in \mathbb{N}} 1_i) \\ {\scriptstyle pr_2} \downarrow & & \downarrow {\scriptstyle pr_1} \\ 1_{i_0} & \xrightarrow{i_{i_0}} & \coprod_{i \in \mathbb{N}} 1_i \\ \| & & \\ 1 & & \end{array}$$

$$\begin{array}{ccc} 1_{i_0} \times 1_{j_0} & \xrightarrow{\alpha_{i_0 j_0}} & (\coprod_{i \in \mathbb{N}} 1_i) \times (\coprod_{i \in \mathbb{N}} 1_i) \\ {\scriptstyle pr_2} \downarrow & & \downarrow {\scriptstyle pr_2} \\ 1_{j_0} & & \\ \| & & \\ 1 & \xrightarrow{i_{j_0}} & (\coprod_{i \in \mathbb{N}} 1_i) \end{array}$$

But, generally speaking, the morphism α is not an isomorphism. In order to get a counterexample, consider the category $\mathscr{C} = (\text{Rg comm})^\circ$, the dual of the category of commutative unitary rings. The canonical morphism

$$\alpha : (\coprod_{i \in \mathbb{N}} Z_i) \otimes_{\mathbb{Z}} (\coprod_{i \in \mathbb{N}} Z_i) \to \coprod_{(i, j) \in \mathbb{N} \times \mathbb{N}} (Z_i \otimes_{\mathbb{Z}} Z_j),$$

where

$$Z_i = \mathbb{Z}, \ \forall \, i \in \mathbb{N},$$

is not an isomorphism.

An extremely important problem of category theory is the following: given a category $\mathscr{C}$, determine another category $\mathscr{C}'$, having reasonable properties, such that a fully truthful functor

$$F : \mathscr{C} \to \mathscr{C}'$$

exists.

Obviously, a fully truthful functor

$$F: \mathscr{C} \to \text{Funct}(\mathscr{C}^\circ, \text{Ens})$$

always exists and, by Yonneda's theorem, it is given by

$$F(X) = h_X.$$

But the properties of the category Funct ($\mathscr{C}^\circ$, Ens) are generally speaking not close to the ones of $\mathscr{C}$.

The theory of topoi allows to find categories $\mathscr{C}'$, in particular topoi, and fully truthful functor F, such that the properties of $\mathscr{C}'$ be very close to those of $\mathscr{C}$.

§ 2. *Some definitions*

DEFINITION 2.1. Let X, Y be two objects of the category $\mathscr{C}$, whose direct sum $S = X \coprod Y$ is assumed to exist, and let $i_x: X \to S$, $i_y: Y \to S$ be the canonical inclusions. We say that the component X, Y of the direct sum S are disjoint if the fibre product associated with the diagram

$$\begin{array}{ccc} & & Y \\ & & \downarrow{\scriptstyle i_Y} \\ X & \xrightarrow{i_x} & S \end{array}$$

exists and is an initial object of $\mathscr{C}$:

$$X \prod_S Y = \emptyset$$

Obviously, most "good" categories (Ens, the Abelian categories, etc.) enjoy the property stated in Definition 2.1.

Nevertheless, there exist categories which do not satisfy this condition. Take for instance the category Rg comm of commutative unitary rings. Consider the objects $\mathbb{Q}$, $\mathbb{Z}/2\mathbb{Z}$ of this category.

As it is well known (cf. I. Bucur–A. Deleanu, *Introduction to the Theory of Categories and Functors)*, we have

$$\mathbb{Q} \coprod \mathbb{Z}/2Z = \mathbb{Q} \otimes_{\mathbb{Z}} \mathbb{Z}/2\mathbb{Z} = 0,$$

consequently the fibre product $\mathbb{Q} \times_0 \mathbb{Z}/2\mathbb{Z}$ associated with the diagram

$$\begin{array}{ccc} & & \mathbb{Q} \\ & & \downarrow i_Q \\ \mathbb{Z}/2\mathbb{Z} & \longrightarrow & \mathbb{Q} \otimes_{\mathbb{Z}} \mathbb{Z}/2\mathbb{Z} = 0 \end{array}$$

coincides with the direct product $\mathbb{Q} \times \mathbb{Z}/2\mathbb{Z}$ which is distinct from the initial object $\mathbb{Z}$ of Rg Comm.

Let $\mathscr{C}$ be a category having fibre products and arbitrary inductive limits, and consider a small category $\mathscr{I}$ and a functor

$$F: \mathscr{I} \to \mathscr{C} \tag{2.1.1}$$

Assume that a morphism in $\mathscr{C}$

$$m: Y \to X$$

and a functorial morphism

$$\varphi: F \to C_X, \tag{2.1.2}$$

are also given, where by C_X we denoted the constant functor

$$C_X: \mathscr{I} \to \mathscr{C}$$

defined by

$$C_X(i) = X, \ \forall\, i \in \mathrm{Ob}(\mathscr{C}),$$

$$C_X(u) = 1_X,$$

for any morphism u of the category $\mathscr{I}$.

Giving the morphism $\mathscr{S}$ is actually the same as giving for every object $i \in \mathrm{Ob}(\mathscr{C})$ the morphism

$$F(i) \xrightarrow{\varphi_i} X$$

such that for every $u \in \mathrm{Hom}_{\mathscr{I}}(i, j)$, the diagram

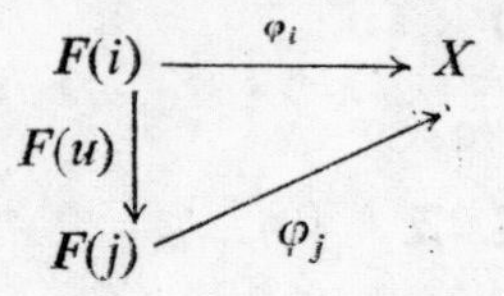

be commutative.

Under these conditions we can form the commutative diagrams

$$\begin{array}{ccc} F(i)\coprod_X Y & \longrightarrow & Y \\ \downarrow & & \downarrow m \\ F(i) & \xrightarrow{\varphi_i} & X \end{array} \tag{2.1.3}$$

$$\begin{array}{ccc} \varinjlim_i F(i)\coprod_X Y) & \longrightarrow & Y \\ \downarrow & & \downarrow m \\ \varinjlim_i F & \xrightarrow{\varinjlim \varphi_i} & X \end{array} \tag{2.1.4}$$

$$\begin{array}{ccc} (\varinjlim_i F)\coprod_X Y & \longrightarrow & Y \\ \downarrow & & \downarrow m \\ \varinjlim F & \xrightarrow{\varinjlim \varphi_i} & X \end{array} \tag{2.1.5}$$

The diagram (2.1.3)–(2.1.5) are Cartesian, hence we have a canonical morphism in the category $\mathscr{C}$

$$\varinjlim (F(i)\coprod_X Y) \to (\varinjlim F)\coprod_X Y. \tag{2.1.6}$$

DEFINITION 2.2. Let $\mathscr{C}$ be a category having arbitrary inductive limits and finite projective limits.

We say that in $\mathscr{C}$ *the inductive limits are universal* if for every little category $\mathscr{I}$, every functor $F: \mathscr{I} \to \mathscr{C}$ and every morphism $m: Y \to X$, the canonical morphism (2.1.6) is an isomorphism.

Categories exist which do not satisfy the condition from Definition 2.2; to see this it is obviously sufficient to consider a category L associated with a lattice $\underline{L}$ which is not infinitely distributive.

Such a lattice can be obtained, for instance, in the following way: Let

$$\mathbb{N} = \{0, 1, 2, \ldots, n, \ldots\}$$

be the set of natural numbers, ordered by means of the divisibility relation:

$$n \leqslant m \Leftrightarrow n \text{ divides } m.$$

In particular, we have

$$n \leqslant 0, \ \forall n \in \mathbb{N}$$

$$1 \leqslant n \ \forall n \in \mathbb{N}.$$

In order to get the required counter-example, we take $\mathscr{C} = \mathbb{N}$, $X = 0$ $Y = 2$, $\mathscr{I} = \{0, 1, 2, 3, \ldots, k, \ldots\}$ having only identity morphisms, $F(k) = 2k + 1$, $\forall k \in \mathrm{Ob}(\mathscr{I})$. The following relations are obvious:

$$\varinjlim F = \coprod_k (2k + 1) = 0,$$

$$F(k) \prod\nolimits_0 2 = (2k + 1) \prod\nolimits_0 2 = 1, \ \forall k \in \mathrm{Ob}(\mathscr{I}),$$

$$(\varinjlim F) \prod\nolimits_0 2 = 0 \prod\nolimits_0 2 = 0 \wedge 2 = 2,$$

$$\varinjlim (F(k) \prod\nolimits_0 2) = 1.$$

Consequently,

$$\varinjlim (F(k) \prod\nolimits_0 2 \neq (\varinjlim F) \prod\nolimits_0 2, \ (1 \neq 2).$$

In writing these relations, we have used the obvious fact that the inductive limit in the category $\mathbb{N}$ is the least common multiple, and the projective limit in this category is the greatest common divisor.

Example 2.2.1. The category $\mathscr{C} = \mathscr{F}\text{asc}(T, \text{Ens})$ of sheaves of sets on the topological space T has arbitrary inductive and projective limits, and the inductive limits are universal.

DEFINITION 2.3. Let $\mathscr{C}$ be a category and consider a family $(A_i)_{i\in I}$ of objects of $\mathscr{C}$.

The family $(A_i)_{i\in I}$ is said to be a family of generators of the category $\mathscr{C}$, if for every monomorphism

$$u: X \hookrightarrow Y$$

the following properties are equivalent:

(α) u is an isomorphism;

(β) For every $i \in I$, the map

$$\text{Hom}_{\mathscr{C}}(A_i, X) \to \text{Hom}_{\mathscr{C}}(A_i, Y)$$

is bijective.

Example 2.3.1. If $\mathscr{C} = \text{Ens}$, then the family (1) defined by the final object 1 is obviously a family of generators for Ens.

If $\mathscr{C} = \text{Ab}$, then $(\mathbb{Z})$ is a family of generators for Ab.

Counterexample 2.3.2. (N. Popescu). Let $\mathscr{A}$ be a class which is not a set and let $(X_\alpha)_{\alpha\in\mathscr{A}}, (Y_\alpha)_{\alpha\in\mathscr{A}}$ be two classes of objects such that $X_\alpha \neq Y_\beta$ for every $\alpha, \beta \in \mathscr{A}$ and $X_\alpha \neq X_\beta$, $Y_\alpha \neq Y_\beta$ for every $\alpha, \beta \in \mathscr{A}$, $\alpha \neq \beta$.

Consider the category $\mathscr{C}$ whose class of objects consists of the objects X_α, Y_β and whose morphisms are defined by

$$\text{Hom}_{\mathscr{C}}(X_\alpha, Y_\alpha) = \{\alpha\}, \ \text{Hom}_{\mathscr{C}}(Y_\alpha, X_\alpha) = \emptyset,$$

$$\text{Hom}_{\mathscr{C}}(X_\alpha, Y_\beta) = \text{Hom}_{\mathscr{C}}(Y_\beta, X_\alpha) = \emptyset \ \text{if}\ \alpha \neq \beta,$$

$$\text{Hom}_{\mathscr{C}}(X_\alpha, X_\alpha) = \{1_{X_\alpha}\},$$

$$\text{Hom}_{\mathscr{C}}(Y_\alpha, Y_\alpha) = \{1_{Y_\alpha}\}.$$

This category has no family of generators. In fact, if $(A_i)_{i\in I}$ would be a family of generators for $\mathscr{C}$, then an $\alpha_0 \in \mathscr{A}$ would exist such that $X_{\alpha_0} \neq A_i$, $\forall\, i \in I$ and $Y_{\alpha_0} \neq A_i$, $\forall\, i \in I$. In this case, the monomorphism

$$X_{\alpha_0} \xrightarrow{\alpha_0} Y_{\alpha_0}$$

satisfies condition β since for any $i \in I$, we have

$$\mathrm{Hom}_{\mathscr{C}}(A_i, X_{\alpha_0}) = \emptyset = \mathrm{Hom}_{\mathscr{C}}(A_i, Y_{\alpha_0}),$$

$$\mathrm{Hom}_{\mathscr{C}}(A_i, \alpha_0) = \mathrm{id}\, \emptyset,$$

On the other hand, α_0 is not an isomorphism, hence condition β is not satisfied.

Example 2.3.3. If X is a topological space, then for every open set V of X one can consider an object $\mathscr{E}_V$ of the category $\mathscr{F}\mathrm{asc}\,(X, \mathrm{Ens})$ of sheaves of sets over X, defined by

$$\mathscr{E}_V(U) = \begin{cases} \{\emptyset\} & \text{if } U \subset V \\ \emptyset & \text{if } U \not\subset V \end{cases}$$

The family $(\mathscr{E}_V)$ is a family of generators for the category $\mathscr{F}\mathrm{asc}(X, \mathrm{Ens})$.

DEFINITION 2.4. Let X be an object of the category $\mathscr{C}$ and assume that the product $X \times X$ exists. An *equivalence relation* on X is a subobject R of $X \times X$,

$$\rho: R \hookrightarrow X \times X$$

such that for every object U of $\mathscr{C}$ the subset $\mathrm{Hom}_{\mathscr{C}}(U, R)$ of the set $\mathrm{Hom}_{\mathscr{C}}(U, X) \times \mathrm{Hom}_{\mathscr{C}}(U, X)$ defines an equivalence relation on $\mathrm{Hom}_{\mathscr{C}}(U, X)$. (We identify $\mathrm{Hom}_{\mathscr{C}}(U, R)$ with its image through the injective map $\mathrm{Hom}_{\mathscr{C}}(U, \rho)\colon \mathrm{Hom}_{\mathscr{C}}(U, R) \hookrightarrow \mathrm{Hom}_{\mathscr{C}}(U, X) \times \mathrm{Hom}_{\mathscr{C}}(U, X)$.

This condition can also be expressed by requiring that the monomorphism ρ should satisfy the following conditions:

(α) *Reflexivity:* There exists a morphism $X \to R$, such that the diagram

$$
(2.4.1)\qquad
\begin{array}{ccc}
X & \xrightarrow{\ \Delta\ } & X \times X \\
 & \searrow \quad \nearrow{\scriptstyle \rho} & \\
 & R &
\end{array}
$$

be commutative, where X is the diagonal morphism.

(β) *Symmetry:* The diagram

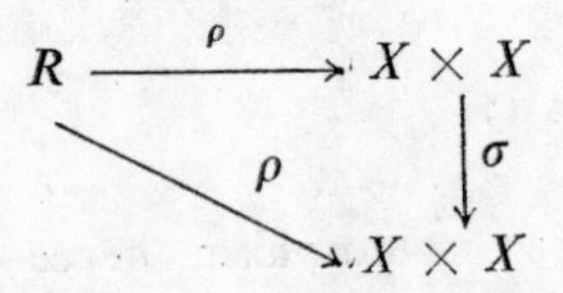

is commutative, where σ, the morphism which permutes factors, is defined by the condition of making commutative the diagram

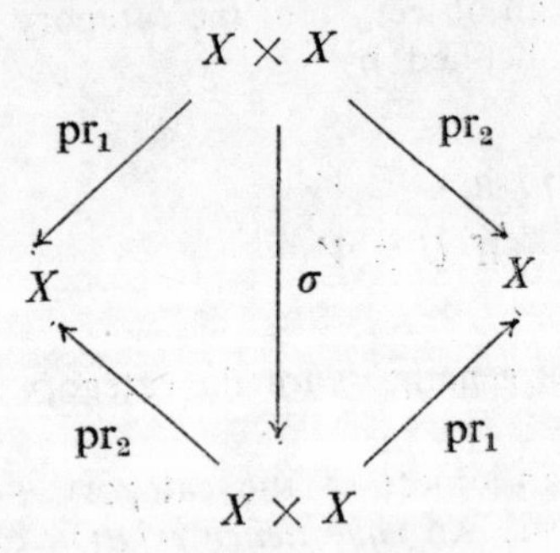

(γ) *Transitivity.* The diagram

$$
\begin{array}{ccc}
R \times_X R & \longrightarrow & X \times X \times X \times X \\
\downarrow & & \downarrow \\
R & \xrightarrow{\ \rho\ } & X \times X
\end{array}
$$

is commutative, where the morphism

$$R \times_X R \to (X \times X) \times (X \times X)$$

is defined by its components

$$R \times X^R \xrightarrow{\mathrm{pr}_i} R \overset{\rho}{\hookrightarrow} X \times X, \quad i = 1, 2$$

and the morphism

$$X \times X \times X \times X \to X \times X$$

is defined by

$$X \times X \times X \times X \xrightarrow{\mathrm{pr}_1} X,$$

$$X \times X \times X \times X \xrightarrow{\mathrm{pr}_4} X.$$

Example 2.4.2. Let $\mathscr{C} = \mathscr{F}\mathrm{asc}(T, \mathrm{Ens})$ be the category of sheaves of sets over the topological space T, let $\mathscr{F}$ be an object of $\mathscr{C}$ and consider a subobject $\mathscr{R} \hookrightarrow \mathscr{F} \times \mathscr{F}$ of $\mathscr{F} \times \mathscr{F}$. The relation $\mathscr{R} \hookrightarrow \mathscr{F} \times \mathscr{F}$ is an equivalence relation on $\mathscr{F}$ if and only if for every point $t \in T$, the fibre $\mathscr{R}_t$ of $\mathscr{R}$ at the point t defines an equivalence relation on the fibre $\mathscr{F}_t$ of $\mathscr{F}$ at the same point.

If $R \overset{\rho}{\hookrightarrow} X \times X$ is an equivalence relation on X, we denote by p_i, $i = 1, 2$, the morphism

$$R \overset{\rho}{\hookrightarrow} X \times X \xrightarrow{\mathrm{pr}_i} X.$$

DEFINITION 2.5. A cokernel of the pair of morphism (p_1, p_2) is called a *quotient object* of X by the equivalence relation R and is denoted by X/R.

Example 2.5.1. In the category of ringed spaces, every equivalence relation has a quotient object.
Indeed, this category has cokernels:
If

$$(X, O_X) \underset{f_2}{\overset{f_1}{\rightrightarrows}} (Y, O_Y)$$

$$f_i = (\varphi_i, \psi_i),\ \varphi_i \colon X \to Y,\ \psi_i \colon O \to \varphi_{i*}(O_X),\ i = 1, 2$$

is a pair of morphisms of ringed spaces, let (Z, π) be the cokernel in the category Top of the continuous maps φ_1, φ_2:

$$X \overset{\varphi_1}{\underset{\varphi_2}{\rightrightarrows}} Y \xrightarrow{\pi} Z$$

It remains to construct a suitable sheaf of rings O_Z on the space Z. To this end, consider an open set U of Z. We put by definition

$$O_Z(U) = O_Y(\pi^{-1}(U)).$$

The morphism $\psi: O_Z \to \pi_*(O_Y)$ is defined in a straightforward way, such that we obtain a morphism $(\pi, \psi): (Y, \mathbb{O}_Y) \to (Z, \mathbb{O}_Z)$ of ringed spaces, yielding a cokernel of the morphisms f_1, f_2.

Definition 2.6. Let R be an equivalence relation on X and let $u: X' \to X$ be a morphism. The equivalence relation R' on X' defined by the monomorphism

$$R' \to X' \times X'$$

such that the diagram

$$\begin{array}{ccc} R' & \longrightarrow & R \\ \downarrow & & \downarrow \\ X' \times X' & & X \times X \end{array}$$

be commutative is called the *inverse image* by u of the equivalence relation R on X.

Example 2.7. *Equivalence relation defined by a group acting freely on an object.* Let G be a group acting on a set E. We recall that the group G is said to act freely on the set E if the following condition is fulfilled:

$$\forall g \,\forall x((g \in G) \wedge (x \in E) \wedge (gx = x) \Rightarrow g = e). \tag{2.7.1}$$

Consider now a group G in the category $\mathscr{C}$, acting on the object X of the category $\mathscr{C}$. The group G is said to act freely on X if for every

$U \in \mathrm{Ob}(\mathscr{C})$, the group $G(U) = \mathrm{Hom}_{\mathscr{C}}(U, G)$ acts freely on the set $X(U) = \mathrm{Hom}_{\mathscr{C}}(U, X)$.

Assume that the action of G on X is given by the morphism

$$\mu: G \times X \to X.$$

Consider the morphism

$$\rho: G \times X \to X \times X \tag{2.7.2}$$

defined by the morphisms

$$\mu: G \times X \to X$$

$$\mathrm{pr}_X: G \times X \to X.$$

As it is readily seen, G acts freely on X if and only if the morphism ρ is a monomorphism. In this case, an equivalence relation on X can be defined taking $R = G \times X$ and considering the morphism $\rho: R \hookrightarrow X \times X$ given by (2.7.2).

Let $\mathscr{C}$ be a category having fibre products and cokernels.

If $R \hookrightarrow X \times X$ is an equivalence relation on X and if $Y = X/R$ is a quotient object of X with respect to R, the following commutative diagram can be considered:

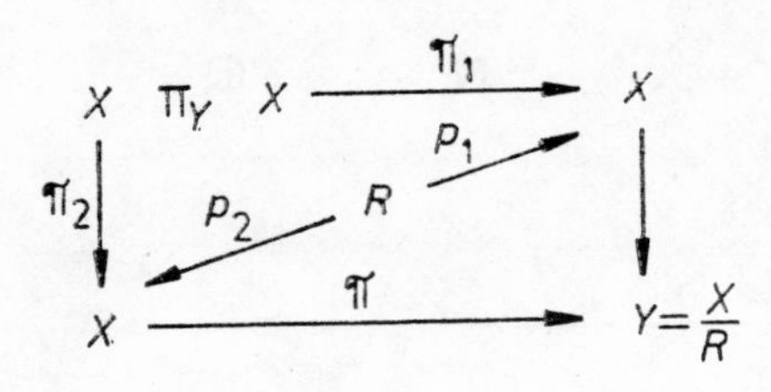

This yields a unique morphism

$$\sigma: R \to X \textstyle\prod_Y X \tag{2.7.3}$$

such that the diagram

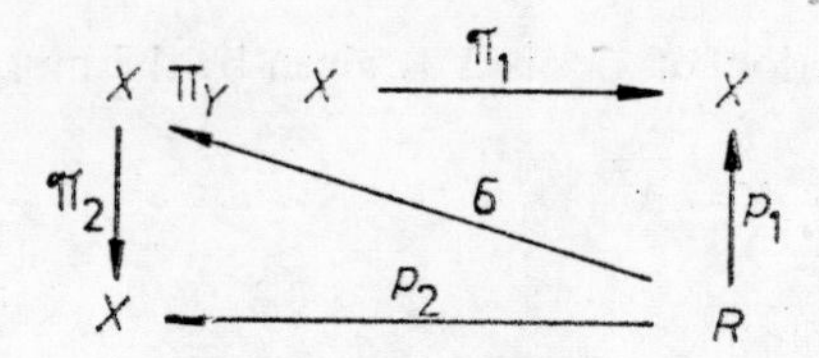

commutes.

DEFINITION 2.8. The equivalence relation R on X is said to be *effective* if the morphism (2.7.3) is an isomorphism.

The equivalence relation R on X is said to be *universally effective* if for every morphism

$$\alpha: X' \to X$$

the inverse image R' of R by α is an effective equivalence relation.

The quotient object X/R is also said to be a *good quotient object* if the morphism σ (2.7.3) is an isomorphism, or if the relation R is an effective one.

In "good" categories (Ens, Abelian categories, etc.), every equivalence relation is effective. Nevertheless, categories with non-effective equivalence relations exist. In order to have an example, consider the category Ab(Top sep) of separated topological Abelian groups, let $\mathbb{R}$ be the additive group of real numbers with the usual topology and consider on $\mathbb{R}$ the equivalence relation $R \hookrightarrow \mathbb{R} \times \mathbb{R}$ defined by the subgroup $\mathbb{Q} \subset \mathbb{R}$ of rational numbers, i.e.

$$R = \{(x, y) | x, y \in \mathbb{R}, x - y \in \mathbb{Q}\}.$$

It is readily seen that

$$\mathbb{R}/R = 0$$

consequently

$$\mathbb{R} \times_0 \mathbb{R} = \mathbb{R} \times \mathbb{R}$$

which cannot be isomorphic to $\mathbb{R}$.

2.8.1. *Another example of non-effective equivalence relation.* We intend to show that in the category Ann of ringed spaces there exist non-effective equivalence relations. This fact is of remarkable principial importance, since from a geometrical point of view, Ann seems nowadays to be the main working category.

In order to get the required example, let $(\mathbb{C}^2, \mathcal{O}_{\mathbb{C}^2})$ be the complex plane with the ringed space structure corresponding to its usual algebraic variety structure over $\mathbb{C}$, let R_1, R_2 be the algebraic subvarieties of $(\mathbb{C}^2, \mathcal{O}_{\mathbb{C}^2})$ represented in Figs. 2.8.1.1 and 2.8.1.2 (R_2 is obtained from R_1 by removing the origin of $\mathbb{C}^2$), and let R be their direct sum. An equivalence relation

$$R \overset{p_1}{\underset{p_2}{\rightrightarrows}} X$$

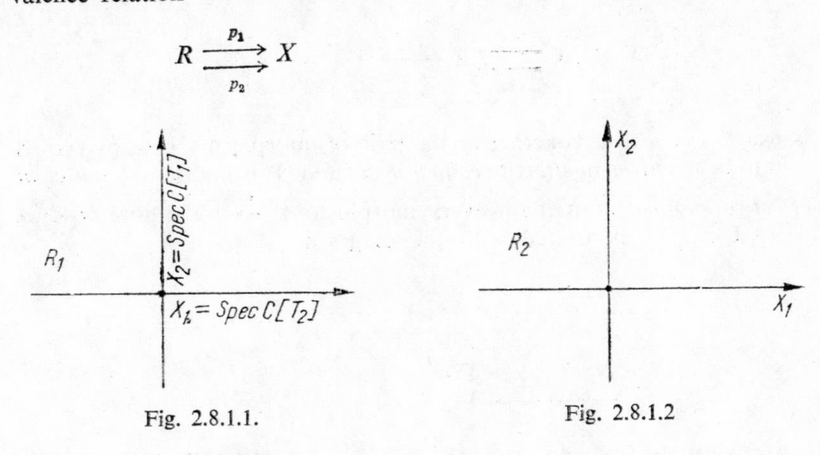

Fig. 2.8.1.1. Fig. 2.8.1.2

on $X = R_1$ can be defined by taking

$$p_1/R_1 = p_2/R_1 = \mathrm{id}_{R_1};$$

$$p_1/R_2 = \text{the canonical inclusion of } R_2 \text{ in } R_1;$$

$$p_2((0, x)) = (x, 0)$$

$$p_2((x, 0)) = (0, x)$$

The quotient $X/R = Y$ is easily described. This quotient coincides with the affine line but for the fibre at the origin of the structure sheaf.

On the other hand, $X \times_Y X \neq R$ as it is readily seen by comparing the fibres of these two ringed spaces at the origin.

DEFINITION 2.9. A morphism $u: X \to Y$ is called an *effective epimorphism* or, equivalently, Y is said to be an *effective quotient* of X if the fibre product $X \times_Y X$ associated with the diagram

$$\begin{array}{ccc} & & X \\ & & \downarrow u \\ X & \xrightarrow{u} & Y \end{array}$$

exists and if the sequence

$$X \times_Y X \underset{\mathrm{pr}_2}{\overset{\mathrm{pr}_1}{\rightrightarrows}} X \xrightarrow{u} Y$$

is exact, i.e. u is a cokernel of the pair of morphisms $(\mathrm{pr}_1, \mathrm{pr}_2)$. u is said to be a *universal effective epimorphism* and Y is said to be a *universal effective quotient* of X if for every morphism $Y' \xrightarrow{v} Y$ the fibre product $X' = X \times_Y Y'$ associated with the diagram

$$\begin{array}{ccc} & & Y' \\ & & \downarrow v \\ X & \xrightarrow{u} & Y \end{array}$$

exists and the morphism $u': X' \to Y'$ is an effective epimorphism.

A monomorphism which is at the same time an effective epimorphism is readily seen to be an isomorphism. This shows that a bijection which is not an isomorphism is a non-effective epimorphism.

Example 2.9.1. In the category Sch, the dual of the category of commutative unitary rings, there exist non-effective epimorphisms.

For instance, the epimorphism

$$\mathrm{Spec}(k[T]) \to \mathrm{Spec}(k[T^3, T^5])$$

induced by the ring morphism

$$k[T^3, T^5] \to k[T]$$

is not effective.

Indeed, the sequence of ring morphisms

$$k[T^3, T^5] \to k[T] \underset{i_2}{\overset{i_1}{\rightrightarrows}} k[T] \otimes_{K[T_3,\, T_3]} K[T]$$

is not exact, since we have

$$i_1(T^7) = T^7 \otimes 1 = T^2 \otimes T^5$$

$$i_2(T^7) = 1 \otimes T^7 = T^5 \otimes T^2 = T^2 \otimes T^5$$

i.e.

$$i_1(T^7) = i_2(T^7)$$

but $T^7 \in k[T^3, T^5]$.

PROPOSITION 2.10. *A one-to-one "Galois" correspondence exists between effective (respectively universally effective) equivalence relations on X and effective (respectively universally effective) quotients of X.*

Proof. With any effective equivalence relation $R \rightrightarrows X$ on X, we associate the effective cokernel Coker (p_1, p_2).

Conversely, with any effective quotient $X \xrightarrow{u} Y$ of X, we associate the effective equivalence relation on X $X \coprod X \underset{\mathrm{pr}_2}{\overset{\mathrm{pr}_1}{\rightrightarrows}} X$.

This is a Galois correspondence in the sense that it realizes an antiisomorphism between the ordered class of effective equivalence relations on X and the ordered class of effective quotients of X.

§ 3. *Elements of descent theory*

Now we intend to explain the utility of the concepts of effective equivalence relation and effective epimorphism. To this end, we have to recall some concepts and preliminary results of descent theory.

DEFINITION 3.1. A fibre category $\mathscr{F}$ is given by:

(I) a category $\mathscr{C}$ called the basis of $\mathscr{F}$;

(II) a family $(\mathscr{F}_X)_{X \in \mathrm{Ob}(\mathscr{C})}$ of categories indexed by the class $\mathrm{Ob}(\mathscr{C})$ of the objects of $\mathscr{C}$; the category $\mathscr{F}_X$ is called the fibre of $\mathscr{F}$ at X;

(III) a map α assigning to every morphism $f: X \to Y$ of a functor

(IV) a map c, assigning to every couple (f, g) of composable morphisms of $\mathscr{C}$:

$$X \xrightarrow{f} Y \xrightarrow{g} Z$$

an isomorphism of functors

$$c_{f,g}: (g,f)^{\alpha} \to f^{\alpha} g^{\alpha}.$$

These data have to satisfy the following conditions:

(α) $(\mathrm{id})^{\alpha} = \text{identity}$;

(β) $c_{f,\mathrm{id}} = \mathrm{id}_{f^{\alpha}}$, $c_{\mathrm{id},g} = \mathrm{id}_{g^{\alpha}}$;

(γ) for every triplet (f, g, h) of morphisms of $\mathscr{C}$,

$$X \xrightarrow{f} Y \xrightarrow{g} Z \xrightarrow{h} T$$

the following diagram commutes:

$$\begin{array}{ccc} ((hg)f)^{\alpha} = (h(gf))^{\alpha} & \xrightarrow{c_{gf,h}} & (gf)^{\alpha} h^{\alpha} \\ \downarrow c_{f,hg} & & \downarrow c_{f,g} h^{\alpha} \\ f^{\alpha}(hg)^{\alpha} \xrightarrow{f c^{\alpha}_{gh}} f^{\alpha}(g^{\alpha} h^{\alpha}) = & & (f^{\alpha} g^{\alpha}) h^{\alpha} \end{array}$$

The concept of cofibre category is obtained by duality (taking the category $\mathscr{C}^{\circ}$ instead of $\mathscr{C}$).

3.2. *Examples* 1. Consider $\mathscr{C} = \mathrm{Ann}$, the category of ringed spaces, and for every $(X, O_X) \in \mathscr{C}$, let $\mathscr{F}_X$ be the category $\mathrm{Mod}(O_X)$ of O_X-modules. If

$$f: (X, O_X) \to (Y, O_Y)$$

is a morphism of ringed spaces, we denote by

$$f^*: \mathscr{F}_Y \to \mathscr{F}_X$$

the inverse image functor corresponding to f, and if f, g are two composable morphisms of ringed spaces, let

$$c_{f,g}: (gf)^* \to f^*g^*$$

be the obvious isomorphism.

One obtains in this way a fibered category, whose basis is the category Ann.

2. The previous example can be varied by starting with the category $\mathscr{C} = \text{Top}$ and by putting for every topological space X, $\mathscr{F}_X = \mathscr{F}\text{asc}(X, \text{Ens})$, the category of sheaves of sets on the space X.

One obtains in this way a fibre category, whose basis is the category Top.

3. Let $\mathscr{C}$ be a category having fiber products and for every $S \in \mathscr{C}$, consider $\mathscr{F}_S = \mathscr{C}/S$, the category of S-objects of $\mathscr{C}$. If $f: S' \to S$ is a morphism of $\mathscr{C}$, denote by $f^\alpha: \mathscr{C}/S \to \mathscr{C}/S'$ the functor

$$\begin{pmatrix} X \\ \downarrow \\ S \end{pmatrix} \longrightarrow \begin{pmatrix} X \times {}_S S' \\ \downarrow \\ S' \end{pmatrix}.$$

If f, g are composable morphisms in $\mathscr{C}$,

$$S \xrightarrow{g} S' \xrightarrow{f} S'',$$

then we have a canonical functor isomorphism (cf. I. Bucur—A. Deleanu, *Introduction to the Theory of Categories and Functors)*,

$$X \times {}_S S'' \xrightarrow{\sim} (X \times {}_S S') \times {}_S S''$$

allowing to define the isomorphism $c_{f,g}$.

One obtains a fibre category whose basis is the category $\mathscr{C}$.

3.3. Let $\mathscr{F} = (\mathscr{C}, (\mathscr{F}_X)_{X \in \text{Ob}(\mathscr{C})}(f \to f^\alpha), ((f, g) \to c_{f,g}))$ be a fibre category and let $S' \xrightarrow{u} S$ be a morphism in $\mathscr{C}$.

Assume that the fibre product $S' \times {}_S S'$ exists and let $p_1, p_2: S' \times {}_S S' \to S'$ be the canonical projections.

A category $\mathscr{R}ec_{\mathscr{F}}(u)$ can be defined as follows:

$$- \mathrm{Ob}(\mathscr{R}ec_{\mathscr{F}}(u)) = \{(X', c) \mid X' \in \mathrm{Ob}(\mathscr{F}_{S'}),\ c : p_1^{\alpha}(X') \xrightarrow{\sim}$$

$$\xrightarrow{\sim} p_2^{\alpha}(X')\}.$$

(Such a couple (X', c) is called a system of local data on X' with respect to the morphism u).

$$- \mathrm{Hom}_{\mathscr{R}ec_{\mathscr{F}}}(u)((X', c), (Y', d)) = \{\xi' \in \mathrm{Hom}_{\mathscr{F}_{S'}}(X', Y')$$

ξ' is such that the diagram

$$\begin{array}{ccc} p_1^{\alpha}(X') & \xrightarrow{\sim} & p_2^{\alpha}(X') \\ \downarrow p_1^{\alpha}(\xi') & & \downarrow p_2^{\alpha}(\xi') \\ p_1^{\alpha}(Y') & \xrightarrow{\sim} & p_2^{\alpha}(Y') \end{array}$$

is commutative.}

– The composition law of the morphisms is the obvious one.

Consider the functors

$$M_u : \mathscr{F}_S \to \mathscr{R}ec_{\mathscr{F}}(u)$$

$$I : \mathscr{R}ec_{\mathscr{F}}(u) \to \mathscr{F}_{S'}$$

defined by

$$M_u(X) = (u^{\alpha}(X),\ C_{p_2, u} \circ C_{p_1, u}^{-1}),$$

$$I(X', c) = X'.$$

For the sake of simplicity, we shall use below the following notation:

Let $\varphi : I \to J$ be a map of sets. For an arbitrary set L and an object X of $\mathscr{C}$, we denote

$$X^L = \prod_{l \in I} X_1,\ X_l = X,\ \forall l \in L.$$

We also denote by pr_φ the morphism

$$\mathrm{pr}_\varphi : X^J \to X^I$$

such that for every $i \in I$, the following diagram is commutative:

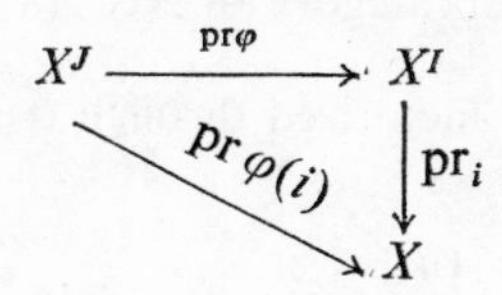

It is easily seen that

$$\mathrm{pr}_{\psi\circ\varphi} = \mathrm{pr}_\varphi \circ \mathrm{pr}_\psi$$

For instance if $I = \{1, 2\}$, $J = \{1, 2, 3\}$ and $\varphi : I \to J$ is the map for which $\varphi(1) = i$, $\varphi(2) = j$, then instead of pr_φ, we shall use the notation p_{ij}.

DEFINITION 3.3.1. A system of local data (X', c) is said to be a *descent data system* if the following diagram commutes:

$$\begin{array}{ccc}
(p_{21}^\alpha \circ p_1^\alpha)(X') & \xrightarrow{p_{21}(c)} & (p_{21}^\alpha \circ p_2^\alpha)(X') \\
\uparrow c_{p_{21}, p_1} & & \uparrow c_{p_{21}, p_2} \\
(p_1 \circ p_{21})^\alpha (X') & & (p_2 \circ p_{21})^\alpha (X') \\
\| & & \| \\
(p_2 \circ p_{32})^\alpha (X') & & (p_2 \circ p_{31})^\alpha (X') \\
\downarrow c_{p_{32}, p_2} & & \downarrow c_{p_{31}, p_2} \\
(p_{32}^\alpha, p_2^\alpha)(X') & & (p_{31}^\alpha \circ p_2^\alpha)(X') \\
\uparrow p_{32}^\alpha(c) & & \uparrow \\
& & (p_{31}^\alpha \circ p_1^\alpha)(X') \\
& & \uparrow \\
(p_{32}^\alpha \circ p_1^\alpha)(X') & \xleftarrow{c_{p_{32}, p_1}} & (p_2 \circ p_{32})^\alpha (X') = (p_1 \circ p_{31})^\alpha (X')
\end{array} \tag{3.3.1.1}$$

Neglecting the isomorphisms of the form $c_{f,g}$, the commutativity of the diagram 3.3.1.1 takes the form

$$p_{21}^{\alpha}(c)\, p_{32}^{\alpha}(c) = p_{31}^{\alpha}(c), \tag{3.3.1.2}$$

relation which is called the "cocycle condition".

We denote by $\mathscr{D}\mathrm{es}_{\mathscr{F}}(u)$ the full subcategory of $\mathscr{R}\mathrm{ec}_{\mathscr{F}}(u)$ generated by the systems of descent data.

The functor M_u can be obviously factorized through $\mathscr{D}\mathrm{es}_{\mathscr{F}}(u)$

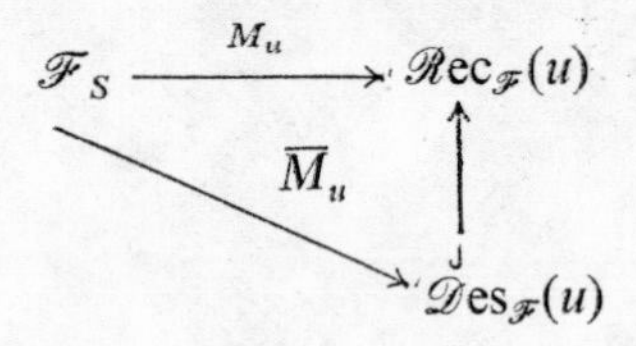

DEFINITION 3.3.2. The morphism u is said to be a *morphism of $\mathscr{F}$-descent* (respectively an *effective morphism of $\mathscr{F}$-descent)* if the functor

$$\overline{M}_u \colon \mathscr{F}_S \to \mathscr{D}\mathrm{es}_{\mathscr{F}}(u)$$

is fully truthful (respectively realizes a category equivalence).

Example 3.3.3. Let $\mathscr{F}$ be the fibre category (example 3.2, 2) whose basis is the category Top and whose fibre $\mathscr{F}_X$ at a topological space X is the category $\mathscr{F}\mathrm{asc}(X, \mathrm{Ens})$ of sheaves of sets on X.

Consider a topological space S and an open covering $(U_i)_{i \in I}$ of S. If $S' = \coprod_{i \in I} U_i$, we have a canonical continuous map (i.e. a morphism in the category Top):

$$S' \xrightarrow{u} S.$$

It is easy to see that

$$S' \times_S S' = \coprod_{(i,j) \in I \times I} (U_i \cap U_j),\; S' \times_S S' \times_S S' =$$

$$= \coprod_{(i,j,k) \in I^3} (U_i \cap U_j \cap U_k)$$

An object of $\mathscr{F}_{S'}$ is a family $(\mathscr{F}_i)_{i\in I}$ such that $\mathscr{F}_i$ is a sheaf of sets on U_i.

A system of local data on such an object is a family of isomorphisms

$$\theta_{\lambda\mu}: \mathscr{F}_\mu| (U_\lambda \cap U_\mu) \xrightarrow{\sim} \mathscr{F}_\lambda| (U_\lambda \cap U_\mu),\ \lambda, \mu \in I.$$

Such a system of local data is a system of descent data if for every triplet (λ, μ, ν) we have

$$\theta'_{\lambda\nu} = \theta'_{\lambda\mu} \circ \theta'_{\mu\nu}$$

where $\theta'_{\lambda\mu}, \theta'_{\mu\nu}, \theta'_{\lambda\nu}$ denote the restrictions of $\theta_{\lambda\mu}, \theta_{\mu\nu}, \theta_{\lambda\nu}$ to $U_\lambda \cap U_\mu \cap U_\nu$.

Under these conditions, there exists a sheaf $\mathscr{F}$ of sets on S and for every $i \in I$ we have an isomorphism

$$\eta_i: \mathscr{F}_{|U_i} \to \mathscr{F}_i$$

such that for every pair of indices $i, j \in I$, the following diagram commutes:

$$\begin{array}{ccc} & \xrightarrow{\eta'_i} & \mathscr{F}_i| U_i \cap U_j \\ \mathscr{F}| U_i \cap U_j & & \big\downarrow \theta_{ji} \\ & \xrightarrow{\eta'_j} & \mathscr{F}_j| U_i \cap U_j \end{array}$$

Indeed, we put $\mathscr{F}(U) = \{(s_k)_{k\in I}|\ s_K \in \mathscr{F}_k(U \cap U_k)$ such that

$$\theta_{l,k}(U \cap U_k \cap U_l)(\mathscr{F}^{U\cap U_k}_{U \cup U_k \cap U_l}(s_k)) =$$

$$= \mathscr{F}^{U\cap U_l}_{U\cap U_k \cap U_l}(s_l) \quad \forall\, k, l \in I$$

$$\begin{array}{ccc} \mathscr{F}_k(U \cap U_k) & \to & \mathscr{F}_k(U \cap U_k \cap U_l) \\ & & \big\downarrow \theta_{1,k}(U \cap U_k \cap U_l) \\ \mathscr{F}_l(U \cap U_l) & \to & \mathscr{F}_l(U \cap U_k \cap U_l) \end{array}$$

The sheaf $\mathscr{F}$ is uniquely determined – up to an isomorphism – by the above conditions.

These conditions can be realized as follows:

Let $(V_i, O_i)_{i \in I}$ be a family of ringed spaces and assume that for every pair $(i,j) \in I \times I$ an open set $V_{i,j} \subset V_i$ and an isomorphism

$$(V_{i,j}, O_i | V_{i,j}) \xrightarrow{\varphi_{i,j}} (V_{j,i}, O_j | V_{j,i})$$

of ringed spaces are given, such that the following conditions be fulfilled:

If $i, j, k \in I$, then φ_{ij} yields a homeomorphism

$$\varphi'_{i,j}: V_{i,j} \cap V_{i,k} \to V_{j,i} \cap V_{k,j}$$

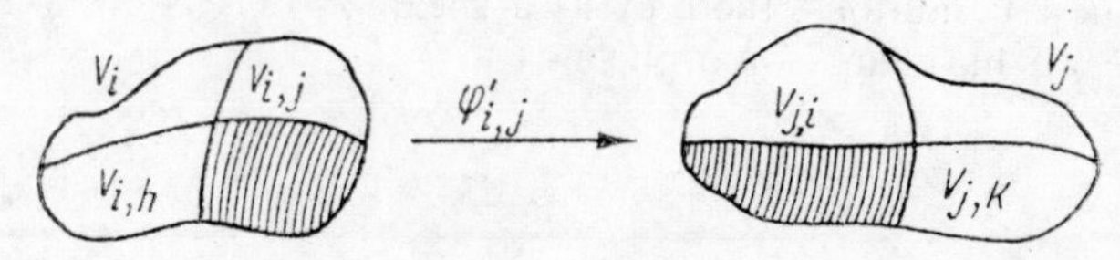

Fig. 3.3.3.1

between $V_{i,j} \cap V_{i,k}$ and $V_{j,i} \cap V_{j,k}$ and in addition we have

$$\varphi'_{i,k} = \varphi'_{j,k} \circ \varphi'_{i,j}.$$

Indeed, it is sufficient to consider on the sum of topological spaces

$$\coprod_{i \in I} V_i$$

the equivalence relation defined as follows:

$$x \sim y \Leftrightarrow \begin{cases} x \sim x, \text{ or} \\ x \in V_{i,j},\ y \in V_{j,i} \text{ and } \varphi_{i,j}(x) = y. \end{cases}$$

Let X be the quotient topological space and consider the canonical projection $\pi: \coprod_{i \in I} V_i \to X$. This map induces for every $i \in I$ a homeomorphism

$$\pi_i: V_i \to U_i$$

such that

$$X = \bigcup_{i \in I} U_i,$$

$$\pi_i(V_{i,j}) = U_i \cap U_j.$$

The homeomorphism π_i defines by structure transport the sheaf $\mathscr{F}_i$ on U_i:

$$\mathscr{F}_i(U) = O_i(\pi_i^{-1}(U)).$$

The isomorphisms $\varphi_{i,j}$ induce the isomorphisms

$$\theta_{j,i}: \mathscr{F}_i|_{U_i \cap U_j} \to \mathscr{F}_j|_{U_i \cap U_j}.$$

Remark. We note that, generally speaking, the space X is not separated, even if all the spaces V_i are so (Fig. 3.3.3.2)

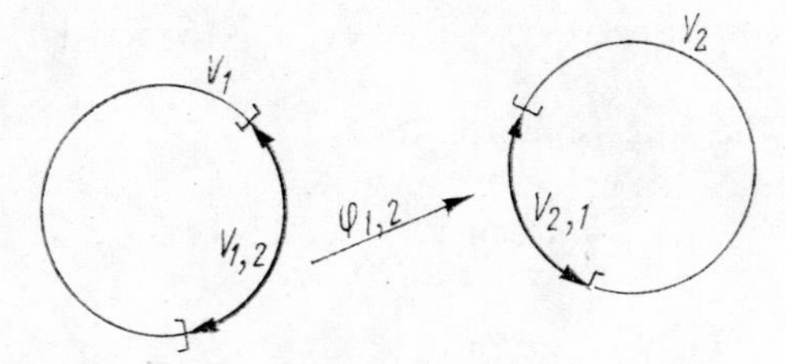

Fig. 3.3.3.2

($V_{1,2}$, $V_{2,1}$ are open arcs)

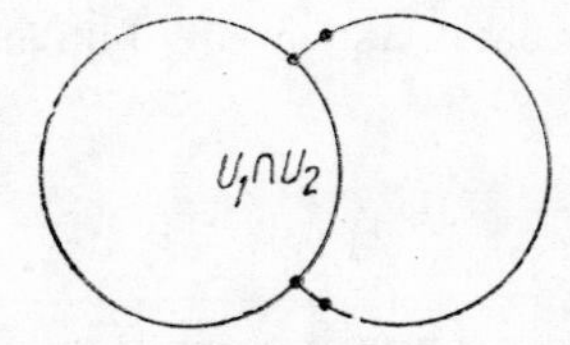

Fig. 3.3.3.3

The space X is obviously a non-separated differentiable variety. Of course, in Fig. 3.3.3.3. the space X has not the subspace topology inherited from the plane.

This fact explains why supplementary conditions are to be imposed to the equivalence relation in order to obtain separated spaces using the above method (See for details Donald Knutson, *Algebraic Spaces*, Lecture Notes 203, Chap. 2).

Examples. P^n schemes. Consider the schemes U_i, $i = 0, 1, 2, \ldots, n$:

$$U_0 = (\mathrm{Spec}(A_0), \tilde{A}_0),\ A_0 = \mathbb{Z}[X_{1,0}, X_{2,0}, \ldots, X_{n,0}]$$

$$U_i = (\mathrm{Spec}(A_i), \tilde{A}_i),\ A_i = \mathbb{Z}[X_{0,i}, X_{1,i}, \ldots, X_{i-1,i},$$

$$X_{i+1,i}, \ldots, X_{n,i}],\ i \geqslant 1$$

$$U_{i,j} = (\mathrm{Spec}(A_i)_{X_{j,i}}, (\tilde{A}_i)_{X_{j,i}}).$$

The isomorphisms

$$\varphi_{i,j}: (\mathrm{Spec}((\tilde{A}_i)_{X_{j,i}}), (\tilde{A}_i)_{X_{j,i}}) \to (\mathrm{Spec}((A_j)_{X_{i,j}}, (\tilde{A}_j)_{X_{i,j}}))$$

correspond to the ring isomorphism

$$u_{i,j}: (A_j)_{X_{i,j}} \to (A_i)_{X_{j,i}}$$

given by

$$u_{i,j}(X_{1,i}|X_{j,i}) - X_{1,i}|X_{j,i}\quad 1 \neq i,\ 1 \neq j$$

$$u_{ij}(X_{i,j}) = 1/X_{j,i}$$

The conditions of a data system are fulfilled since there exists a unique isomorphism

$$(A_j)_{X_{i,j}X_{k,j}} \to (A_i)_{X_{j,i}X_{k,i}}$$

such that the following diagram commutes:

$$\begin{array}{ccc} (A_j)_{X_{i,j}} & \longrightarrow & (A_i)_{X_{j,i}} \\ \downarrow & & \downarrow \\ (A_j)_{X_{i,j}X_{k,j}} & \longrightarrow & (A_i)_{X_{j,i}X_{k,i}} \end{array}$$

The motivation of this definition can be seen if we study the geometric points of the scheme P^n so defined.

To this end, let K be a commutative field. The set $P^n(K)$ of the geometric points of P^n which are rational over K is described as follows:

Consider the disjoint union

$$U = \bigcup_{i=0}^{n} U_i(K).$$

We obviously have a surjective map

$$\pi: \bigcup_{i=0}^{n} U_i(K) \to P^n(K),$$

$$\pi_i = \pi|_{U_i(K)}$$

assigning to every element of $U_i(K)$ the corresponding geometric point of P^n localized at U_i.

On the other hand, the elements of $U_i(K)$ can be identified with the points of the affine space $A^n(K)$:

$$U_i(K) = \{\xi_{0,i}, \xi_{1,i}, \ldots, \xi_{i-1,i}, \xi_{i+1,i}, \ldots, \xi_{n,i}\}.$$

Two points $\xi_i \in U_i(K)$, $\xi_j \in U_j(K)$ are mapped by π onto the same point if the following conditions are fulfilled:

1. ξ_i is localized at $U_{i,j}$, hence $\xi_{j,i} \neq 0$.
2. ξ_j is localized at $U_{j,i}$, hence $\xi_{i,j} \neq 0$.
3. The following diagram commutes:

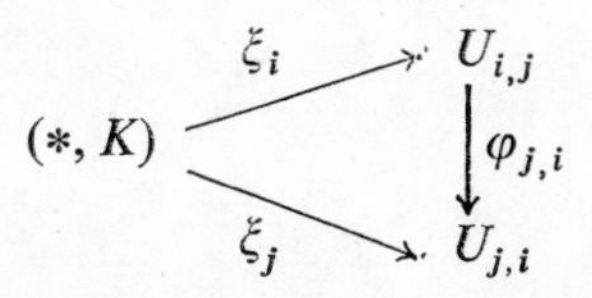

In view of the definition of $\varphi_{j,i}$, it follows that the diagram

$$\begin{array}{ccc} (A_j)_{X_{i,j}} & \xrightarrow{U_{i,j}} & (A_i)_{X_{j,i}} \\ & \searrow^{\xi_j} & \downarrow \xi_i \\ & & K \end{array}$$

also commutes, where

$$X_{1,j} \to X_{1,i}/X_{j,i}$$

$$X_{i,j} \to 1/X_{j,i}$$

$$\xi_i(X_{1,i}) = \xi_{1,i} 1 \neq i$$

$$\xi_j(X_{m,j}) = \xi_{m,j} \; m \neq j$$

consequently the following relations hold:

$$\xi_{1,j} = \xi_{1,i}/\xi_{j,i}, \; 1 \neq i, \; 1 \neq j$$

$$\xi_{i,j} = 1/\xi_{j,i}.$$

Now consider the points of the classical projective space P_K^n over K:

$$P_K^n = \{\overline{(\xi_0, \xi_1, \ldots, \xi_n)}^{\bullet} \mid \xi_i \in K, \; i=0, 1, \ldots, n, \; (\xi_0, \ldots, \xi_n) \neq$$

$$\neq (0, \ldots, 0), \; (\eta_0, \ldots, \eta_n) \in \overline{(\xi_0, \ldots, \xi_n)}^{\bullet} \Leftrightarrow$$

$$\Leftrightarrow \lambda \in K, \; \lambda \neq 0 \; \eta_i = \lambda \xi_i, \; i = 0, \ldots, n\}.$$

We have the maps:

$$\sigma_i \colon U_i(K) \to P_K^n$$

$$(\xi_{0,i}, \xi_{1,i}, \ldots, \xi_{i-1,i}, \xi_{i+1,i}, \ldots, \xi_{ni}) \to$$

$$\to \overline{(\xi_{0,i}, \xi_{1,i}, \ldots, \xi_{i-1,i}, 1, \xi_{i+1,i}, \ldots, \xi_{n,i})}^{\bullet}.$$

The equalities

$$\sigma_i(\xi_i) = \sigma_j(\xi_j), \; i \neq j$$

or

$$\overline{(\xi_{0,i}, \xi_{1,i}, \ldots, \xi_{i-1,i}, 1, \xi_{i+1,i}, \ldots, \xi_{n,i})}^{\bullet} =$$

$$= \overline{(\xi_{0,j}, \xi_{1,j}, \ldots, \xi_{i-1,j}, 1, \xi_{j+1,j}, \ldots, \xi_{n,j})}^{\bullet}$$

are equivalent with the relations

$$\xi_{1,i} = \lambda\xi_{1,j},\ 1 \neq i, j$$

$$1 = \lambda\xi_{i,j}$$

$$\xi_{j,i} = \lambda 1$$

or

$$\xi_{1,j} = \xi_{1,i}|\xi_{j,i},$$

$$\xi_{i,j} = 1|\xi_{j,i}.$$

Consequently, P_K^n can be naturally identified with $P^n(K)$.

PROPOSITION 3.3.4. *Let $\mathscr{C}$ be a category having fibre products and let $\mathscr{F}$ be the fibre category whose basis is the category $\mathscr{C}$ and whose fibre $\mathscr{F}_X$ for an object X of $\mathscr{C}$ is the category $\mathscr{C}/X$ of the X-objects of $\mathscr{C}$* (cf. 3.2.2).

A morphism

$$u: S' \to S$$

of $\mathscr{C}$ is a morphism of $\mathscr{F}$-descent if and only if it is a universal effective epimorphism.

Proof. Assume that u is a universal effective morphism. By making explicit in the case of the fibre category $\mathscr{F}$ the condition in order that u be a morphism of $\mathscr{F}$-descent we find that it is equivalent to the condition that for every two S-objects X, Y of $\mathscr{C}$, the following sequence of sets

$$\text{(3.3.4.1)} \qquad \mathrm{Hom}_S(X, Y) \to \mathrm{Hom}_{S'}(X', Y') \overset{p_1}{\underset{p_2}{\rightrightarrows}} \mathrm{Hom}(X'', Y'')$$

is exact ($X' = X \times_S S', \ldots, X'' = X \times_S S'', \ldots$).

But we have the canonical bijections

$$\mathrm{Hom}_{S'}(X', Y) \simeq \mathrm{Hom}_S(X', Y),$$

$$\mathrm{Hom}_{S''}(X'', Y) \simeq \mathrm{Hom}_S(X'', Y).$$

Consequently, the exactness of 3.3.4.1 is equivalent to that of the sequence

$$\text{(3.3.4.2)} \qquad \mathrm{Hom}_S(X, Y) \to \mathrm{Hom}_S(X, Y) \rightrightarrows \mathrm{Hom}_S(X'', Y)$$

which follows from the exactness of the sequences

$$X'' \rightrightarrows X' \to X,$$

$$S'' \rightrightarrows S' \to S.$$

In this way, we have proved the implication
u is a universal effective morphism $\Rightarrow$ u is a morphism of $\mathscr{F}$-descent.

Now assume that u is a morphism of $\mathscr{F}$-descent. In this case the sequence (3.3.4.1) is exact, and in view of the existence in the category $\mathscr{C}$ of fibre products, this implies that u is a universal effective morphism.

§ 4. *Grothendieck topologies*

DEFINITION 4.1. A category $\mathscr{C}$ is said to be a topos if the following conditions are fulfilled:

(a) $\mathscr{C}$ has finite projective limits and arbitrary inductive limits;

(b) the components of the direct sums in $\mathscr{C}$ are disjoint (Definition 2.1);

(c) the inductive limits in $\mathscr{C}$ are universal (Definition 2.2);

(d) $\mathscr{C}$ has a family of generators;

(e) every equivalence relation on an object X of $\mathscr{C}$ is a universal effective relation.

The motivation for calling it a "topos" consists in the fact that every such category can be realized as a category of sheaves of sets with respect to a more general topology than the classical one, which has been introduced by Grothendieck.

Some details in this direction are the following:

DEFINITION 4.1.1. We say that on the category $\mathscr{C}$ a *Grothendieck topology* τ is given if for every object X of $\mathscr{C}$ a set $\mathrm{Cov}_\tau(X)$ of families $(X_i \xrightarrow{\varphi_i} X)_{i\in I}$ of morphisms in $\mathscr{C}$, called the τ-coverings of the object X, are given such that the following conditions be fulfilled:

1. If $\varphi: Y \to X$ is an isomorphism, then $(\varphi) \in \mathrm{Cov}_\tau(X)$

2. If $(X_i \xrightarrow{\varphi_i} X)_{i\in I} \in \mathrm{Cov}_\tau(X)$ and $(Y_{ij} \xrightarrow{\psi_{ij}} Y_i)_{j\in J} \in \mathrm{Cov}_\tau(Y_i)$ for every $i \in I$, then $(Y_{ij} \xrightarrow{\varphi_i\psi_{ij}} X)_{i\in I, j\in J_i} \in \mathrm{Cov}_\tau(X)$.

3. If $(X_i \xrightarrow{\varphi_i} X) \in \mathrm{Cov}_\tau(X)$ and if $Y \to X$ is any morphism in $\mathscr{C}$, then the fibre product $X_i \times {}_X Y$ associated with the diagram

$$\begin{array}{ccc} & & Y \\ & & \downarrow \\ X_i & \xrightarrow{\varphi_i} & X \end{array}$$

exists for every $i \in I$ and $(X_i \times {}_X Y \xrightarrow{v_i} Y)_{i\in I} \in \mathrm{Cov}_\tau(Y)$, where v_i is the canonical projection.

DEFINITION 4.1.2. Let $\mathscr{C}$ be a category endowed with a Grothendieck topology and let $\mathscr{E}$ be a category with infinite products.

Un $\mathscr{E}$-valued presheaf on $\mathscr{C}$ is a contravariant $\mathscr{E}$-valued functor F defined on $\mathscr{C}$

$$F: \mathscr{C} \to \mathscr{E}.$$

An $\mathscr{E}$-valued sheaf on $\mathscr{C}$ is a presheaf satisfying the following condition:

(F) If $(X_i \xrightarrow{\varphi_i} X)_{i\in I} \in \mathrm{Cov}_\tau(X)$, then the sequence

$$F(X) \xrightarrow{u} \prod_{i\in I} F(X_i) \underset{w}{\overset{v}{\rightrightarrows}} \prod_{i,j\in I2} F(X_i, \times {}_X X_j),$$

is exact, where the morphisms v, w are determined by the commutativity of the diagrams

$$\begin{array}{ccc} \prod_{i \in I} F(X_i) & \xrightarrow{v} & \prod_{(i,j) \in I^2} F(X_i \times_X X_j) \\ \downarrow p_\alpha & & \downarrow p_{(\alpha,\beta)} \\ F(X_\alpha) & \xrightarrow{F(\pi_\alpha)} & F(X_\alpha \times_X X_\beta) \end{array}$$

$$\begin{array}{ccc} \prod_{i \in I} F(X_i) & \xrightarrow{w} & \prod_{(i,j) \in I^2} F(X_i \times_X X_j) \\ \downarrow & & \downarrow p_{(\alpha,\beta)} \\ F(X_\beta) & \xrightarrow{F(\pi_\beta)} & F(X_\alpha \times_X X_\beta) \end{array}$$

in which $p_\alpha, p_\beta, p(\alpha, \beta)$ are canonical projections.

Example 4.1.2.1. Let T be a topological space and let $\mathscr{D}_T$ be the category associated with the partially ordered set of open subsets of T. A Grothendieck topology τ_T on $\mathscr{D}_T$ can be obtained as follows:

$$\mathrm{Cov}_\tau(U) = \{(U_i \xrightarrow{\varphi_i} U)_{i \in I} | \varphi_i \text{ inclusion and } \bigcup_{i \in I} U_i = U\}.$$

A sheaf on $\mathscr{D}_T$ is a sheaf on the topological space T in the classical sense (cf. for instance R. Godement, *Théorie des faisceaux)*.

DEFINITION 4.1.2.2. Let $\mathscr{C}$ be a category endowed with a Grothendieck topology τ. We denote by $\mathscr{F}\mathrm{asc}(\tau, \mathrm{Ens})$ the full subcategory of the category $\mathscr{F}\mathrm{onct}(\mathscr{C}^\circ, \mathrm{Ens})$ generated by the sheaves defined on $\mathscr{C}$ with values in the category Ens with respect to the topology τ.

PROPOSITION 4.1.3. *The category* $\mathscr{F}\mathrm{asc}(\tau, \mathrm{Ens})$ *is a topos.*

The proof of this proposition is relatively long and intricate. In fact, it extends by suitable adaptations the proof of the particular case corresponding to the category $\mathscr{F}\mathrm{asc}(X, \mathrm{Ens})$ of the sheaves of sets on the topological space X. Details can be found in SGA (4), *Théorie des topos et cohomologie étale des schémas*, Lecture Notes, Springer, vol. 269.

In the following we give some hints related to the extension of some concepts and results concerning the category $\mathscr{F}\mathrm{asc}(X, \mathrm{Ens})$ to the case of the category $\mathscr{F}\mathrm{asc}(\tau, \mathrm{Ens})$.

4.1.4. Let $\mathscr{C}$ be a category endowed with a Grothendieck topology τ. The inclusion functor

$$\mathscr{F}\mathrm{asc}(\tau, \mathrm{Ens}) \xrightarrow{i} \mathscr{F}\mathrm{onct}(\mathscr{C}^\circ, \mathrm{Ens})$$

has a right adjoint

$$\mathscr{F}\mathrm{onct}(\mathscr{C}^\circ, \mathrm{Ens}) \xrightarrow{a} \mathscr{F}\mathrm{asc}(\tau, \mathrm{Ens})$$

i.e. for every sheaf $\mathscr{F}$ and for every presheaf $\mathscr{G}$, we have a functorial bijection

$$\mathrm{Hom}_{\mathscr{F}\mathrm{asc}(\tau, \mathrm{Ens})}(a(\mathscr{F}), \mathscr{G}) \to \mathrm{Hom}_{\mathscr{F}\mathrm{onct}(\mathscr{C}^\circ, \mathrm{Ens})}(\mathscr{F}, i(\mathscr{G})).$$

The sheaf $a(\mathscr{F})$ is called the sheaf associated with the presheaf $\mathscr{F}$.

This result allows a simple description of cokernels in the category $\mathscr{F}\mathrm{asc}(\tau, \mathrm{Ens})$.

Indeed, in order to form cokernels in the category $\mathscr{F}\mathrm{asc}(\tau, \mathrm{Ens})$ it is sufficient to form them first in the category $\mathscr{F}\mathrm{onct}(\mathscr{C}^\circ, \mathrm{Ens})$, an operation which is realized on components in Ens, and then to pass to the associated sheaf.

In the following we give some ideas about the definition of the functor a. To this end, let X be an object of $\mathscr{C}$ and let $\mathrm{Cov}_\tau(X)$ be the set of the τ-coverings of X.

A preorder relation on the set $\mathrm{Cov}_\tau(X)$ can be defined as follows: Let

$$\alpha = (X_a \xrightarrow{f_a} X)_{a \in A'}, \ \beta = (X_b \xrightarrow{g_b} X)_{b \in B}$$

be two elements of the set $\mathrm{Cov}_\tau(X)$.

By definition we say that $\alpha \geqslant \beta$ if and only if a map $\varphi: A \to B$, and for every $a \in A$, a morphism

$$h_a: X_a \to X_{\varphi(a)}$$

exist such that the following diagram be commutative:

$$\begin{array}{ccc} X_a & \xrightarrow{f_a} & X \\ h_a \downarrow & \nearrow_{g_{\varphi(a)}} & \\ X_{\varphi(a)} & & \end{array}$$

This preorder relation on $\mathrm{Cov}_\tau(X)$ is a filtering one. Indeed, let

$$\alpha' = (X_{a'} \xrightarrow{f'_{a'}} X)_{a' \in'}$$

$$\alpha'' = (X_{a''} \xrightarrow{f''_{a''}} X)_{a'' \in A''}$$

be two elements of $\mathrm{Cov}_\tau(X)$ and let

$$\alpha = (X_{a''} \coprod_X X_{a'} \xrightarrow{f'_{a'} \circ \mathrm{pr}_{X_{a'}}} X)_{(a'', a') \in A'' \times A'}$$

be the family of morphisms having the address X, where $\mathrm{pr}_{X_{a'}}$ is the morphism from the cartesian square

$$(4.1.4.1) \qquad \begin{array}{ccc} X_{a''} \coprod_X X_{a'} & \xrightarrow{\mathrm{pr}_{X_{a'}}} & X_{a'} \\ \mathrm{pr}_{X_{a''}} \uparrow & & \downarrow \\ X_{a''} & \longrightarrow & X \end{array}$$

The element α belongs to the set $\mathrm{Cov}_\tau(X)$. Indeed, it is sufficient first to remark that for every element $a' \in A'$, the family of morphisms

$$(X_{a''} \coprod_X X_{a'} \to X_{a'})_{a'' \in A''}$$

is an element of the set $\mathrm{Cov}_\tau(X_{a'})$.

The diagram (4.1.4.1) shows at the same time that $\alpha \geqslant \alpha'$, $\alpha \geqslant \alpha''$. Putting

$$(LF)(X) = \varinjlim_{\alpha \in \mathrm{Cov}_\tau(X)} \mathrm{Ker}(\prod_{a \in A} F(X_a) \rightrightarrows \prod_{(a, b) \in A^2} F(X_a \times_X X_b))$$

we get a presheaf such that $L(LF)$ is a sheaf. We can take $a(F) = L(LF)$

We obviously have a functorial morphism

$$F \to (aF).$$

The concept of sheaf associated with a presheaf allows among other things to make explicit the definition in the category $\mathscr{F}$asc(τ, Ens) of the functor which represents the cokernel of a couple of morphisms

$$R \underset{\pi_1}{\overset{\pi_2}{\rightrightarrows}} X.$$

This cokernel is obtained as follows:
First consider the presheaf cokernel

$$P \to \mathscr{C}^\circ \underset{\pi_1}{\overset{\pi_2}{\rightrightarrows}} \text{Ens}$$

having a purely set-theoretic definition, i.e. for an arbitrary object T of the category $\mathscr{C}$, consider on the set $X(T)$ the equivalence relation generated by

$$t_1, t_2 \in X(T),\ t_1 \sim t_2 : \exists t \in R(T),\ \pi_{1,T}(t) = t_1,\ \pi_{2,T}(t) = t_2,$$

and take for $P(T)$ the quotient set of $X(T)$ with respect to this relation. In fact, P is the cokernel of the couple of morphisms $R \underset{\pi_1}{\overset{\pi_2}{\rightrightarrows}} X$ in the category $\mathscr{F}$onct($\mathscr{C}^\circ$, Ens).

In order to obtain the cokernel Y in the category $\mathscr{F}$asc(τ, Ens), of the couple of morphisms $R \underset{\pi_1}{\overset{\pi_2}{\rightrightarrows}} X$, we take the associated sheaf $a(P)$ of the presheaf P.

4.1.5. Let $\mathscr{C}$ be a category endowed with a Grothendieck topology such that every representable functor be a sheaf in the τ-topology and let $R \underset{\pi_1}{\overset{\pi_2}{\rightrightarrows}} X$ be an equivalence relation on X. If the quotient Y in the topos $\mathscr{F}$asc(τ, Ens) is representable, then obviously Y is a quotient of X and even a good quotient of X with respect to the equivalence relation R in the category $\mathscr{F}\mathrm{asc}(\tau, \mathrm{Ens})$.

It is accordingly very important to have manageable criteria for the representability of a quotient in the topos $\mathscr{F}\mathrm{asc}(\tau, \mathrm{Ens})$. Of course, a quotient in $\mathscr{C}$ is not necessarily a quotient in $\mathscr{F}\mathrm{asc}(\tau, \mathrm{Ens})$, as it is readily seen by considering a quotient which is not effective. (One uses the remark that considering the diagram in $\mathscr{C}$

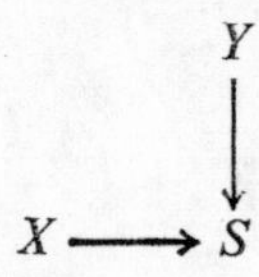

we have

$$(h_X \times_{h_S} h_Y)(U) = h_X(U) \times_{h_S(U)} h_Y(U) = h_{X \times_S Y}(U),$$

where the fibre product $X \times_S Y$ is taken in $\mathscr{C}$).

4.1.6. *The "étale" topology.* For the sake of simplicity, consider an algebraically closed field k, and denote by $\mathscr{S}ch_k$ the category of separated schemes of finite type over k.

DEFINITION 4.1.6.1. Let X, Y be two schemes of finite type over an algebraically closed field k.

A morphism of schemes $f: X \to Y$ is said to be étale if for every point $y \in Y$, the set $f^{-1}(y)$ is finite and if, in addition, for every point $x \in f^{-1}(y)$, the homomorphism of local rings

$$(4.1.6.2) \qquad f_x^*: \mathcal{O}_y \to \mathcal{O}_x$$

induced by f yields an isomorphism

$$\hat{f}_x^*: \hat{\mathcal{O}}_y \to \hat{\mathcal{O}}_x$$

of the completed rings (with respect to the $\underline{m}$-adic topology).

The condition for f to be étale is equivalent to the following two conditions:

(a) the morphism f is flat, i.e. O_x is a flat O_y-module (with respect to the O_y-module structure of O_x induced by the morphism f_x^*).

(b) For every $y \in Y$, the set $f^{-1}(y)$ is finite and, in addition, we have

$$\underline{m}_x = f_x^*(\underline{m}_y)\mathcal{O}_x$$

where $\underline{m}_x$ (or $\underline{m}_y$) is the unique maximal ideal of O_x (or O_y).

DEFINITION 4.1.6.2. The étale topology on the category $\mathcal{S}ch_k$ is the Grothendieck topology γ on $\mathcal{S}ch_k$ defined as follows:

$$\mathrm{Cov}_\gamma(X) = \{(X_i \xrightarrow{\varphi_i} X)_{i\in I} \mid \varphi_i \text{ étale morphism and}$$

$$\bigcup_{i\in I} \varphi_i(X_i) = X\}.$$

It is useful to associate with every object X of the category $\mathcal{S}ch_k$ the category Ét/X, the full subcategory of $\mathcal{S}ch_k$ generated by the étale morphisms $X' \to X$.

A morphism in Ét/X between the objects $U \xrightarrow{p} X$, $V \xrightarrow{q} Y$ is by definition a commutative diagram

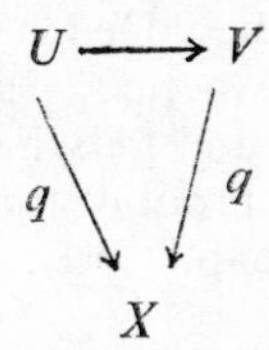

A Grothendieck topology on Ét/X, induced by the étale topology of $\mathcal{S}ch_k$ and called the étale topology on Et/X, can be defined as follows:

A family of morphisms

$$\begin{array}{ccc} U_\alpha & \xrightarrow{u_\alpha} & U \\ & {\scriptstyle p_\alpha}\searrow \quad \swarrow{\scriptstyle q} & \\ & X & \end{array} \qquad (\alpha \in A)$$

is said to be a covering of the object $U \xrightarrow{p} X$ of the category Ét/X in the sense of the étale topology if and only if the following relation is fulfilled:

$$U = \bigcup_{\alpha \in A} u_\alpha(U_\alpha).$$

Let $\mathscr{F} \in \mathrm{Ab}(\mathscr{F}\mathrm{asc}(\gamma_X, \mathrm{Ens}))$ be an Abelian group in the category $\mathscr{F}\mathrm{asc}(\gamma_X, \mathrm{Ens})$. Since the category $\mathrm{Ab}(\mathscr{F}\mathrm{asc}(\gamma_X, \mathrm{Ens})$ is Abelian and has sufficient injective objects, it follows that the right derived functors of the functor

$$\Gamma_X \colon \mathrm{Ab}(\mathscr{F}\mathrm{asc}(\gamma_X, \mathrm{Ens})) \to \mathrm{Ab},$$

$$\mathscr{F} \mapsto \mathscr{F}(X \xrightarrow{1_X} X).$$

can be defined.

In particular, we consider the groups

$$R^i\Gamma_X(\mathscr{F}) = H^i(X_{\mathrm{et}}, \mathscr{F})$$

which are usually called the étale cohomology groups of the scheme X.

For this theory, Grothendieck, Artin, Verdier, Deligne have proved the reasonable results which were to be expected – finiteness and Poincaré duality theorems, Künneth formulae and Lefschetz fixed point formulae, which in particular have allowed the proof of the Diophantine conjectures of Weil (see, for some details, Chapter VI).

4.1.7. *The "Zariski topology".*

DEFINITION 4.1.7.1. The *Zariski topology* on $\mathscr{S}ch_k$ is the Grothendieck topology on $\mathscr{S}ch_k$ defined as follows:

$$\mathrm{Cov}_{\mathrm{Zar}}(X) = \{(X_i \xrightarrow{\varphi_i} X) \mid \varphi_i \text{ open immersions and } \bigcup_{i \in I} \varphi_i(X_i) = X\}.$$

As in the case of the étale topology, with every element X of $\mathscr{S}ch_k$ it is useful to associate the full subcategory of $\mathscr{S}ch_k/X$ generated by the morphisms $X' \to X$ which are open immersions. The category $\mathscr{S}ch_k/X$ can obviously be endowed with a Grothendieck topology Zar_X induced by the Zariski topology on $\mathscr{S}ch_k$.

It is readily seen that an equivalence can be established between the category $\mathscr{F}\mathrm{asc}(\mathrm{Zar}_X, \mathrm{Ens})$ and the category of sheaves of sets (in the classical sense) defined on X.

This category has been studied in detail by J. P. Serre in his paper *Faisceaux algébriques cohérents*, Ann. Math., 1955.

4.1.8. Let $\mathscr{C}$ be a category. We denote by τ_c the topology on $\mathscr{C}$ defined as follows:

$$\mathrm{Cov}_{\tau c}(X) = \{(Y \xrightarrow{u} X) | \, u \text{ universal effective epimorphism}\}$$

PROPOSITION 4.1.8.1. *Every representable functor is a sheaf with respect to the topology* τ_c

The functor

$$\mathscr{C} \to \mathscr{F}\mathrm{asc}(\tau_c, \mathrm{Ens})$$

$$A \rightsquigarrow h_A = \mathrm{Hom}_{\mathscr{C}}(-, A)$$

defines an equivalence between $\mathscr{C}$ *and a full subcategory of* $\mathscr{F}\mathrm{asc}(\tau_c, \mathrm{Ens})$.

Proof. All we have to prove is the fact that the representable functors are sheaves with respect to the topology τ_c. But if

$$(Y \to X) \in \mathrm{Cov}_{\tau_c}(X),$$

then the following sequence of sets

$$(4.1.8.2) \qquad \mathrm{Hom}_{\mathscr{C}}(X, A) \xrightarrow{\mathrm{Hom}_{\mathscr{C}}(u,\, A)} \mathrm{Hom}_{\mathscr{C}}(Y, A) \underset{\mathrm{Hom}_{\mathscr{C}}(\mathrm{pr}_2 A)}{\overset{\mathrm{Hom}_{\mathscr{C}}(\mathrm{pr}_1 A)}{\rightrightarrows}} \mathrm{Hom}_{\mathscr{C}}(Y \times {}_X Y, A)$$

is exact for every object A of $\mathscr{C}$, where pr_1, pr_2 are the canonical projections

$$Y \times {}_X Y \underset{\mathrm{pr}_2}{\overset{\mathrm{pr}_1}{\rightrightarrows}} Y.$$

The exactness of the sequence (4.1.8.2) results from the fact that the sequence

$$Y \times {}_X Y \rightrightarrows Y \xrightarrow{u} X$$

is exact by hypothesis. But, according to Definition 4.1.2, the exactness of the sequence (4.1.8.2) expresses the fact that the functor h_A is a sheaf.

4.1.9. The "*canonical*" *topology* c on $\mathscr{C}$ can in fact be defined as the finest topology for which the representable functors are sheaves.

In order to define the topology c, consider first a family of morphisms

$$(X_i \xrightarrow{\varphi_i} X)_{i \in I}.$$

We say that this family is *strictly epimorphic* if the following conditions are fulfilled:

(1) For every couple $(\alpha, \beta) \in I \times I$, the fibre product $X_\alpha \times_X X_\beta$ associated with the diagram

$$\begin{array}{ccc} & & X_\beta \\ & & \downarrow \\ X_\alpha & \longrightarrow & X \end{array}$$

exists.

(2) For every object Z in $\mathscr{C}$, the sequence of sets

$$\operatorname{Hom}_{\mathscr{C}}(X, Z) \to \prod_{i \in I} \operatorname{Hom}_{\mathscr{C}}(X_i, Z) \rightrightarrows$$

$$\rightrightarrows \prod_{(i,j) \in I^2} \operatorname{Hom}_{\mathscr{C}}(X_i \textstyle\prod_X X_j, Z)$$

is exact.

$(X_i \to X)_{i \in I}$ is a *strictly epimorphic universal family* if for every morphism $Y \xrightarrow{u} X$, the fibre product $X_i \times_X Y$ associated with the diagram

$$\begin{array}{ccc} & & Y \\ & & \downarrow u \\ X_i & \xrightarrow{\varphi_i} & X \end{array}$$

exists, and the family $(X_i \times_X Y \xrightarrow{\mathrm{pr}_i^Y} Y)$ is strictly epimorphic.

The canonical topology c is defined as follows: $\mathrm{Cov}_c(X)$ = the set of strictly epimorphic universal families $(X_i \xrightarrow{\varphi_i} X)_{i \in I}$.

Obviously, we have a canonical imbedding

$$\mathscr{C} \to \mathscr{F}\mathrm{asc}(c, \mathrm{Ens}).$$

Example 4.1.9.1. On a (sufficiently general) category $\mathscr{C}$, there exist topologies with respect to which not all representable functors are sheaves. Such a topology can be defined as follows. Let $\mathscr{C}$ be a little category having fibre products, in which there exists a morphism which is not an effective epimorphism.

We introduce on $\mathscr{C}$ the topology $\mathscr{X}$ defined by

$$\mathrm{Cov}_{\mathscr{X}}(X) = \{(Y \to X)\}.$$

We see that there exist representable functors which are not sheaves, since otherwise every morphism would be an effective epimorphism.

4.1.10. Proposition 4.1.3 has a remarkable reciprocal, which justifies once more the "topos" terminology.

PROPOSITION 4.1.11. *Let $\mathscr{T}$ be a topos. A category $\mathscr{C}$ and a Grothendieck topology τ on $\mathscr{C}$ exist such that $\mathscr{T}$ is equivalent to the category* $\mathscr{F}\mathrm{asc}(\tau, \mathrm{Ens})$.

The proof of this proposition can be found in A. Grothendieck, *Théory des topos et cohomologie étale des schémas*, Lecture Notes in Mathematics, Springer-Verlag, 269. We merely remark that this proof essentially consists in showing that if we introduce on $\mathscr{T}$ the canonical topology (4.1.9), then every sheaf of sets on $\mathscr{T}$ with respect to c is representable.

4.1.12. A remarkable immediate consequence of Proposition 4.1.11 is the following:

PROPOSITION 4.1.13. *Every topos has arbitrary projective limits.*

Proof. In view of condition a from Definition 4.1, it is sufficient to show that every topos has arbitrary direct products (cf. I. Bucur–A. Deleanu, *Introduction to the Theory of Categories and Functors)*. But if $(F_\alpha)_{\alpha \in A}$ is a family of sheaves, i.e. for every covering of X

$$(X_i \to X)_{i \in I},$$

the sequence

$$F_\alpha(X) \to \prod_{i \in I} F_\alpha(X_i) \rightrightarrows \prod_{(i,j) \in I^2} F_\alpha(X_i \textstyle\prod_X X_j)$$

is exact, then the sequence

$$\prod_{\alpha \in A} F_\alpha(X) \to \prod_{i \in I} \left(\prod_{\alpha \in A} F_\alpha(X_i)\right) \rightrightarrows \prod_{(i,j) \in I^2} \left(\prod_{\alpha \in A} F_\alpha(X_i \textstyle\prod_X X_j)\right)$$

is exact too.

II. THEORY OF LAWVERE-TIERNEY TOPOI

1. As mentioned in the previous section, the concept of topos has been introduced by Grothendieck's school as a generalization of the concept of topological space.

In the present conception of this school, general topology is the study of topoi.

In the study and development of the theory of topoi, Grothendieck's school had in mind applications to algebraic geometry. A special attenion has been paid to the so-called étale topology (cf. 4.1.6) and its applications to algebraic geometry, in particular for formulating and proving the remarkable conjectures of A. Weil.

F. W. Lawvere and M. Tierney have been led to consider a class of categories similar to that of topoi, starting from the idea of generalizing the category of sets, more precisely, of finding the categories in which the concepts of a higher degree language could be interpreted – i.e. in which a natural category – theoretical model for a "universe of discourse" could be defined.

A natural problem which arose in this way was: given an object A of a category $\mathscr{C}$, define the object $\mathscr{P}(A)$ of all subobjects of A, viewed as an object of $\mathscr{C}$, and more generally, define the object B^A of the morphisms of A in B. One also had to assume the existence of a "characteristic morphism" associated to a subobject X'.

Definition 1.1. Let $\mathscr{C}$ be a category with finite direct products. We say that $\mathscr{C}$ has an *internal hom* if for every object A of $\mathscr{C}$, the functor

$$- \textstyle\prod A : \mathscr{C} \to \mathscr{C}$$

$$X \rightsquigarrow X \textstyle\prod A$$

has a right adjoint

$$\underline{\mathrm{Hom}}_{\mathscr{C}}(A, -): \mathscr{C} \to \mathscr{C},$$

$$Y \rightsquigarrow \underline{\mathrm{Hom}}_{\mathscr{C}}(A, Y) = Y^A.$$

in other words, if we have a functorial bijection

$$\mathrm{Hom}_{\mathscr{C}}(X \textstyle\prod A, Y) \simeq \mathrm{Hom}_{\mathscr{C}}(X, Y^A).$$

If 1 is a final object of the category $\mathscr{C}$, for every couple Y, A of objects in $\mathscr{C}$ we have a canonical bijection

$$\mathrm{Hom}_{\mathscr{C}}(1, Y^A) \simeq \mathrm{Hom}_{\mathscr{C}}(A, Y).$$

Example 1.1.2. Every topos has an internal hom.

We shall merely prove this fact for the case of a topos $\mathscr{T}$ of the form $\mathscr{T} = \mathscr{F}\mathrm{asc}(X, \mathrm{Ens})$, where X is a topological space.

To this end, consider two sheaves of sets F, G on the space X. A sheaf $\underline{\mathrm{Hom}}(F, G)$ on X can be defined by taking for every open set U of X

$$\underline{\mathrm{Hom}}(F, G)(U) = \mathrm{Hom}_{\mathscr{F}\mathrm{asc}(U, \mathrm{Ens})}(F_{|U}, G_{|U}).$$

If H is a sheaf on X, $H_{|U}$ denotes as usual the sheaf on U defined by $(H_{|U})(V) = H(V)$ if V is open and $V \subset U$.

Example 1.1.3. Let L be a lattice with a first and a last element and let $\underline{L}$ be the associated category. L is a Heyting algebra (Chapter II, Example 2.12) if and only if the category $\underline{L}$ has an internal hom.

If $a, b \in L$, one writes $a \Rightarrow b$ instead of $\underline{\mathrm{Hom}}_L(a, b)$. Consequently,

$$a \wedge b \leqslant c \text{ if and only if } a \leqslant (b \Rightarrow c).$$

Example 1.1.4. The category Ens/S has an internal hom.

If $X \xrightarrow{u} S$, $Y \xrightarrow{v} S$ are two objects of Ens/S, then $\underline{Hom}(u, v)$ is constructed fibre by fibre as follows:

The underlying set of $\underline{Hom}(u, v)$ is

$$\bigcup_{s \in S} \mathrm{Hom}_{\mathrm{Ens}}(u^{-1}(s), v^{-1}(s)).$$

and the morphism into S maps the set $\mathrm{Hom}_{\mathrm{Ens}}(u^{-1}(s), v^{-1}(s))$ onto the point $s \in S$.

DEFINITION 1.2. Let $\mathscr{C}$ be a category with an internal hom and let A, X be objects of $\mathscr{C}$. We denote by ev_A the morphism

$$\mathrm{ev}_A : X^A \times A \to X$$

called the *evaluation morphism*, which corresponds by the bijection (1.1.1)

$$\mathrm{Hom}_{\mathscr{C}}(X^A \times A, X) \simeq \mathrm{Hom}_{\mathscr{C}}(X^A, X^A)$$

to the identical morphism of X^A.

DEFINITION 1.3. Let $\mathscr{C}$ be a category with fibre products and let 1 be a final object of $\mathscr{C}$. We say that $\mathscr{C}$ has a *classifying object* for its subobjects if there exists a morphism $t: 1 \to \Omega$ such that for every subobject

$$\xi : X' \hookrightarrow X$$

of X, a unique morphism

$$\chi_\xi : X \to \Omega$$

called the "characteristic morphism" of the subobject ξ, exists such that the diagram

$$\begin{array}{ccc} X' & \longrightarrow & 1 \\ {\scriptstyle \xi}\downarrow & & \downarrow{\scriptstyle t} \\ X & \longrightarrow & \Omega \end{array}$$

be Cartesian.

(This condition obviously determines the morphism t up to an isomorphism).

PROPOSITION 1.4. *Every topos (in the sense of Grothendieck) has a classifying object for its subobjects.*

Proof. We consider the particular case of a topos of the form $\mathscr{F}\mathrm{asc}(X, \mathrm{Ens})$.

Let Ω, 1 be sheaves of sets on the topological space X, defined as follows:

$\Omega(U) = \mathscr{D}(U) =$ the set of the open subsets of X which are included in U

$1(U) = x =$ the final object of the category

Let $t: 1 \to \Omega$ be the monomorphism of sheaves defined by $t(U)(x) = U$ for every $U \in \mathscr{D}(X) = \Omega(X)$.

With any monomorphism $\xi: \mathscr{F}' \to \mathscr{F}$ of sheaves of sets on X we can associate a morphism

$$\chi_\xi: \mathscr{F} \to \Omega.$$

In order to define this morphism, we first remark that for every point $x \in X$, the fibre Ω_x contains two privileged elements: the class of the empty set and the class of the total space. Consequently, the morphism χ_X can be defined on each fibre by taking

$$\chi_{\xi,x}: \mathscr{F}_x \to \Omega_x,$$

$\chi_{\xi,x} = \chi_{\mathscr{F}'_x} =$ the characteristic function (with respect to the set of the two privileged elements of Ω_x) of the subset $\mathscr{F}'_x$ of $\mathscr{F}_x$.

The family of morphisms $(\chi_{\xi,x})_{x \in X}$ defines a morphism of sheaves

$$\chi_\xi: \mathscr{F} \to \Omega.$$

Indeed, if U is an open subset of X and if $s \in \mathscr{F}(U)$, then the subset

$$S = \{y \in U | \, s(y) \in \mathscr{F}'_x\}$$

is open and

$$\chi_{\xi,y}(s, (y)) = s(y), \quad \forall y \in U.$$

The diagram

$$\begin{array}{ccc} \mathscr{F}' & \longrightarrow & 1 \\ \xi \downarrow & & \downarrow t \\ \mathscr{F} & \xrightarrow{\chi_\xi} & \Omega \end{array}$$

is obviously commutative and even Cartesian, while χ_ξ is clearly uniquely determined by this property.

Example 1.4.1. The category Ens/S has a classifying object for its subobjects.

Such an object is readily seen to be

$$S \times \{0, 1\} \xrightarrow{\mathrm{pr}_s} S.$$

DEFINITION 1.5. A category $\mathscr{C}$ is called a *topos in the sense of Lawvere–Tierney*, or briefly an L–T topos, if the following conditions are satisfied:

(a) $\mathscr{C}$ has finite inductive limits;
(b) $\mathscr{C}$ has finite projective limits;
(c) $\mathscr{C}$ has an internal hom (Definition 1.1);
(d) $\mathscr{C}$ has a classifying object for its subobjects (Definition 1.3).

Example 1.5.1. Every Grothendieck topos is a Lawvere–Tierney topos.

Example 1.5.2. The category Ens_f of finite sets is a Lawvere–Tierney topos but it is not a Grothendieck topos.

PROPOSITION 1.6. *Let $\mathscr{C}$ be an L–T topos and let Ω be a classifying object for the subobjects of $\mathscr{C}$. For every object A of $\mathscr{C}$, there exists a monomorphism*

$$\{\cdot\}: A \to \Omega^A.$$

Proof. In order to define a morphism

$$A \to \Omega^A,$$

it is sufficient to point out a morphism

$$A \times A \to \Omega$$

or, equivalently, a subobject of $A \times A$. In the present case, we take the diagonal morphism

$$\Delta: A \to A \times A$$

and we define $\{\cdot\}$ as being the morphism which corresponds through the canonical bijection

$$\mathrm{Hom}_{\mathscr{C}}(A, \Omega^A) \rightrightarrows \mathrm{Hom}_{\mathscr{C}}(A \times A, \Omega)$$

to the morphism μ from the cartesian diagram

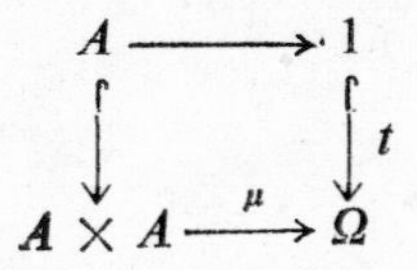

We still have to prove that $\{\cdot\}$ is a monomorphism. To this end, first we remark that for every morphism $X \xrightarrow{u} A$, the diagram

$$\begin{array}{ccc} X & \longrightarrow & A \\ {\scriptstyle (1_X, u)} \downarrow & & \downarrow {\scriptstyle \Delta} \\ X \times A & \longrightarrow & A \times A \end{array}$$

is Cartesian.

Now consider $\xi_1, \xi_2 \colon X \to A$ such that

$$\{\cdot\}\xi_1 = \{\cdot\}\,\xi_2.$$

It follows that in the diagram

$$\begin{array}{ccccc} X & \longrightarrow & A & \longrightarrow & L \\ {\scriptstyle (1_X, \xi_i)} \downarrow & & \downarrow & & \downarrow {\scriptstyle t} \\ X \times X & \xrightarrow{\xi_i \times 1_A} & A \times A & \xrightarrow{\mu} & \Omega \end{array} \qquad (i = 1, 2)$$

all three rectangles are cartesian.

By the uniqueness property of Ω, this shows that the subobject $(1_X, \xi_1)$, of $X \times X$ coincides with the subobject $(1_X, \xi_2)$, hence $\xi_1 = \xi_2$.

Remark. In the case of the category Ens, the map $\{\cdot\}$ is defined by

$$\{\cdot\} \colon A \to \mathscr{P}(A) = 2^A,$$

$$a \mapsto \{a\}.$$

PROPOSITION 1.7. *In an $I-T$ topos, every morphism $f: A \to B$ can be uniquely (up to isomorphisms) factorized as follows:*

$$f = vu,$$

where u is an epimorphism and v is a monomorphism.

Proof. Consider the cartesian diagram

$$\begin{array}{ccc} A \times_B A & \xrightarrow{p_1} & A \\ {\scriptstyle p_2}\downarrow & & \downarrow{\scriptstyle f} \\ A & \xrightarrow{f} & B \end{array}$$

Let (Q, q) be the cokernel of the couple of morphisms (p_1, p_2). The sequence of morphisms

$$A \times_B A \underset{p_2}{\overset{p_1}{\rightrightarrows}} A \xrightarrow{q} Q$$

being exact, the commutativity of (1.7.1) yields the required factorization:

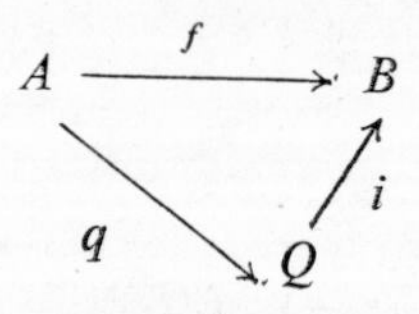

(The proof of the uniqueness of the factorization and the fact that it is a monomorphism are left to the reader. Details can be found in A. Koch and G. C. Wraith, *Elementary Toposes*, Aarhus Universitet, Lecture Notes Series No. 30, 1971, Prop. 1.1.9, p. 22).

DEFINITION 1.8. The subobject of B defined by the monomorphism v is called the *image* of f and is denoted by $\operatorname{Im} f$. We state without proof the following results:

PROPOSITION 1.9. *In an $I-T$ topos the factors of every direct sum are disjoint.*

Cf. A. Koch and G. C. Wraith, loc. cit. Prop. 1.24, p. 28.

PROPOSITION 1.10. *In an I—T topos, every equivalence relation is effective.*

Cf. A. Koch and G. C. Wraith, loc. cit., Prop. 1.21, p. 24.

PROPOSITION 1.11. *In an L—T topos, the object Ω classifying subobjects has a structure of Heyting algebra.*

Proof: First we define the following morphisms:

I. $1 \xrightarrow{f} \Omega$,

II. $\Omega \times \Omega \xrightarrow{\cap} \Omega$,

III. $\Omega \times \Omega \xrightarrow{\cup} \Omega$,

IV. $\Omega \xrightarrow{\neg} \Omega$.

f is defined by the condition that the following diagram be Cartesian

$$\begin{array}{ccc} 0 & \hookrightarrow & 1 \\ \downarrow & & \downarrow t \\ 1 & \xrightarrow{f} & \Omega \end{array}$$

i.e. f is the characteristic morphism of the zero subobject of the final object 1.

$\cap$ is defined as the characteristic morphism of the subobject of $\Omega \times \Omega$ defined by the monomorphism

$$1 = 1 \times 1 \xrightarrow{(t,t)} \Omega \times \Omega$$

i.e. the following diagram is Cartesian:

$$\begin{array}{ccc} 1 & \longrightarrow & 1 \\ (t,t) \downarrow & & \downarrow t \\ \Omega \times \Omega & \xrightarrow{\cap} & \Omega \end{array}$$

In order to define U, we first consider the morphism

$$\Omega \amalg \Omega \xrightarrow{\mu} \Omega \amalg \Omega$$

defined by the matrix

$$\begin{pmatrix} \mathrm{id}_\Omega & t_\Omega \\ t_\Omega & \mathrm{id}_\Omega \end{pmatrix}$$

where t_Ω is the morphism

$$\Omega \to 1 \xrightarrow{t} \Omega$$

Let $|m\mu$ be the image of μ (Definition 1.8.1). The morphism U is by definition the characteristic morphism of the subobject $|m\mu$ of $\Omega \times \Omega$. In particular, the following diagram is Cartesian:

$$\begin{array}{ccc} |m\mu & \longrightarrow & 1 \\ \downarrow & & \downarrow \\ \Omega \times \Omega & \xrightarrow{U} & \Omega \end{array}$$

The morphism $\neg$ is defined by the condition that the following diagram be Cartesian:

$$\begin{array}{ccc} 1 & \longrightarrow & 1 \\ f \downarrow & & \downarrow t \\ \Omega & \xrightarrow{\neg} & \Omega \end{array}$$

In order to define a structure of Heyting algebra on Ω, we consider on Ω the order relation whose subobject R of $\Omega \times \Omega$ is the kernel of the sequence

$$\Omega \times \Omega \underset{\cap}{\overset{\mathrm{pr}_1}{\rightrightarrows}} \Omega.$$

Now we have to define the morphism

$$\Rightarrow : \Omega \times \Omega \to \Omega.$$

By definition, $\Rightarrow$ is the characteristic morphism of the subobject R, i.e. the following diagram:

$$\begin{array}{ccc} R & \longrightarrow & 1 \\ \downarrow & & \downarrow t \\ \Omega \times \Omega & \xrightarrow{\Rightarrow} & \Omega \end{array}$$

is Cartesian.

The morphisms t, f, $\leqslant$ are readily seen to define on Ω a lattice structure inducing on the set $\mathscr{P}(X)$ of all subobjects of X the usual lattice structure. In order to show that this lattice is actually a Heyting algebra, one can proceed in two different ways:

(1) one shows that $\mathscr{P}(X)$ has a structure of Heyting algebra for every object X;

(2) one considers the subobjects S, T of $\Omega \times \Omega \times \Omega$ defined by the condition that the following diagrams be commutative:

$$\begin{array}{ccc} S & \longrightarrow & R \\ \sigma \downarrow & & \downarrow \rho \\ \Omega \times \Omega \times \Omega & \xrightarrow{(\cap,\, \mathrm{id}_\Omega)} & \Omega \times \Omega \end{array}$$

$$\begin{array}{ccc} T & \longrightarrow & R \\ \tau \downarrow & & \downarrow \rho \\ \Omega \times \Omega \times \Omega & \xrightarrow{(\leqslant,\, \mathrm{id}_\Omega)} & \Omega \times \Omega \end{array}$$

One shows that there exists a (necessarily unique) morphism

$$S \to T$$

making commutative the diagram

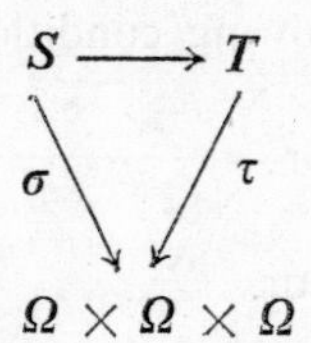

The details are left to the reader (cf. A. Koch and G. C. Wraith, *Elementary Toposes*, and P. Freyd, *Aspects of Topoi*).

DEFINITION 1.12. A *Boole topos* is a topos in which the following relation is fulfilled:

$$\neg \cdot \neg = \mathrm{id}_\Omega.$$

PROPOSITION 1.13. *For every topos $\mathscr{C}$, the following assertions are equivalent:*

(I) *is a Boole topos.*

(II) *The morphism*

$$1 \coprod 1 \xrightarrow{(t, f)} \Omega$$

is an isomorphism.

The proof is left to the reader.

2. A remarkable property of $I-T$ topoi is the fact that in the framework of these categories one can interpret every relation (proposition) of a higher order language.

In the following, given an object A, we show how to interpret a relation of a second order language. As usual, the possibility of interpreting such a relation is proved inductively.

Consider first a relation R of the form

$$R = P(\tau_1(x_{11}, \ldots, \ldots), \tau_2(x_{21}, \ldots, \ldots), \ldots,$$

$$\ldots, \tau_n(x_{n1}, \ldots, \ldots))$$

in which P is a fixed n-ary predicate.

Suppose we are given a partition

$$\{1, 2, \ldots, n\} = L \cup F,$$

$$L = \{i_1, i_2, \ldots, i_l\}, \ F = \{j_1, j_2, \ldots, j_f\}$$

of the set $\{1, 2, \ldots, n\}$ such that the following condition be fulfilled:

$$i \in L \Rightarrow \tau_i \text{ is a free operator;}$$

$$j \in F \Rightarrow \tau_j \text{ is a fixed operator.}$$

We introduce the following notation:

I, for the set of the variables occurring in R;

I_{i_r}, for the set of the variables occurring in the term

$$\tau_{i_r}(x_{i_{r1}}, \ldots);$$

A^M, for the product of the family of subobjects of the considered topos, whose indexation is

$$m \mapsto A, \ \forall\, m \in M.$$

With these notation, we assign to the relation R a subobject $\overline{R}$ of the object

$$A^I \times (A^1)^{A^{I_{i_1}}} \times A^{I_{i_2}} \times \ldots \times A^{I_{i_l}}$$

defined in the following way:

First we consider a subobject P of A^n, which yields in particular a morphism

$$A^n \to \Omega.$$

For every fixed term τ_{j_s} $(x_{j_s}1, \ldots,)$ we also consider a morphism

$$\overline{\tau}_{j_s}\colon A^{I^j_s} \to A, \qquad s = 1, 2, \ldots, f.$$

In order to define the required subobject, it is sufficient to define a morphism

$$A^I \times (A^1)\, A^{I_{i_1}} \times A^{I_{i_2}} \times \ldots \times A^{I_i}\colon \to \Omega.$$

This morphism is obtained as follows: first we consider the morphism

$$\overline{\tau}_{j_1} \times \overline{\tau}_{j_2} \times \ldots \times \overline{\tau}_{j_f}\colon A^{I_{j_1}} \times A^{I_{j_2}} \times \ldots \times A^{I_{j_f}} \to A^f$$

Taking also into account the canonical morphisms

$$(A^{I_{i_1}} \times \ldots \times A^{I_{i_1}}) \times (A^1)^{A^{I_{i_1}} \times \ldots \times A^{I_{i_l}}} \to A^1$$

$$A^I \to A^{I_{j_1}} \times A^{I_{j_2}} \times \ldots \times A^{I_{i_f}} \times A^{I_{i_1}} \times \ldots \times A^{I_{i_l}}$$

we get the required morphism

$$A^I \times (A^l)^{A^{I_{i_1}} \times \ldots \times A^{I_{i_l}}} \to A^{I_{j_1}} \times \ldots \times A^{I_{j_f}} \times A^{I_{i_1}} \times$$

$$\times \ldots \times A^{I_{i_2}} \times (A^l)^{A^{I_{i_1}} \times \ldots A^{I_{i_l}}} \to A^f \times A^l \simeq A^n \to \Omega.$$

Now consider a relation R of the form

$$R = P(\tau_1(x_{11}, \ldots, \ldots), \tau_2(x_{21}, \ldots, \ldots), \ldots$$

$$\ldots, \tau_n(x_{n1}, \ldots))$$

where P is a variable n-ary predicate.

With the same notation as above, we show how to assign to the relation R a subobject $\bar{R}$ of the object

$$A^I \times (A^1)^{A^{I_{i_1}} \times A^{I_{i_2}} \times \ldots A^{I_{i_1}}} \times \Omega^{A^n}.$$

Arguing as above, we get the morphism

$$A^I \times (A^1)^{A^{I_{i_1}} \times \ldots A^{I_1}} \to A^n.$$

The canonical morphism

$$A^n \times \Omega^{A^n} \to \Omega$$

together with the previous one leads to the required morphism

$$A^I \times (A^1)^{A^{I_{i_1}} \times \ldots \times A^{I_{i_1}}} \times \Omega^{A^n} \to A^n \times \Omega^{A^n} \to \Omega.$$

Now suppose that to the relation R_1 a subobject $\bar{R}_1$ of the object

$$A^{I_1} \times (A^{l_1})^{A^{I_{i_1}} \times A^{I_{i_2}}} \times \ldots \times A^{I_{i_{l_1}}} \times \Omega^{A^{n_1}} \times$$

$$\times \Omega^{A^{n_2}} \times \ldots \times \Omega^{A^{n_r}}.$$

has been assigned. If we are given in addition a relation R_2 to which a subobject $\bar{R}_2$ of the object

$$A^{I_2} \times (A^{l_2})^{A^{I}j_1 \times \ldots \times A^{Ij}l_2} \times \Omega^{A^{m_1}} \times \Omega^{A^{m_2}} \times \ldots \times \Omega^{A^{m_r}}$$

has been assigned, then for interpreting the relations

$$R_1 \vee R_2,\ R_1 \wedge R_2,\ R_1 \Rightarrow R_2$$

we proceed as follows:

Using suitable projections and taking inverse images by these projections, we lift the subobjects $\bar{R}_1, \bar{R}_2$ to subobjects of

$$A^{I} \times (A^{l})^{A^{I}k_1 \times \ldots \times A^{I}k} \times \Omega^{Ap_1} \times \Omega^{Ap_2} \times \ldots \times \Omega^{Ap_r}$$

where $I, l, \ldots$ are determined by the relations $R_1 \vee R_2, \ldots$

Thereafter, we make use of the fact that in an $L\text{—}T$ topos the set of the subobjects of an object forms a Heyting algebra (Example 2.12, Chap. II).

The case of the relation $\neg R$ is dealt with in a similar manner.

Now we show how to interpret a relation of the form

$$\exists \rho(R), \quad \forall \rho(R)$$

assuming that the interpretation of the relation R is known.

To this end, we use the following proposition, whose proof is left to the reader (cf. P. Freyd, *Aspects of Topoi*, p. 50, Prop. 4.11).

PROPOSITION 2.1. *Consider a morphism $u: M \to N$ and a subobject M' of M. There exists a maximal subobject N' of N such that $u^{-1}(N')$ be a subobject of M'. This subobject is denoted by $V_u(M')$.*

Going back to the interpretation of the relation $\exists \rho(R)$, it is sufficient to remark that by the corresponding projection the element ρ determines a morphism of the object

$$A^{I} \times (A^{l})^{A^{I}i_1 \times \ldots A^{I}i} \times \Omega^{A^{n_1}} \times \ldots \times \Omega^{A^{n_r}}$$

into an object of the same form, having one factor in minus. In order to get the interpretation of the relation $\exists \rho(R)$ (or $\forall \rho(R)$), one takes the image of the subobject R through this projection.

3. The usual axioms of set theory can be formulated in an $L-T$ topos. To this end, we could use the above results concerning the interpretation of a higher degree logical language. Nevertheless we choose a different approach, holding in more general categories.

3.1. *The axiom of the infinity* can be formulated by simply requiring the existence in the topos under consideration of an arithmetical structure (Definition 1.1 of section I).

3.2. *The axiom of choice.* A category $\mathscr{C}$ is said to satisfy the axiom of choice if every epimorphism is splittable i.e. for every epimorphism

$$A \xrightarrow{u} B$$

there exists a morphism

$$B \xrightarrow{v} A$$

such that

$$uv = 1_B.$$

Example 3.2.1. There exist topoi which do not satisfy the axiom of choice. Consider for instance the topological space X defined by the interior of a circle of the complex plane from which the origin has been removed, viewed as a subspace of the plane from Fig. 3.2.1.1.

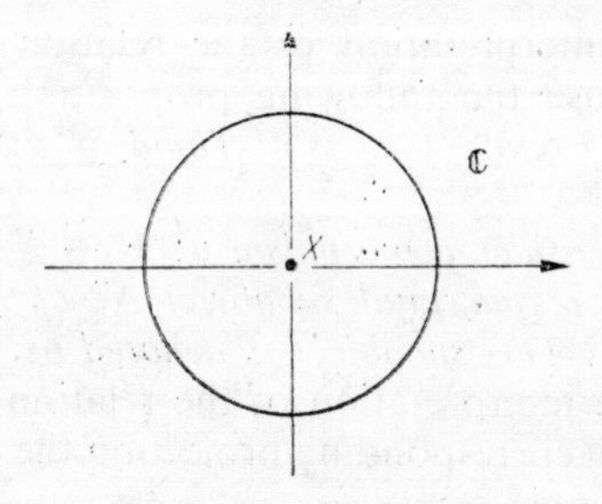

Fig. 3.2.1.1.

Consider the topos $\mathscr{F}$asc (X, Ens) of the sheaves of sets on X. Let $A = O_X$ be the sheaf of germs of holomorphic functions on X and let $B = O_X^*$ be the sheaf of germs of holomorphic functions which are nowhere vanishing.

The morphism

$$O_X \xrightarrow{e^{\cdot}} O_X^*$$

$$f \to e^f$$

is an epimorphism, since each of its fibres is so ($e^{\cdot}$ can be locally inverted by lg) but is not splittable.

Indeed, if this morphism were splittable, then the induced map

$$O_X(X) \to O_X^*(X)$$

would be an epimorphism of sets, which is absurd since the section $f \in O_X^*(X)$ defined by

$$f(x) = x, \quad \forall\, x \in I$$

cannot be the image of any global section of O_X by the map $e^{\cdot}$.

Remark 3.2.2. Simpler examples of epimorphisms which are not splittable can be pointed out in categories which are not topoi.

For instance: In Top

$$A = [0, 1],\ B = S^1 = \{z \mid |z| = 1\}$$

$$u\colon [0, 1) \to S^1$$

$$x \to e^{2\pi i x}$$

In $\mathscr{G}$r: Every group epimorphism

$$u\colon G \to H$$

where G is free and H is not.

One can also formulate the "continuum hypothesis": If X is an object such that

$$w \hookrightarrow X \hookrightarrow 2^w,$$

then X is isomorphic either to w or to 2^w.

4. *The theory of real numbers in an L—T topos.* In section I we have seen how an arithmetical theory can be initiated in a suitable category. In the following we show that such a theory can be adequately developed in a topos. Even a theory of real numbers can be initiated in an *I—T* topos, a fact of special importance since recent researches of logic and analysis have pointed out ordered fields more adapted to the study of some problems than is the standard model of real numbers. To be more precise, we refer to the following two concrete situations:

4.1. *The field of constructive real numbers.* It is easy to see the existence of real numbers, written for instance in the decimal formal

$$\alpha = 0, a_1, a_2, \ldots, a_n \ldots a_i \in \{0, 1, 2, \ldots, 9\},$$

$$i \in \{1, 2, \ldots, n, \ldots\}$$

such that no algorithm is available, allowing, for a given n, to compute the value of the decimal a_n of α.

Consider for instance a function

$$f\colon \mathbb{N} \to \{0, 1, 2, \ldots, 9\}$$

which is not recursive and let α be the real number defined by the convergent sequence of rational numbers $(r_n)_{n \geqslant 1}$ determined by the relation

$$r_n = \frac{f(1)}{10} + \frac{f(2)}{10^2} + \ldots + \frac{f(n)}{10^n}.$$

Indeed, the decimal a_n of the real number α defined by the sequence (r_n) is precisely $f(n)$.

Of course, the use of such real numbers is to be avoided in problems of constructive analysis, in particular in problems which are to be solved by means of computers. This leads to considering a subfield of the field of usual real numbers, called the field of constructive real numbers, with the property that every decimal of a given number is effectively (algorithmically) computable.

4.2. *Fields of "non-standard" real numbers.* In the previous example, the field $\mathbb{R}$ of real numbers has turned out to be too rich with respect to computability requirements. From another point of view, it turns out to be too poor in order to allow the natural treatment of some problems. It is well known for instance that the fathers of mathematical analysis used to define its basic concepts as continuity, derivability, etc. by means of "infinitesimal elements" for which they took for granted a calculus similar to that from the field of real numbers. Such a calculus cannot be justified in the framework of the standard field of real numbers. Nevertheless, recent researches of logic and analysis have shown that "non-standard" fields of real numbers can be defined in which the concepts of "infinitesimal elements" as well as a calculus with these elements could be introduced in such a way that all concepts, arguments and results of the fathers of mathematical analysis become a full justification. Moreover, these researches have shown how open problems of classical analysis could be dealt with and successfully solved by means of these concepts.

It is hoped that the topoi will yield a unitary frame for various theories of real numbers. This hope is sustained by the fact that a field of real numbers can be defined in every topos $\mathscr{C}$ with the infinity axiom, the particular case of $\mathbb{R}$ corresponding to $\mathscr{C} = \text{Ens}$.

4.3. In this section we intend to sketch the general lines of a theory of real numbers in a Lawvere — Tierney topos which satisfies the infinity axiom.

4.3.1. Let w be an object of a Lawvere—Tierney topos $\mathscr{C}$ and assume that

$$1 \xrightarrow{0} w \xrightarrow{s} w.$$

is an arithmetical structure on $\mathscr{C}$. We show that two algebraic operations, the addition σ and the multiplication μ, can be defined on $\mathscr{C}$ as follows:

Definition of σ. First of all, every morphism of

$$A \xrightarrow{f} B$$

obviously induces a morphism of functors

$$\coprod A \xrightarrow{\pi(f)} \coprod B$$

such that for every $X \in \mathscr{C}$ we have

$$\pi(f)(X) = (\text{id}_X, f) \colon X \coprod A \to X \coprod B.$$

This yields a morphism of functors

$$\underline{\mathrm{Hom}_{\mathscr{C}}(B, -)} \xrightarrow{\mathrm{Hom}_{\mathscr{C}}(f)} \underline{\mathrm{Hom}_{\mathscr{C}}(A, -)},\ Y^B \overset{Y^f}{\longmapsto} Y^A$$

and in particular a morphism

$$w^w \xrightarrow{w^s} w^w.$$

By definition, σ is the unique morphism for which the following diagram commutes:

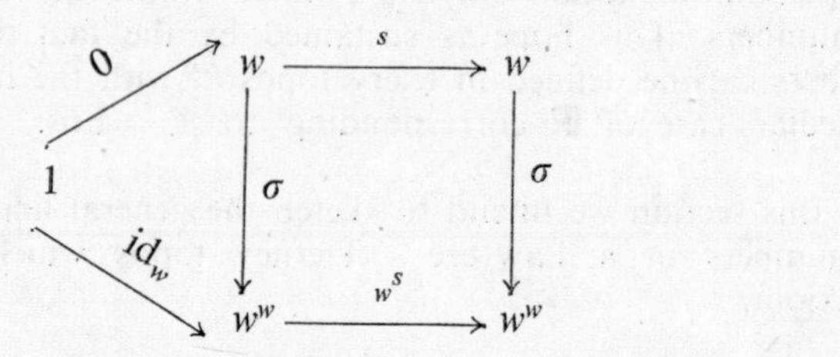

(Here and in the following, we use the canonical identifications

$$\mathrm{Hom}_{\mathscr{C}}(1, w^w) = \mathrm{Hom}_{\mathscr{C}}(w, w),$$

$$\mathrm{Hom}_{\mathscr{C}}(w, w) = \mathrm{Hom}_{\mathscr{C}}(w \times w, w)).$$

DEFINITION of μ. For every couple A, B of objects in $\mathscr{C}$, consider the "evaluation morphism" introduced in Definition 1.2:

$$\mathrm{ev}: A^B \times B \to A.$$

Denote by τ the composed morphism

$$w^w \times w \xrightarrow{(\mathrm{ev},\, \mathrm{pr}_2)} w \times w \to w.$$

This yields a unique morphism, still denoted by σ:

$$w^w \xrightarrow{\tau} w^w.$$

By definition, μ is the unique morphism for which the following diagram commutes:

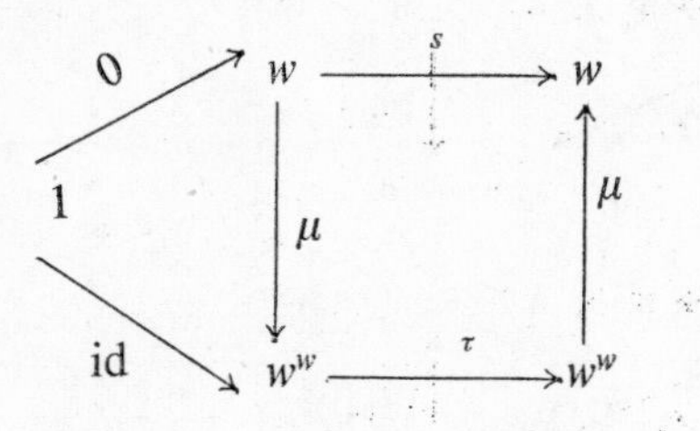

4.3.2. *Construction of the object* $\mathbb{Q}^+_{\mathscr{C}}$ *of "positive rational numbers" in a topos* $\mathscr{C}$. The classical definition will be adapted to this general case.

Consider the image (Definition 1.8.1) w^+ of the morphism

$$w \xrightarrow{s} w$$

and let

$$w \times w^+ \times w \times w^+ \underset{a_2}{\overset{a_1}{\rightrightarrows}} w$$

be the composed morphisms

$$w \times w^+ \times w \times w^+ \xrightarrow{\text{pr}_{14}} w \times w^+ \xrightarrow{\mu} w \qquad (a_1)$$

$$w \times w^+ \times w \times w^+ \xrightarrow{\text{pr}_{32}} w \times w^+ \xrightarrow{\mu} w \qquad (a_2)$$

Denote by P the subobject of $w \times w^+ \times w \times w^+$ which is the kernel of the couple of morphisms (a_1, a_2):

$$P = \text{Ker}(a_1, a_2),$$

and consider the morphisms

$$P \underset{p_2}{\overset{p_1}{\rightrightarrows}} w \times w^+$$

for which the following diagrams commute:

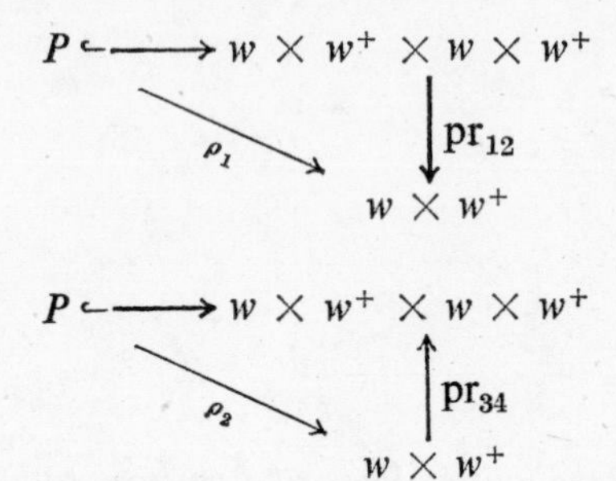

The morphisms ρ_1, ρ_2 are readily seen to define an equivalence relation on $w \times w^+$.

In view of the classical definition of positive rational numbers, we put

$$\mathbf{Q}_{\mathscr{C}}^{+} = \operatorname{Coker}(\rho_1, \rho_2)$$

4.3.3. *The order structure on* $\mathbf{Q}_{\mathscr{C}}^{+}$. First of all, an order structure on w can be defined by considering the subobject

$$0 \hookrightarrow w \times w$$

associated with the relation

$$\exists \xi (y = \sigma(x, \xi))$$

from the first order language, in which $=$ is a fixed second order predicate, naturally interpreted by the diagonal of the product $w \times w$, and σ is a fixed second-order operator, interpreted by the algebraic operation (cf. 4.3.1):

$$\sigma : w \times w \to w.$$

The composition of the morphisms

$$w \to w \times 1 \xrightarrow{\mathrm{id}_w \times 0} w \times w \xrightarrow{\mathrm{id}_w \times s} w \times w^+ \to \mathbf{Q}_{\mathscr{C}}^{+}.$$

is readily seen to yield a monomorphism

$$i\colon w \to \mathbb{Q}^+_{\mathscr{C}}$$

Moreover, the morphism $s\colon w \to w$ can be extended to a morphism $S\colon \mathbb{Q}^+_{\mathscr{C}} \to \mathbb{Q}^+_{\mathscr{C}}$ such that the following diagram commutes:

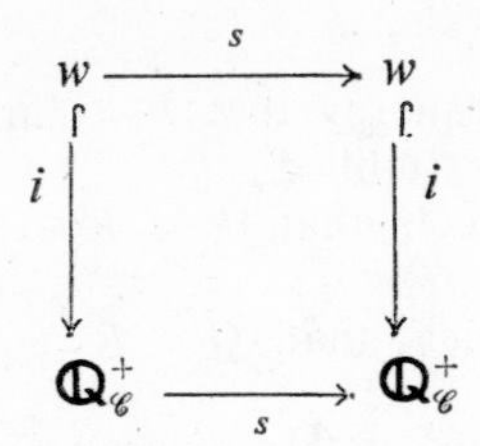

The algebraic operations σ, μ can also be extended to $\mathbb{Q}^+_{\mathscr{C}}$ allowing to define on $\mathbb{Q}^+_{\mathscr{C}}$ an order relation in just the same way as in the case of w.

4.3.4. *Definition of the object* $\mathbb{R}^+_{\mathscr{C}}$ *of "positive real numbers" in a topos* $\mathscr{C}$. The classical method of "Dedekind cuts" can be adapted in order to define the concept of "object of positive real numbers in the topos $\mathscr{C}$". To this end, we recall that a cut in the set of positive rational numbers $\mathbb{Q}^+$ is defined by a non-empty subset P of $\mathbb{Q}^+$, such that

$$(4.3.4.1) \qquad (\forall \xi\, \forall \eta(((\xi \leqslant \eta) \wedge (\xi \in P)) \Rightarrow \eta \in P).$$

The subset P is usually called the upper subset of the corresponding cut of $\mathbb{Q}^+$. As it is well known, the first step in defining classical real numbers consists in considering the set $\mathscr{T}$ of all cuts of the set $\mathbb{Q}^+$.

On the other hand, an object which corresponds to the set $\mathscr{T}$ can be introduced in an arbitrary topos $\mathscr{C}$. In view of 4.3.4.1, we consider in the second-order language the relation

$$(4.3.4.2) \qquad (\forall \xi\, \forall \eta(((\xi \leqslant \eta) \wedge P(\xi)) \Rightarrow P(\eta)))) \wedge \exists \xi\, P(\xi)$$

in which $\leqslant$ is a fixed second-order predicate and P is a variable first-order predicate. This relation defines a subobject $\mathscr{T}_{\mathscr{C}}$ of the object $\Omega^{\mathbb{Q}^+_{\mathscr{C}}}$.

In order to conclude the definition of the object $\mathbb{R}^+_{\mathscr{C}}$ of positive real numbers, we have to define on $\mathscr{T}_{\mathscr{C}}$ an equivalence relation whose quotient object is $\mathbb{R}^+_{\mathscr{C}}$.

To this end, recall that two cuts (A, B), (C, D) of the set $\mathbb{Q}^+$ are said to be equivalent if they define the same rational number, i.e. if a unique rational number r exists such that

$$(r \in A \cap D) \vee (r \in B \cap C).$$

In terms of upper subsets, we can say that P, Q are equivalent if one of the following properties is fulfilled:

– $P \subseteq Q$ and if R is a cut such that $P \subseteq R \subseteq Q$, then either $P = R$, or $Q = R$;

– $Q \subseteq P$ and if R is a cut such that $Q \subset R \subset P$, then either $Q = R$ or $P = R$.

The set $\mathscr{T}$ being a set of subsets of $\mathbb{Q}^+$, it is naturally endowed with an order relation $\subseteq$ which defines on $\mathscr{T}$ a second-order predicate. This fact suggests the following relation in the first-order language:

$$\forall R(((P \subseteq Q) \wedge (P \subseteq R) \wedge (R \subseteq Q)) \vee$$

$$\vee ((Q \subseteq P) \wedge (Q \subseteq R) \wedge (R \subseteq P))) \Rightarrow$$

$$\Rightarrow (P = R) \wedge (Q = R)).$$

This relation involves two free variables P, Q and a fixed predicate $\subseteq$. If the predicate $\subseteq$ on $\mathscr{T}$ is interpreted in the usual way, the above relation defines a subset of the product $\mathscr{T} \times \mathscr{T}$ which yields precisely the graph of the required equivalence relation.

In order to extend these concepts to the general case of the object $\mathscr{T}_{\mathscr{C}}$ of an arbitrary topos $\mathscr{C}$, we first remark that the kernel of the couple of morphisms

$$\Omega \times \Omega \underset{\text{pr}_1}{\overset{\cap}{\rightrightarrows}} \Omega,$$

which will be denoted by c, defines for every object A a subobject of $\Omega^A \times \Omega^A$. In particular, if P is a subobject of Ω^A, this yields a subobject of $P \times P$. This will be applied in the particular case of the object $\mathscr{T}_{\mathscr{C}}$.

In order to define the required equivalence relation, we consider the relation

$$\forall \xi (((x \subseteq y) \wedge (x \subseteq \zeta) \wedge (\zeta \subseteq y)) \vee (y \subseteq x) \wedge$$

$$\wedge (y \subseteq \zeta) \wedge (\zeta \subseteq x))) \Rightarrow (x = \zeta) \vee (y = \zeta))$$

together with the subobject $R_{\mathscr{C}}$ of $\mathscr{T}_{\mathscr{C}} \times \mathscr{T}_{\mathscr{C}}$ defined by this relation, where the predicate $\subseteq$ is interpreted as above.

By composing the canonical monomorphism

$$R_{\mathscr{C}} \hookrightarrow \mathscr{T}_{\mathscr{C}} \times \mathscr{T}_{\mathscr{C}}$$

with the two projections, we get the morphisms

$$R_{\mathscr{C}} \underset{\pi_2}{\overset{\pi_1}{\rightrightarrows}} \mathscr{T}_{\mathscr{C}}.$$

We put by definition $\mathbb{R}^+_{\mathscr{C}} = \mathrm{Coker}(\pi_1, \pi_2)$

4.3.5. The definition of the object $\mathbb{R}_{\mathscr{C}}$ of real numbers can now be completed by following the classical way of passing from the set $\mathbb{N}$ of natural numbers to the set $\mathbb{Z}$ of integers. Details are left to the reader.

4.3.6. *Examples.* (a) In order to develop the theory of "non-standard" real numbers (cf. 4.2) one can use a category of the following type:

Let M be a partially ordered set, let $\widetilde{M}$ be the category canonically associated with M and let $\mathscr{C}$ be the category of contravariant functors from $\widetilde{M}$ to the category of sets. $\mathscr{C}_M$ is a topos.

(b) In order to develop the theory of constructive real numbers (cf. 4.1) one can use the category $\mathscr{E}$ defined as follows:

The objects of $\mathscr{E}$ are surjections

$$\mathbb{N} \xrightarrow{\nu} S$$

where $\mathbb{N}$ is the set of natural numbers and S is a set. Such a surjection is called an "enumeration" of the set S.

– A morphism between the surjections

$$\mathbb{N} \xrightarrow{\nu} S$$

$$\mathbb{N} \xrightarrow{\mu} T$$

in the category $\mathscr{E}$ is a map $f\colon S \to T$ for which a recursive function $g\colon N \to N$ exists such that the following diagram commutes

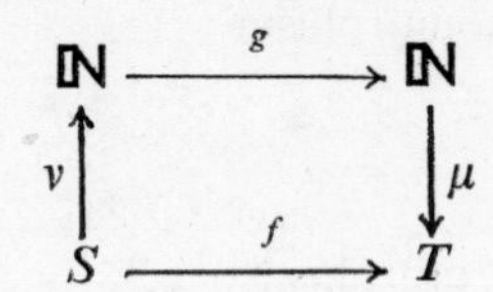

Chapter V

Elements of the theory of elliptic curves

I. ELLIPTIC CURVES DEFINED OVER THE COMPLEX FIELD

DEFINITION 1. A subgroup of the underlying Abelian group of $\mathbb{R}^n$ is said to be a *lattice* if it is generated by a set of vectors which are linearly independent over $\mathbb{R}$. A *complete lattice* is a lattice generated by a basis of $\mathbb{R}^n$.

PROPOSITION 2. *The underlying set of every lattice is discrete in the $\mathbb{R}^n$-space (with regard to the usual topology).*

Proof. We can restrict ourselves to the case of complete lattices generated by the canonical basis $\{e_1, \ldots, e_n\}$ of $\mathbb{R}^n$. Given a point $\alpha = (\alpha_1, \ldots, \alpha_n)$ of $\mathbb{R}^n$ and a cubical neighbourhood of α

$$V = \{(\xi_1, \ldots, \xi_n) \mid |\xi_i - \alpha_i| < t_i, \ i = 1, 2, \ldots, n\},$$

the set of points from L belonging to V coincides with the set

$$\{(m_1, \ldots, m_n) \mid m_i \in \mathbb{Z}, \ |m_i - \alpha_i| < t_i, \ i = 1, \ldots, n\}$$

which is obviously finite.

DEFINITION 3. Let G be a group operating on a set M. A subset $E \subset M$ is said to be a *fundamental domain* for G provided that:

(*i*) every point of M is a congruent modulo G to some point of E.

(*ii*) no pair of points from E are congruent modulo G.

From these conditions, it follows that every point of M is congruent with a single point of E.

PROPOSITION 4. *Let L be a complete lattice of $\mathbb{R}^n$, generated by the linearly independent vectors $f_1, \ldots, f_n$, and let P_2 be the subset of $\mathbb{R}^n$ defined by*

$$P_L = \{\alpha_1 f_1 + \ldots + \alpha_n f_n | 0 \leqslant \alpha_i \leqslant 1,\ i = 1, \ldots, n\},$$

and called the fundamental parallelipiped of the lattice L corresponding to the vectors $f_1, \ldots, f_n$. P_L is a fundamental domain for the group L operating in the natural way on the $\mathbb{R}^n$-space.

Proof. Let $x = \xi_1 e_1 + \ldots + \xi_n e_n$ be an arbitrary point of $\mathbb{R}^n$. For every real number ξ_i, we can find an entire k_i such that

$$\xi_i = k_i + \alpha_i, \quad 0 \leqslant \alpha_i < 1,\ i = 1, \ldots, n.$$

From this we get

$$x = k_1 e_1 + \ldots + h_n e_n + \alpha_i e_1 + \ldots + \alpha_n e_n,$$

which yields condition i from Definition 3.

Condition ii is trivially fulfilled.

PROPOSITION 5. *A subgroup L of $\mathbb{R}^n$ is a lattice iff L is a discrete subset of $\mathbb{R}^n$.*

Proof. By Proposition 2, we have only to show that every discrete subgroup L of $\mathbb{R}^n$ is lattice, i.e. it is generated by a set of linearly independent vectors of $\mathbb{R}^n$. Let V be the vector subspace generated by L. Let $(f_1, \ldots, f_m)$ be a basis of V with elements from L and let L_0 be the lattice generated by this basis. We clearly have $L_0 \subset L \subset V$.

We show that the index of the L_0 subgroup of L is finite. In fact, if P_{L_0} is the fundamental parallelepiped of the lattice L_0, then every element of L is a congruent modulo L_0 to some point from P_{L_0} which belongs to L, since $L_0 \subset L$. In this way, we get an injection of the factor group L/L_0 in the set of points of L which lie in P_L. But as P_{L_0} is bounded and L is, by assumption, a discrete subset of $\mathbb{R}^n$, it follows that the set of points from L lying in P_{L_0} is finite, hence $L|L_0$ is finite too. Let j be the cardinal of L/L_0. For every $x \in L$, we have $jx = 0 \pmod{L_0}$, i.e. $jx \in L_0$.

Let L^* be the lattice generated by $\left(\frac{1}{j} f_1, \ldots, \frac{1}{j} f_m\right)$; clearly, $L \subset L^*$.

By a classical theorem from the theory of Abelian groups, it follows that L is a free group, spanned by a basis extracted, up to multiplication by some entire factors, from a basis of L^*. Now, take into account the fact that every basis of the group L^* necessarily consists of linearly independent vectors of $\mathbb{R}^n$, since $\left(\frac{1}{j} f_1, \ldots, \frac{1}{j} f_m\right)$ is such a basis.

Definition 6. Let L be a lattice of the complex field $\mathbb{C}$, identified as a real vector space with $\mathbb{R}^2$. A complex function f meromorphic on $\mathbb{C}$ is said to be an *elliptic function with respect to the lattice L* (or *L-elliptic*), if the relation

$$f(z+w) = f(z) \tag{6.1}$$

holds for every $w \in L$, provided that one of the members from (6.1) be defined.

The set of all L-elliptic functions forms a field ξ_L, which is an extension of $\mathbb{C}$.

Proposition 7. *For every lattice* L *of* $\mathbb{C}$, *there is an* L-*elliptic non-constant function* $\mathfrak{p}$ *such that*

(a) $\mathfrak{p}$ *is an even function;*

(b) $\mathfrak{p}'$ *is an odd function;*

(c) *the following relation holds:*

$$(\mathfrak{p}')^2 = 4\mathfrak{p}^3 - g_2\mathfrak{p} - g_3,$$

the complex numbers g_2, g_3 *being defined by the formulae*

$$g_2 = 60 \sum_{w \in L-0} \frac{1}{w^4}$$

$$g_3 = 140 \sum_{w \in L-0} \frac{1}{w^6}$$

and satisfying the additional condition

$$g_2^3 - 27g_3^2 \neq 0;$$

(d) *every even L-elliptic function is a rational function of* $\mathfrak{p}$.

Proof. The $\mathfrak{p}$ function has been introduced and studied by Weierstrass and is defined by the formula

$$\mathfrak{p}(z) = \frac{1}{z^2} + \sum_{w \in L-0} \left(\frac{1}{(z-w)^2} - \frac{1}{w^2} \right).$$

The properties stated in Proposition 8 can be found in every book on functions theory. (See, for instance, S. Stoilow, *Theory of Functions of a Complex Variable*, vol. I, Chap. VI, § II, 24, Chap. VII, 12).

PROPOSITION 8. The extension $\xi_L \supset \mathbb{C}$ is of transcendence degree one. More specifically, in the above assumptions, we have the isomorphism

$$\xi_L \simeq \mathbb{C}(X)[Y]/(Y^2 - 4X^3 + g_2X + g_3).$$

Proof. First of all, Proposition 7 implies the existence of an element belonging to ξ_L which is transcendent over $\mathbb{C}$. In fact, $\mathfrak{p}$ is transcendent over $\mathbb{C}$ because $\mathfrak{p}$ is non-constant and $\mathbb{C}$ is algebraically closed. On the other side, by property (d) from the same proposition, the set of even elliptical functions forms a subfield of ξ_L, which is isomorphic to $\mathbb{C}(X)$:

$$\text{(8.2)} \qquad \mathbb{C}(X) \simeq \mathbb{C}(\mathfrak{p}), \quad \mathfrak{p} \in \xi_L.$$

To complete the proof of Proposition 8, we show that ξ_L is an algebraic extension of degree two of $\mathbb{C}(\mathfrak{p})$. More specifically, we prove that every $f \in \xi_L$ can be represented in a unique way as

$$\text{(8.3)} \qquad f = f_1 + f_2\mathfrak{p}',$$

where f_1, f_2 are even elliptical functions. In fact, it will suffice to set

$$f_1(z) = \frac{1}{2}(f(z) + f(-z)), \qquad \forall z \in \mathbb{C}$$

$$f_2 = \frac{f - f_1}{\mathfrak{p}'}.$$

The uniqueness of representation (8.3) is straightforward since the function $\mathfrak{p}'$ is odd, nonzero, and a function which is both even and odd must be zero.

We conclude the proof using properties b and c from Proposition 7 and the following purely algebraic remark:

Given an extension $K \subset L$, and an element $w \in L$ such that $\{1, w, \ldots, w^n\}$ is a basis of L as a K-vector space and w annihilates an irreducible polynomial $P \in K[Y]$, there exists an isomorphism

$$L \xrightarrow{\sim} K[Y]/P.$$

PROPOSITION 9. *The algebraic affine curve defined by the polynomial*

$$P = Y^2 - 4X^3 + g_2X + g_3$$

has no singular points.

Proof. Assuming that (ξ, η) would be a singular point of this curve, we should have

$$\frac{\partial P}{\partial X}(\xi, \eta) = -12\xi^2 + g_2 = 0,$$

$$\frac{\partial P}{\partial Y}(\xi, \eta) = 2\eta = 0$$

$$P(\xi, \eta) = \eta^2 - 4\xi^3 + g_2\xi + g_3 = 0.$$

But the first two relations yield

$$\eta = 0, \quad \xi = \pm \sqrt{\frac{g_2}{12}}\,.$$

Consider for instance the case $\xi = \sqrt{\frac{g_2}{12}}$. We show that the point $\left(\sqrt{\frac{g_2}{12}}, 0\right)$ does not belong to the given curve, i.e. it does not satisfy the equation $P(\xi, \eta) = 0$. In fact, we should have

$$4\frac{g_2}{12}\sqrt{\frac{g_2}{12}} - g_2\sqrt{\frac{g_2}{12}} + g_3 = 0,$$

$$g_2\sqrt{\frac{g_2}{12}}\left(\frac{1}{3} - 1\right) = -g_3, \quad \frac{2}{3}g_2\sqrt{\frac{g_2}{12}} = g_3,$$

$$\frac{4g_2^3}{9} = 12\,g_3^2,$$

in contradiction with the inequality $g_2^3 - 27\,g_3^2 \neq 0$ from Proposition 7,c).

Remark 9.1. If $K \supset \mathbb{C}$ is an extension of finite type, then any algebraic variety defined over C and having K as rational functions field is said to be a *model* of the field K. In his paper *Resolution of singularities of an algebraic variety over a field of characteristic zero* (*Ann. Math.*, **79**, 1964), Hironaka showed that every field K, as above, admits a projective and non-singular model.

Propositions 8 and 9 show that every elliptic functions field admits an affine non-singular model. It is enough to observe the existence of a canonical isomorphism

$$\mathrm{Quot}\,(\mathbb{C}[X, Y]_{,}(Y^2 - X^3 + g_2X + g_3) \rightleftarrows$$

$$\rightleftarrows \mathbb{C}(X)[Y]/(Y^2 - X^3 + g_2X + g_3).$$

Example 10. The curve $Y^2 = X^2 + X^3$ is a model for the field $k(T)$. In fact, there is an isomorphism between $\mathrm{Quot}(k[X, Y])/(Y^2 - X^2 - X^3)$ and $k(T)$.

For every point $P \in \mathbb{C}$ and every meromorphic function $f (f \not\equiv 0$ in the neighbourhood of $P)$, we denote by $v_P(f)$ the number defined as follows

$$v_P(f) = \begin{cases} n \text{ if } P \text{ is a zero of order } n \text{ for } f, \\ -n \text{ if } P \text{ is a pole of order } n \text{ for } f, \\ 0 \text{ if } P \text{ is neither a zero nor a pole for } f. \end{cases}$$

By convention, we set $v_p(0) = \infty$. The above definition extends in a straightforward manner to every one-dimensional complex variety. The following formulae hold true:

(*i*) $v_P(f, g) = v_P(f) + v_P(g)$

(*ii*) $v_P(f + g) \geqslant \min(v_P(f), v_P(g))$.

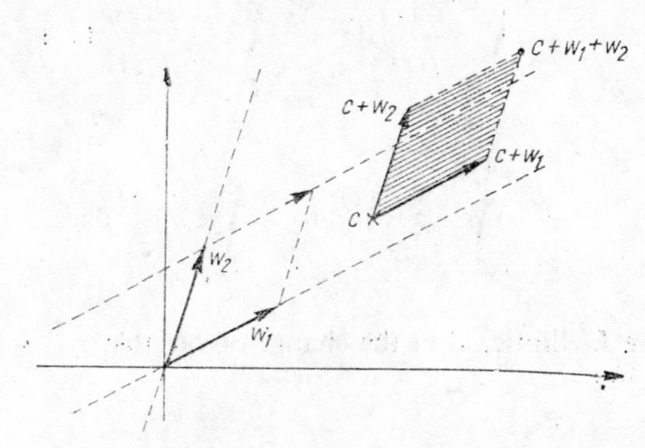

Fig. 10.1

In other words, v_P defines a valuation on the field of meromorphic functions in the complex plane.

If $L \subset \mathbb{C}$ is a complete lattice generated by the vectors w_1, w_2 and $c \in \mathbb{C}$, we denote by $P_{L,c}$ the fundamental domain defined as follows:

$$P_{L,c} = \{z \mid z = c + \alpha_1 w_1 + \alpha_2 w_2,\ 0 \leqslant \alpha_1,\ \alpha_2 < 1\}$$

Given a meromorphic function f, there clearly exist fundamental domains $P_{L,f}$ of the $P_{L,c}$-type, the frontier of which includes no zero and no pole of f.

PROPOSITION 11. *If f is an L-elliptic function and $P_{L,f}$ a fundamental domain containing no pole and no zero of f on its frontier, then*

$$\sum_{z \in P_{L,f}} v_z(f), \quad z \in L.$$

Proof. By a well-known theorem of the complex functions theory, we have

$$2\pi i\left(\sum_{z\in P_{L,f}} v_z(f)\cdot z\right)=\int_{\mathrm{Front}P_{L,f}} z\frac{f'(z)}{f(z)}\,\mathrm{d}z=$$

$$=\int_{c}^{c+w_1} z\frac{f'(z)}{f(z)}\,\mathrm{d}z+\int_{c+w_1}^{c+w_1+w_2} z\frac{f'(z)}{f(z)}\,\mathrm{d}z+ \tag{11.1}$$

$$+\int_{c+w_1+w_2}^{c+w_2} z\frac{f'(z)}{f(z)}\,\mathrm{d}z+\int_{c+w_2}^{c} z\frac{f'(z)}{f(z)}\,\mathrm{d}z.$$

But f being L-elliptic, after the change of variable $z=u+w_2$, we get

$$\int_{c+w_1+w_2}^{c+w_2} z\frac{f'(z)}{f(z)}\,\mathrm{d}z=-\int_{c}^{c+w_1}(u+w_2)\frac{f'(u)}{f(u)}\,\mathrm{d}u=$$

$$=-\int_{c}^{c+w_1} z\frac{f'(z)}{f(z)}\,\mathrm{d}z-w_2\int_{c}^{c+w_1}\frac{f'(z)}{f(z)}\,\mathrm{d}z,$$

while the change of variable $z=u+w_1$ yields

$$\int_{c+w_1}^{c+w_1+w_2} z\frac{f'(z)}{f(z)}\,\mathrm{d}z=\int_{c}^{c+w_2}(u+w_1)\frac{f'(u)}{f(u)}\,\mathrm{d}u=$$

$$=\int_{c}^{c+w_2} z\frac{f'(z)}{f(z)}\,\mathrm{d}z+w_1\int_{c}^{c+w_2}\frac{f'(z)}{f(z)}\,\mathrm{d}z.$$

Taking into account formula (11.1), we obtain

$$2\pi i\left(\sum_{z\in P_{L,f}} v_z(f)\cdot z\right) = w_1 \int_c^{c+w_2} \frac{f'(z)}{f(z)}\,dz - w_2 \int_c^{c+w_1} \frac{f'(z)}{f(z)}\,dz.$$

But

$$\int_c^{c+w_2} \frac{f'(z)}{f(z)}\,dz = \int_c^{c+w_2} \frac{d}{dz}(\lg f(z))\,dz = \lg f(z)\Big|_c^{c+w_2} =$$

$$= 2\pi i n_1,\ n_1 \in \mathbb{Z}$$

and similarly

$$\int_c^{c+w_1} \frac{f'(z)}{f(z)}\,dz = 2\pi i n_2.$$

Hence,

$$\sum_{z\in P_{L,f}} v_z(f)\cdot z = n_1 w_1 - n_2 w_2$$

which gives the desired property.

PROPOSITION 12. *With the same notation as in Proposition* 11, *we have*

$$\sum_{z\in P_{L,f}} \mathrm{Res}_z(f) = 0, \tag{12.1}$$

where $\mathrm{Res}_z(f)$ *designates the residue of* f *in* z.

Proof. By Cauchy's theorem, we have

$$2\pi i \int_{\mathrm{Fr}(P_{L,f})} f(z)\,dz = \sum_{z\in P_{L,f}} \mathrm{Res}_z(f),$$

from which (12.1) *follows by a straightforward calculation, taking into account the periodicity of* f.

COROLLARY 13. The sum of the multiplicity orders of the zeros of an L-elliptic function f, which are contained in a $P_{L,f}$-domain, is at least 2.

Proof. In fact, this sum is equal to that corresponding to the poles of f, and no elliptical function can have in $P_{L,f}$ a unique pole with multiplicity one, since this would contradict formula 12.1.

II. DIVISORS

Let L be a complete lattice of $\mathbb{C}$ viewed as a vector space over $\mathbb{R}$.

By definition, an element of the free Abelian group $\mathbb{Z}[\mathbb{C}/L]$ having the set $\mathbb{C}/L$ for basis is called a divisor of the elliptic curve $X = \mathbb{C}/L$. The group $\mathbb{Z}[\mathbb{C}/L]$ is called the group of divisors of the curve X and is sometimes denoted by $\mathrm{Div}(X)$. Consequently, a divisor on X is a formal linear combination of elements from X with integer coefficients:

$$D \in \mathrm{Div}(X) \Leftrightarrow D = \sum_{P \in X} n_P \cdot P, \; n_P \in \mathbb{Z},$$

where $n_p = 0$ except for a finite set of points $P \in X$.

For every fundamental parallelogram P associated with the lattice L there obviously exists a canonical group isomorphism

$$\mathrm{Div}(X) \simeq \mathbb{Z}[P_L].$$

We have a group homomorphism

$$\deg\colon \mathrm{Div}(X) \to \mathbb{Z}$$

$$\sum_{P \in X} n_P \cdot P \mapsto \sum_{P \in X} n_P.$$

The group $\mathrm{Div}(X)$ has a structure of (partially) ordered group:

$$\sum_{P \in X} n_P \cdot P \geqslant 0 \overset{\mathrm{Def}}{\Leftrightarrow} n_p \geqslant 0, \quad \forall P \in X.$$

Sometimes, a positive divisor is called an effective divisor.

Example 1. The divisor (f) associated to an L-elliptical function f. Let P_f be a fundamental parallelogram whose frontier contains no zero and no pole of f. Such parallelograms obviously exist. If we put:

$$(f) = \sum_{p \in P_f} v_P(f) \cdot P$$

we get a divisor on X. Such a divisor is said to be principal.

From the immediate formulae

$$(f \cdot g) = (f) + (g), \tag{1.2}$$

$$\left(\frac{f}{g}\right) = (f) - (g) \tag{1.3}$$

it follows that the set $P(X)$ of all principal divisors is a subgroup of the group $\mathrm{Div}(X)$. We denote by $C(X)$ the quotient group $\mathrm{Div}(X)/P(X)$.

PROPOSITION 2. *For every L-elliptical function f, we have* $\deg((f)) = 0$.

Proof. Using the well-known formula of function theory

$$\frac{1}{2\pi i} \int_{\mathrm{Fr}(P_f)} \frac{f'(z)}{f(z)} \mathrm{d}z = \deg((f)) \tag{2.1}$$

the required relation is an immediate consequence of the periodicity property of f.

COROLLARY 3. The "degree" function defines a homomorphism

$$\deg\colon C(X) \to \mathbb{Z}.$$

Example 4. There exist divisors of degree zero which are not principal. Indeed, consider two interior points of the parallelogram P_L and form the divisor

$$D = P_1 - P_2.$$

D has the degree zero. Nevertheless, there exists no L-elliptical function f such that $D = (f)$, since such a function would have in the parallelogram P_L a single pole of multiplicity one, contradicting Corollary 13 from section I.

Example 5. $(\mathfrak{p}') = -3.0 + 1.\frac{w_1}{2} + 1.\frac{w_2}{2} + 1.\frac{w_3}{2}$, where $w_3 = w_1 + w_2$.

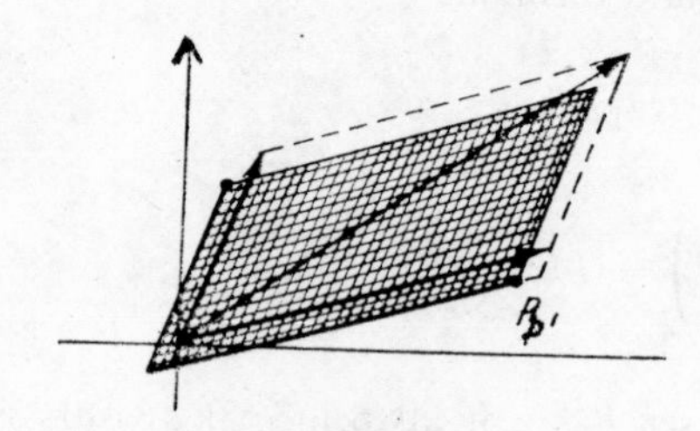

Fig. 5.1

Indeed, it is easy to see that there exists a fundamental parallelogram $P_{\mathfrak{p}}$ containing the origin and the points w_1, w_2, w_3 and having no pole and no zero of $\mathfrak{p}'$ on its frontier (Fig. 5.1).

The single pole of the function p' situated in the parallelogram $P_{\mathfrak{p}}$, is the origin, which is a pole of order three for this function.

In view of Proposition 2, the required formula will follow if we show that $\frac{w_1}{2}, \frac{w_2}{2}, \frac{w_3}{2}$ are zeros of $\mathfrak{p}'$, since each of them will necessarily be a simple zero of $\mathfrak{p}'$. But since the function $\mathfrak{p}'$ is odd, it first follows that every point z satisfying the relation $z = -z \bmod L$ is a zero of $\mathfrak{p}'$. On the other hand, we obviously have:

$$\frac{w_1}{2} \equiv -\frac{w_1}{2} \bmod L$$

$$\frac{w_2}{2} \equiv -\frac{w_2}{2} \bmod L$$

$$\frac{w_1}{2} + \frac{w_2}{2} \equiv -\frac{w_1}{2} - \frac{w_2}{2} \bmod L$$

The vector space $L(D)$ associated with a divisor. For every divisor D, let $L(D)$ be the set of elliptical functions f satisfying the relation:

$$D + (f) \geqslant 0,$$

to which the identically vanishing function is added. In order to show that this is a vector space over $\mathbb{C}$ one uses for instance the fact that $v_{\mathfrak{p}}$ is a valuation.

PROPOSITION 6. *The vector space* $L(D)$ *depends only on the class of* D mod $P(X)$.

Proof. Assume that $D_1 - D = (g)$.
Since

$$D + (f.g) = D + (g) + (f) = D_1 + (f) \geqslant 0,$$

we get a vector space isomorphism

$$L(D_1) \to L(D)$$

$$f \mapsto f \cdot g$$

Riemann–Roch problem. For every divisor D, the vector space $L(D)$ is shown to have a finite dimension, which is denoted by $l(D)$. Riemann–Roch's problem requires to express the natural number $l(D)$ in terms of natural invariants of the curve X and of the divisor D.

PROPOSITION 7. *(Riemann–Roch theorem). There exists a divisor* K *depending only on the lattice* L, *such that for every divisor* D, *we have*

$$l(D) = \deg(D) + l(K - D).$$

8. *System of local divisorial data.* By definition, a system of local divisorial data on a curve X is an open covering $\mathscr{U} = (U_i)_{i \in I}$ of X together with a system $\{f_i\}_{i \in I}$ of meromorphic functions such that on $U_i \cap U_j$ the meromorphic function f_i has no zero and no pole.

Let $D = \sum_{P \in X} n_p P$ be a divisor on X and consider a system $\mathscr{D} = (U_i, f_i)_{i \in I}$ of local divisorial data on X. This divisorial system is said to be associated to the divisor D if for every $i \in I$ and every $P \in U_i$ we have

$$v_p(f_i) = n_P.$$

Example 8.1. Take $D = n.0$ and let $\mathscr{D} = (U_i, f_i)_{1 \leqslant i \leqslant 4}$ be the system of local divisorial data associated to D as represented in fig. 8.1.1.

A property of this system of local divisorial data is the fact that *for every point $z \in X$ there exists U_i, $i \in \{1, 2, 3, 4\}$, such that, denoting by the same symbol z the element associated to this point in the fundamental parallelogram, we have $f_i(z) = z^n$.*

Every divisor can be defined by a system of divisorial data.

If D is the divisor associated to the system of local divisorial data $\mathscr{D} = (U_i, f_i)_{i \in I}$, then obviously $L(D) = \{f \in \mathscr{E}_L | f_i f$ is holomorphic on U_i for every $i \in I\}$.

Morphism associated with a divisor. D being a divisor on X, defined by a system of local divisorial data $\mathscr{D} = (U_i, f_i)_{i \in I}$, consider $r + 1$

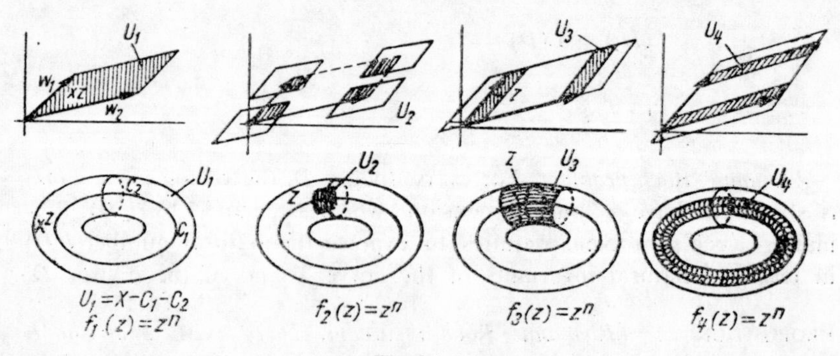

Fig. 8.1.1

elements $s_0, s_1, \ldots, s_r$ of the vector space $L(D)$ and a point $z \in U_i$. Since the functions $f_i s_0, f_i s_1, \ldots, f_i s_r$ are holomorphic on U_i, this yields a point

$$((f_i s_0)(z), (f_i s_1)(z), \ldots, (f_i s)(z)) \in \mathbb{C}^{r+1}$$

This system depends on the open set $U_i \in \mathscr{D}$ containing the point z, as well as on the function f_i.

Now assume that $(f_i s_l)(z) \neq 0$ for at least one $l \in \{1, 2, \ldots, r\}$. In this case, the system of complex numbers $((f_i s_0)(z), (f_i s_1)(z), \ldots, (f_i s_r)(z))$ defines a point of the projective space $P^r(\mathbf{C})$, depending only on z and on the system of elements $s_0, s_1, \ldots, s_r$. Indeed, if $z \in U_i \cap U_j$, the meromorphic function $h_{ij} = f_i/f_j$ has neither zeros nor poles on $U_i \cap U_j$, consequently

$$(f_i s_k)(z) = h_{ij}(z)(f_i s_k)(z), \; k = 0, 1, 2, \ldots, r,$$

$$h_{ij}(z) \neq 0.$$

Denoting by $V_D(s_0, s_1, \ldots, s_r)$ the subset of X consisting of the points z for which there exists an $l \in \{1, 2, \ldots, r\}$ such that $(f_i s_l)(z) \neq 0$, this yields a map

$$\varphi_D(s_0, s_1, \ldots, s_r) \colon V_D(s_0, s_1, \ldots, s_r) \to P^r(\mathbf{C}).$$

The subset $V_D(s_0, s_1, \ldots, s_r)$ is open and the map $\varphi_D(s_0, s_1, \ldots, s_r)$ is obviously analytic. This is an immediate consequence of the fact that for every $i \in I$ and $k \in \{0, 1, \ldots, r\}$ the map

$$f_i s_k \colon U_i \to \mathbf{C}$$

is analytic.

An extremely important problem of algebraic geometry is that of finding classes of divisors D and of systems of functions $s_0, s_1, \ldots, s_r$ such that $V_D(s_0, s_1, \ldots, s_r)$ coincides with X and $\varphi_D(s_0, s_1, \ldots, s_r)$ yields an analytic isomorphism of X into an algebraic subset of $P^r(\mathbf{C})$.

Proposition 9. *Let $D = 3.0$ and $\wp, \wp', 1 \in L(D)$. The map $\varphi_D(\wp, \wp', 1)$ yields an analytic isomorphism of X onto the projective curve in $P^2(\mathbf{C})$ defined by the equation*

$$Y^2 T = 4X^3 - g_2 X T^2 - g_3 T^3. \tag{9.1}$$

Proof. We begin by making some remarks:

— First of all, $\wp, \wp', 1$ obviously belong to $L(D)$ since $\wp$ (or $\wp'$) have a single pole, namely 0, of order 2 (or 3).

— Let $(U_i, f_i)_{1 \leq i \leq 4}$ be a system of local divisorial data of the type described in Example 8.1, whose associated divisor be our divisor D. As already remarked, if we identify a point $z \in X$ with the point canonically associated with it in the fundamental parallelogram, then $f_i(z) = z^3$ for some $i \in \{1, 2, 3, 4)$, hence $\varphi_D(\mathfrak{p}, \mathfrak{p}', 1)$ is defined by the formula

$$\varphi_D(\mathfrak{p}, \mathfrak{p}', 1)(z) = (z^3\mathfrak{p}(z), z^3p'(z), z^3)$$

— The map $\varphi_D(\mathfrak{p}, \mathfrak{p}', 1)$ is defined at every point of X, i.e. $V_D(\mathfrak{p}, \mathfrak{p}', 1) = X$. Indeed, the only point to arise some questions is $z = 0$. But $\mathfrak{p}'$ has at $z = 0$ a pole of order 3, hence $z^3p'(z)$ does not vanish at $z = 0$. Moreover, we have

$$\varphi_D(\mathfrak{p}, \mathfrak{p}', 1)(0) = (0, 1, 0).$$

— The point $(0, 1, 0)$ is the only point of the curve $Y^2T = 4X^3 - g_2XT^2 - g_3T^3$ situated on the line $T = 0$ (the "line at infinity" of the projective plane $P^2(\mathbb{C})$).

— Consequently, it will be sufficient to prove that the map

$$z \mapsto (\mathfrak{p}(z), \mathfrak{p}'(z))$$

yields an analytic bijection between $X \setminus \{0\}$ and the curve defined in the affine space $\mathbb{C}^2$ by the equation

$$(9.3) \qquad Y^2 = 4X^3 - g_2X - g_3.$$

To this end, we remark first that for every $z \neq 0$, the point $(\mathfrak{p}(z), \mathfrak{p}'(z))$ belongs to the above curve (cf. section I, proposition 7, c)).

Now we show that the considered map is surjective. If (ξ, η) is an arbitrary point of curve (9.3), then, since $\mathfrak{p}$ is L-elliptic, there exists a point $z \in P_L$ such that $\xi = \mathfrak{p}(z)$. On the other hand, the point $(\mathfrak{p}(z), \mathfrak{p}'(z))$ satisfies equation (9.3). It follows that $\mathfrak{p}'(z) = \pm \eta$. But $\mathfrak{p}'$ being odd and $\mathfrak{p}$ being even, this yields either $(\mathfrak{p}(z), \mathfrak{p}'(z)) = (\xi, \eta)$, or $(\mathfrak{p}(-z), \mathfrak{p}'(-z)) = (\xi, \eta))$.

— In order to prove that the considered map is injective, let z, u be two complex numbers such that

$$(\mathfrak{p}(z), \mathfrak{p}'(z)) = (\mathfrak{p}(u), \mathfrak{p}'(u)).$$

The following two cases are possible:

(i) $z \not\equiv -z \bmod L$.

(ii) $z \equiv -z \bmod L$.

In the first case, the relation $\mathfrak{p}(z) = \mathfrak{p}(u)$ and the fact that $\mathfrak{p}$ has only two zeros on X show that $u \equiv -z \bmod L$. But this would yield $\mathfrak{p}'(z) = \mathfrak{p}(-z)$, i.e. $\mathfrak{p}'(z) = 0$ since $\mathfrak{p}'$ is odd. This implies the relation $z \equiv -z \bmod L$ (Example 5) which contradicts our assumption.

In order to deal with the second case, we remark the following preliminary facts:

– All the roots of the polynomial $4X^3 - g_2X - g_3$ are simple. Indeed, otherwise the polynomials

$$4X^3 - g_2X - g_3$$

$$12X^2 - g_2$$

would have a common root and the calculation developed in the proof of Proposition 10 from section I shows that this fact contradicts the "non-singularity condition" $g_2^3 - 27g_3^2 \neq 0$. In fact the discriminant δ of the polynomial $4X^3 - g_2X - g_3$ is readily seen to be given by the formula

$$\delta = (e_1 - e_2)^2(e_2 - e_3)^2(e_3 - e_1)^2 = \frac{(g_2^3 - 27g_3^4)}{16},$$

– The roots e_1, e_2, e_3 of the polynomial $4X^3 - g_2X - g_3$ are $\mathfrak{p}\left(\frac{w_1}{2}\right)$, $\mathfrak{p}\left(\frac{w_2}{2}\right)$, $\mathfrak{p}\left(\frac{w_3}{2}\right)$. Indeed, these roots satisfy the relations

$$4X^3 - g_2X - g_3 = 4(X - e_1)(X - e_2)(X - e_3),$$

$$(\mathfrak{p}'(z))^2 = 4(\mathfrak{p}(z) - e_1)(\mathfrak{p}(z) - e_2)(\mathfrak{p}(z) - e_3).$$

If $\xi \in P_{\mathfrak{p}'} = P_{\mathfrak{p}}$ are such that $e_i = \mathfrak{p}(\xi)$, $(i = 1, 2, 3)$, it follows that $\mathfrak{p}'(\xi) = 0$, hence ξ is one of the numbers $\frac{w_1}{2}$, $\frac{w_2}{2}$, $\frac{w_3}{2}$ (cf. Example 5).

– Now we are able to deal with case ii. The hypothesis shows that

$$(p(z), 0) = (p(u), p'(u)).$$

Consequently, the numbers z, u are elements of the set $\left\{\frac{w_1}{2}, \frac{w_2}{2}, \frac{w_3}{2}\right\}$. It follows that z cannot be distinct from u, since otherwise the previous remark would yield $p(z) \neq p(u)$.

— The map $\varphi_D(p, p', 1)$ is actually an analytic isomorphism (with respect to the canonical analytic structures) between $X = \mathbb{C}/L$ and the curve defined by equation (9.1). This is a consequence of the following remark:

If

$$f: X \to Y$$

is an analytic map and if $G \subset Y$ is an analytic subvariety of Y such that

$$f(X) \subset G,$$

then the map

$$X \to G,$$
$$x \mapsto f(x)$$

is analytic with respect to the analytic structure of G induced by Y.

COROLLARY 9.1. *Every complex 1-dimensional torus is analytically isomorphic with a projective plane curve. In particular, it is algebraizable.*

Remark 9.2. There exist 2-dimensional analytic tori which are not algebraizable.

Remark 9.3. The point $(0, 1, 0)$ of the projective curve

$$\text{(E)}\quad Y^2T = 4X^3 - g_2XT^2 - g_3T^3$$

is non-singular. It is an inflexion point of the curve: the intersection multiplicity between the curve and the tangent at this point is 3.

Indeed, consider the affine curve

$$(9.3.1)\qquad T = 4X^3 - g_2XT^2 - g_3T^3,$$

whose origin corresponds to the point $(0, 1, 0)$ on (E). The origin is obviously non-singular on the curve (9.3.1).

The tangent at $(0,1, 0)$ to (E) is the line $T = 0$, as already remarked, $(0, 1, 0)$ is the only point of this line situated on (E). Since the degree of the curve (E) is 3, this shows that $(0, 1, 0)$ is an inflexion point.

We have seen in Proposition 9 that the map $\varphi_D(p, p', 1)$ yields an analytic isomorphism of the curve X onto the projective curve (E). On the other hand, X has an Abelian group structure:

$$X = \mathbb{C}/L.$$

Consequently $\varphi_D(\mathfrak{p}, \mathfrak{p}', 1)$ defines by structure transport an Abelian group structure on (E). We intend to make explicit this Abelian group structure of the curve (E) using elementary geometric considerations.

We have already seen that the null element of this Abelian group coincides with the only point of (E) lying on the line $T = 0$ and is an inflexion point (Remark 9.3). We have in addition the following result:

PROPOSITION 10. *Let P_1, P_2, P_3 be three points of the curve (E). The sum of these points (with respect to the Abelian group structure induced by $\varphi_D(\mathfrak{p}, \mathfrak{p}', 1)$ on (E)) vanishes if and only if these points are colinear.*

Proof. One can assume that two among the three points under consideration do not lay on the "line at infinity" $T = 0$. Hence we suppose that P_1, P_2 lay in the affine plane defined by the variables X, Y and we denote by D the line through these points. We distinguish the following two cases:

(i) D is not parallel to the y-axis, or equivalently, the coordinates $(x_1, y_1), (x_2, y_2)$ of the two points satisfy the condition $x_1 \neq x_2$. In this case, the equation of D writes

$$y = mx + n. \tag{10.1}$$

(ii) D is parallel to the y-axis. In this case, P_3 coincides with the point $(0, 1, 0)$ which is the null element of the Abelian group (E).

Suppose we are in case *i*. Consider the elliptic function $\varphi = \mathfrak{p}' - - m\mathfrak{p} - n$ associated to equation (10.1). This function has at the origin a pole of order 3, which, in a suitable fundamental parallelogram P_φ containing the origin, is the only pole of φ. It follows that the fundamental parallelogram P_φ contains three points z_1, z_2, z_3 such that

$\varphi(z_1) = \varphi(z_2) = \varphi(z_3) = 0$. For every such point z_i, $i = 1, 2, 3$, the point $(\varphi(z_i), \varphi'(z_i))$ obviously is a point of (E) lying on D. Hence we may assume

$$P_i = (p(z_i), p'(z_i)), \quad i = 1, 2, 3.$$

In view of Proposition 11 from section I, this gives the required relation

$$z_1 + z_2 + z_3 \equiv 0 \bmod L.$$

In case *ii*, the equation of D is of the form

$$X - a = 0. \tag{10.2}$$

Consider the elliptic function $\varphi = p - a$. The argument goes on as in the first case, with the remark that this time D intersects our curve also in the "point at infinity" $(0, 1, 0)$.

Now suppose that z_1, z_2, z_3 are points in a fundamental parallelogram such that $z_1 + z_2 + z_3 \equiv 0 \bmod L$. We show that the points P_1, P_2, P_3 defined by

$$P_i = (p(z_i), p'(z_i)), \quad i = 1, 2, 3$$

are colinear. Let D be the line through the points P_1, P_2. The equation of this line has one of the form (10.1), (10.2). Assume its form is (10.1), the case of (10.2) being similar. Consider again the elliptic function $\varphi = p' - mp - n$. It follows that in a suitable parallelogram containing the origin, the points z_1, z_2 are zeros of φ. But the relation $z_1 + z_2 + z_3 \equiv 0 \bmod L$ and Proposition 11 from section I show that z_3 is a third zero of this elliptic function. Consequently, the point $(p(z_3), p'(z_3))$ lies on D and obviously also on (E), which concludes the proof.

COROLLARY 11. *The algebraic operation induced by* $\varphi_D(\mathfrak{p}, \mathfrak{p}', 1)$ *on* (E) *can be defined as follows. Let* P_1, P_2 *be two points of* (E) *and let* D *be the line through these points. Denote by* Q *the third intersection point of* D *and* (E) (Fig. 10.1).

Let I *be the line through* Q *and* $(0, 1, 0)$ *(i.e. the line through* Q*, parallel to the y-axis). The third intersection point* S *of the line* I *with the curve* (E) *is the sum of the points* P_1, P_2:

$$P_1 + P_2 = S.$$

Proposition 10 and Corollary 11 lead to the following result:

PROPOSITION 12. *Let E be a plane, non-singular, absolutely irreducible curve defined over the field of characteristic $\neq 2, 3$ and having an inflexion point P whose coordinates lay in k. An algebraic operation*

$$+_P: E(k) \times E(k) \to E(k)$$

on the set $E(k)$ of k-rational points of the curve (E) can be defined as follows:

If $P_1, P_2 \in E(k)$ and D is the line through these points, let Q be the third intersection point of D with the curve (E). Then $P_1 + P_2$ is the third intersection point between (E) and the line I through P and Q.

This algebraic operation endows $E(k)$ with an Abelian group structure such that the following properties hold:

— P is the null element of the algebraic operation.

— The points P_1, P_2, P_3 are colinear if and only if $P_1 +_P P_2 +_P +_P P_3 = 0$.

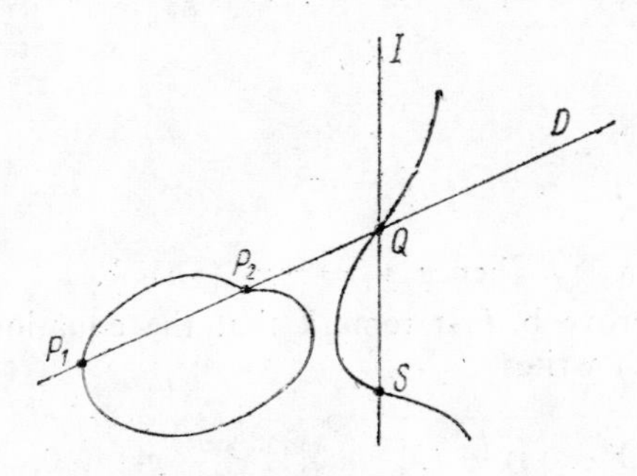

Fig. 10.1.1

Proof. We merely show that the set $E(k)$ is stable with respect to the above algebraic operation. It is sufficient to consider the elliptic curve (E) defined by the equation

$$Y^2T = 4X^3 - g_2 \times T^2 - g_3T^3, \quad g_2, g_3 \in k$$

whose inflexion point is $(0, 1, 0)$. We assume char $(k) \neq 2$.

All we have to do is to prove the following two facts:

(a) Let $P_i(x_i, y_i)$, $i = 1, 2$, be two k-rational points of the affine curve

$$Y^2 = 4X^3 - g_2X - g_3$$

and let D be the line passing through them. The third intersection point between (E) and D is k-rational.

(b) Let T be tangent to (E) at a k-rational point $P(x, y)$. The third intersection point between T and (E) is k-rational.

In order to prove a, we can assume $x_1 \neq x_2$ since otherwise the third intersection point is the inflexion point $(0, 1, 0)$. Let

$$y = mx + n, \quad m, n \in k$$

be the equation of D and denote by (x_3, y_3) the third intersection point between D and (E). *A priori* the coordinates (x_3, y_3) lie in some extension of k.

But since x_3 satisfies the relation

$$(mX + n)^2 = 4X^3 - g_2X - g_3$$

which yields

$$x_1 + x_2 + x_3 = -m^2$$

it follows that $x_3 \in k$, hence $y_3 = mx_3 + n \in k$.

In order to prove b, first remark that the equation of the tangent at $P(x, y)$ to (E) writes

$$2(Y - y)y + (X - x)(12x^2 - g_2) = 0,$$

hence it is of the above type, with

$$m = -\frac{12x^2 - g_2}{2y},$$

$$n = \frac{2y^2 + x(12x^2 - g_2)}{2y}.$$

The rest of the argument runs as above.

COROLLARY 12.1. *The necessary and sufficient condition for point A of the set E(k) to be an inflexion point is*

$$A +_P A +_P A = 0.$$

In particular, the set of these points is a subgroup of the group $(E(k), +_P)$.

Example 13. Let E be the non-singular curve defined by the equation

$$Y^2T = 4X^3 - g_2XT^2 - g_3T^3, \quad g_2, g_3 \in K$$

where $K \supset \mathbb{Q}$, $[K:\mathbb{Q}] < \infty$, is a field of algebraic numbers. The set $E(K)$ is an Abelian group having the point at infinity as null element. Mordell and Weil have shown that this group is finitely generated. An extremely difficult problem is to determine the rank $r = rk(E/k)$ using classical invariants associated with (E).

Remark 13.1. For an elliptic non-singular curve E with complex coefficients:

$$Y^2T = 4X^3 - g_2 \times T^2 - g_3T^3, \quad g_2, g_3 \in \mathbb{C}$$

the structure of the group $E(\mathbb{C})$ is a very simple one:

$$E(\mathbb{C}) \cong S^1 \times S^1 \cong \mathbb{C}/L, \ L = \{n_1w_1 + \\ + n_2w_2 | n_1, n_2 \in \mathbb{Z}, \ w_1/w_2 \notin \mathbb{R}\}$$

If g_2, g_3 belong to a field K of algebraic numbers, $(K \subset Q, [K:Q] < < \infty)$, then, by the Mordell-Weil theorem, $E(K)$ is a group of finite rank. Nevertheless, we have $E(K) \subset E(\mathbb{C})$, and $\mathbb{C}$ being an infinite-dimensional vector space over $\mathbb{Q}$, in $E(\mathbb{C}) = \mathbb{C}/L$ there exist subgroups of any rank.

Conjecture 13.2. Let $\mu: \mathbb{N} \times \mathbb{N} \to \mathbb{N}$ be the function defined by:

$$\mu(n, m) = \begin{cases} 0 \text{ if } n^3 - 27m^2 = 0 \\ \text{the rank of the group } E(Q), \text{ of the elliptic curve} \\ Y^2T = X^3 - nXT^2 - mT^3, \text{ otherwise.} \end{cases}$$

Then μ is a computable function.

Remark 14. In the case of a cubic with a singular point – such a curve can have at most one singular point – the above construction leads to simple groups which are isomorphic either to the additive underlying group k^+, or to the multiplicative group k^*.

DEFINITION 15 (of the concept of elliptic curve). Let K be a field and consider a projective curve X defined over K and a K-rational point P_0 of X. The couple (X, P_0) is said to be an elliptic curve over K if there exists a K-isomorphism between this curve and a non-singular, absolutely irreducible plane cubic E, such that the image of P_0 through this isomorphism be an inflexion point of E.

It follows that the coordinates of this inflexion point are elements of K, consequently the curve X has an Abelian group structure, with P_0 as null element.

Chapter VI

Algebraic varieties over a finite field

§ 1. *The zeta function*

1.1. Let X be an algebraic projective variety defined over the finite field k, let $\bar{k}$ be an algebraic closure of k, let k_n be the unique extension of degree n of k which is contained in $\bar{k}$, and let $X_n = X(k_n)$ be the set of points of X which are rational over k_n. In other words, X_n is the set of the points of X of the form

$$x = (\xi_0, \ldots, \xi_n), \; \xi_i \in K_n, \quad i = 0, \ldots, n.$$

The set X_n is clearly finite; we denote its cardinal by $v_n(X)$. In this way, the diophantine properties of the variety X are concentrated in the formal series with entire coefficients

$$v_1(X)T + \ldots + v_n(X)T^n + \ldots \in \mathbb{Z}[[T]].$$

Experience shows that it is more convenient to introduce another formal series with *integer coefficients*

$$Z_X(T) = 1 + b_1 T + \ldots + b_n T^n + \ldots,$$

called the *zeta-function associated to the variety* V, and defined by

$$\log Z_X(T) = \sum_{n=1}^{\infty} \frac{v_n(X)}{n} T^n \tag{1.1.1}$$

or by

$$T \frac{Z'_X(T)}{Z_X(T)} = \sum_{n=1}^{\infty} v_n(x) T^n. \tag{1.1.2}$$

The coefficients b_i can be determined by recurrence, from the relations

$$(1.1.3) \qquad b_1 = \gamma_1(X)$$

$$(1.1.4) \qquad ib_i = v_i(X) + v_{i-1}(X)b_1 + \ldots + v_1(X)b_{i-1}, \quad (i > 1).$$

To show that the numbers b_i (which *a priori* must be rational) are, in fact, entire, let us denote by X^0 the set of closed points of the Grothendieck scheme associated to the variety X.

If x is a point of X^0, the residual field $k(x)$ is a finite extension of k, of degree, say, $\deg(x)$.

With these notation, we have

$$(1.1.5) \qquad Z_X(T) = \prod_{x \in X^0} \left(\frac{1}{1 - T^{\deg(x)}} \right).$$

In order to prove this equality, it is enough to verify that the series defined by formula (1.1.5) satisfies (1.1.1). In fact, if $\mu_n(X)$ is the number of the points $x \in X^0$ such that $\deg(x) = n$, then we have

$$(1.1.6) \qquad v_n(X) = \sum_{i,n} i\mu_i(X),$$

and the function $Z_X(T)$ defined by formula (1.1.5) can be written:

$$Z_X(T) = \prod_{n \geqslant 1} \left(\frac{1}{1 - T^n} \right) \mu_n(X).$$

In order to verify condition (1.1.1) it suffices to take $\log Z_X(T)$ and use relation (1.1.6).

Example 1.2. Let $X = P^n$ be the projective n-dimensional space over k. We have

$$(1.2.1) \qquad v_S(P^n) = \frac{(q^s)^{n+1} - 1}{q^s - 1} = \sum_{i=0}^{n} q^{is}, \quad q = \mathrm{Card}(k),$$

hence

(1.2.2) $$\log Z_{P^n}(T) = \sum_{s\geq 1}\left(\sum_{i=0}^{n} q^{is}\right)\frac{T^s}{s} = \sum_{i=0}^{n}\sum_{s\geq 1}\frac{(q^iT)^s}{s}.$$

But using the well-known formula

(1.2.2.1) $$\log(1-t) = -t - \frac{t^2}{2} - \ldots - \frac{t^2}{s} - \ldots,$$

we get

$$\sum_{s\geq 1}\frac{(q^iT)^s}{s} = -\log(1-q^iT),$$

so that

$$\log Z_{P^n}(T) = \log\frac{1}{(1-T)(1-qT)\ldots(1-q^nT)},$$

and

$$Z_{P^n}(T) = \frac{1}{(1-T)(1-qT)\ldots(1-q^nT)}.$$

Example 1.3. Let $X = A^n$ be the affine n-dimensional space over k. From the straightforward relation

$$\nu_S(A^n) = q^{s^n},$$

it follows that

$$\log Z_{A^n}(T) = \sum_{s=1}^{\infty}\frac{q^{s^n}}{s}T^n = \sum_{s=1}^{\infty}\frac{(q^nT)^s}{s}.$$

Again, using formula (1.2.2.1), we get

$$\log Z_{A^n}(T) = \log\frac{1}{(1-q^nT)},$$

whence

$$Z_{A^n}(T) = \frac{1}{1 - q^n T}.$$

Example 1.4. Let $V = G_{m,r}$ be the Grassmann variety of r-dimensional subspaces of A^m. We have

$$(1.4.1.1) \qquad v_S(G_{m,r}) = \frac{(q^{sm} - 1)}{(q^{sr} - 1)} \frac{(q^{s(m-1)} - 1)}{(q^{s(r-1)} - 1)} \cdots \frac{(q^{s(m-r+1)} - 1)}{(q^s - 1)}.$$

In order to prove this relation, it is sufficient to consider the particular case $s = 1$ and to make an induction on r. If $r = 1$, then $G_{m,1} = P_{m-1}$ from which (1.4.1.1) follows in this case.

In order to continue the proof we deduce the recurrence formula:

$$(1.4.1.2) \qquad (G_{m,r+1}) = \frac{(q^{m-r} - 1)}{(q^{r+1} - 1)} v_1(G_{m,r}).$$

For the proof of this last formula, we consider the auxiliary variety $D_{m,(r,r+1)}$ of the flags $E_{r+1} \supset E_r$ of the space k^m. This variety has a fibration over $G_{m,r}$:

$$D_{m,(r,r+1)} \xrightarrow{p} G_{m,r},$$

$$(E^{r+1}, E^r) \mapsto E^r,$$

the fibre $p^{-1}(E^r)$ being the set of $r+1$-dimensional subspaces of k^m containing E^r. Consequently, we have

$$v_1(p^{-1}(E^r)) = v_1(P(k^m/E^r)) = v_1(P_{m-r}) = \frac{(q^{m-r} - 1)}{(q - 1)},$$

which yields

$$(1.4.1.3) \qquad v_1(D_{m,(r,r+1)}) = \frac{(q^{m-r} - 1)}{(q - 1)} v_1(G_{m,r}).$$

On the other hand, we obviously have

$$(1.4.1.4)\qquad \nu_1(D_{m(r,r+1)}) = \nu_1(G_{m,r+1})\nu_1(G_{r+1,r}) = = \frac{q^{r+1}-1}{q-1}\nu_1(G_{m,r+1}).$$

The recurrence formula (1.4.1.2) now follows from (1.4.1.3), (1.4.1.4). From (1.4.1.2) we can deduce (1.4.1.1) in the case $s = 1$, for we have

$$\nu_1(G_{m,n}) = \frac{(q^{m-r+1}-1)}{(q^r-1)}\nu_1(G_{m,r-1})$$

$$\nu_1(G_{m,r-1}) = \frac{(q^{m-r+2}-1)}{(q^{r-1}-1)}\nu_1(G_{m,r-2})$$

$$\cdots\cdots\cdots\cdots\cdots\cdots$$

$$\nu_1(G_{m,1}) = \frac{q^m-1}{q-1}$$

consequently

$$\nu_1(G_{m,r}) = \frac{(q^{m-r+1}-1)}{(q^r-1)}\cdot\frac{(q^{m-r+2}-1)}{(q^{r-1}-1)}\cdot \cdot\frac{(q^{m-r+3}-1)}{(q^{r-2}-1)}\cdots\frac{(q^m-1)}{(q-1)}$$

which is the required formula.

1.5. The concept of zeta-function associated with an algebraic variety X can be relativized with respect to a sheaf in the étale topology in the following way (for more details, see A. Grothendieck, *Formule de Lefschetz et rationalité des fonctions* L, Sém. Bourbaki 1964/nr. 279, or *Dix exposés sur la cohomologie des schémas,* pp. 37—40).

Let $\Omega \supset \mathbb{Q}$ be an extension of $\mathbb{Q}$ and let $\mathscr{F}$ be a sheaf (in the étale topology) of finite dimensional vector spaces over Ω, on the scheme X. For every closed point $x \in X^0$ of X, choose a separable closure $\overline{k(x)}$ of the residual field $k(x)$, and denote by $\bar{x}$ the geometric $\overline{k(x)}$-rational point of X defined by the extension $\overline{k(x)} \supset k(x)$. The Galois group $\prod_x \mathrm{Gal}(\overline{k(x)}, k(x))$ has a natural left action on the fibre $\mathscr{F}_{\bar{x}}$ of $\mathscr{F}$ at the geometric point $\bar{x}$ of X, in particular we can consider the Frobenius isomorphism

$$(f_x)_{\mathscr{F}_x} : \mathscr{F}_{\bar{x}} \to \mathscr{F}_{\bar{x}}.$$

Setting

$$(1.5.1) \qquad Z_x(\mathscr{F}, T) = \prod_{x \in X^0} \frac{1}{\det(1 - (f_x)^{-1} \mathscr{F}_{\bar{x}} T^{d(x)})}$$

we get the following equivalent formulae:

$$(1.5.2) \qquad \log Z_X(\mathscr{F}, T) = \sum_{n=1}^{\infty} v_n(\mathscr{F}) \frac{T^n}{n}$$

$$(1.5.3) \qquad T \frac{Z'_X(\mathscr{F}, T)}{Z_X(\mathscr{F}, T)} = \sum_{n=1}^{\infty} v_n(\mathscr{F}) T^n,$$

where

$$(1.5.4) \qquad v_n(\mathscr{F}) = \sum_{x \in X(k_n)} \mathrm{Tr}((f_x)^{-1})\mathscr{F}_{\bar{x}}$$

§ 2. *Cohomology theories*

2.1. Let K be a fixed field of characteristic 0. A Weil cohomology theory is a contravariant functor

$$X \mapsto H^*(X, K) = \bigoplus_{q \geqslant 0} H^q(X, K)$$

from the category of algebraic varieties to the category of commutative graded K-algebras, satisfying the following conditions:

(a) *Finiteness condition:* $\dim_K H^q(X, K) < \infty$, $q = 0, 1, \ldots$

$$H^i(X, K) = 0 \text{ for } i \notin [0, 2 \dim X]$$

(b) *Künneth formula:* The canonical morphism

$$H^*(X, K)\otimes_K H^*(X, K) \to H^*(X\times Y, K)$$

induced by the two projections

$$\mathrm{pr}_X : X\times Y \to X$$

$$\mathrm{pr}_Y : X\times Y \to Y$$

is an isomorphism.

(c) *Poincaré duality:* (I) There exists a canonical isomorphism:

$$H^{2n}(X, K) \simeq K, n = \dim X.$$

(II) The bilinear canonical map

$$H^i(X, K)\times H^{2n-1}(X, K) \to H^{2n}(X, K)$$

induced by the multiplication from $H(X, K)$ (the cup-product) is a perfect duality.

2.2. Assume that the variety X is defined over the finite field K having q elements. In this case, we have the Frobenius morphism

$$F: X \to X$$

$$(x_\alpha) \mapsto (x_\alpha^q)$$

which yields the linear maps:

$$H^i(F) : H^i(\overline{X}, K) \to H^i(X, K)$$

defining the polynomials

$$P_i^X(T) = \det(1 - H^i(F)T), \quad i = 0, 1, 2, \ldots, 2 \dim X. \tag{2.2.1}$$

§ 3. *Weil's diophantine conjectures*

3.1. *First Weil conjecture.* There exists a field of characteristic 0, say K, and a Weil cohomology theory

$$X \mapsto H^*(X, K)$$

such that for every algebraic, projective, non-singular variety X, defined over a finite field k, we have

$$Z_X(T) = \frac{P_1^X(T)\, P_3^X(T) \ldots P_{2n-1}^X(T)}{P_0^X(T)\, P_2^X(T) \ldots P_{2n}^X(T)}, \quad n = \dim X. \tag{3.1.1}$$

Sketch of the proof. We use the following lemma from linear algebra:

LEMMA 3.2. *Let E be a finite dimensional vector space over K and let*

$$U : E \to E$$

be an endomorphism of E.

Write

$$P_u = \det\,(1 - uT), \quad L_u = \frac{1}{P_u}, \tag{3.2.1}$$

where P_u, L_u are considered as formal power series with coefficients in K: $P_u, L_u \in K[[T]]$.

Then the following formula holds:

$$T \frac{L_u'(T)}{L_u(T)} = \mathrm{Tr}(u) T + \mathrm{Tr}(u^2) T^2 + \ldots + \mathrm{Tr}(u^n)\, T^n + \ldots$$

(For any endomorphism $v : E \to E$, $T_r(v)$ denotes the trace of v).

Proof of the lemma. First of all, by an extension of scalars, we obviously can reduce ourselves to the case of an algebraically closed field K.

We begin by considering the case of simple eigenvalues. In this case we are readily reduced to the case of a one-dimensional vector space, whence

$$P_u(T) = 1 - \lambda T$$

consequently

$$T \frac{L'_u(T)}{L_u(T)} = T \frac{\dfrac{\lambda}{(1-\lambda T)^2}}{\dfrac{1}{1-\lambda T}} = T \frac{\lambda}{1 - \lambda T} =$$

$$= \lambda T (1 + \lambda T + \lambda^2 T + \ldots)$$

$$= \sum_{i \geqslant 1} \lambda T^i = \sum_{i \geqslant 1} \mathrm{Tr}(u^i)\, T^i.$$

In order to show that we can always limit ourselves to the case of simple eigenvalues, we first remark that for any endomorphism

$$v : L^n(A) \to L^n(A)$$

of a free A-module $L^n(A)$ of rank n, where A is any commutative ring, we can define the characteristic polynomial $P_v \in A[T]$ and its inverse $A_v \in A[[T]]$.

For any ring homomorphism

$$A \xrightarrow{\varphi} B,$$

the map

$$v \otimes_A \mathrm{id}_B : L^n(B) \to L^n(B),$$

corresponding to the A-module isomorphism

$$L^n(B) \simeq L^n(A) \otimes_A B$$

is readily seen to satisfy the formula

$$(P_v) = P_{v \otimes_A \mathrm{id}_B}$$

Now consider $A = \mathbb{Z}[X_{i,j}]$, $i, j \in \{1, 2, \ldots, n\}$, $n = \dim_K E$ and let

$$\boldsymbol{v}: L^n(A) \to L^n(A)$$

be the linear map whose matrix in the canonical basis is $\|X_{ij}\|$.

Considering the ring homomorphism

$$\varphi: \mathbb{Z}[X_{ij}] \to K$$

$$X_{ij} \mapsto \mu_{ij}$$

where $\|\mu_{ij}\|$ is the matrix of the map $u: E \to E$ with respect to a basis chosen for E, we see that $\varphi \otimes_{\mathbb{Z}[X_{ij}]}$ id K is identical to u. On the other hand, since $Z[X_{ij}]$ is an integrity domain, the scalars of $L^n(A)$ can be extended first to the field k of the quotients of A and then to the algebraic closure $\bar{k}$ of k. All eigenvalues of the linear map w obtained from v by these extensions are simple.

In order to prove (3.1.1) we also use the following result of Grothendieck:

There exists a Weil cohomology theory such that the Lefschetz fixed point theorem be true. In other words, if

$$f: X \to X$$

is an isomorphism of a projective, non-singular algebraic variety, then

(3.2.3) $\qquad \mathrm{Lef}(f) = v(f)$

where

(3.2.4) $\qquad \mathrm{Lef}(f) = \sum_{i \geqslant 0} (-1)^i \, \mathrm{Tr}(H^i(f)).$

(3.2.5) $v(f) =$ the algebraic number of fixed points of f.

Setting

(3.2.6) $\qquad L_X(T) = \dfrac{L_{H^0(F)}(T)\, L_{H^2(F)}(T) \ldots L_{H^{2n}(F)}(T)}{L_{H^1(F)}(T)\, L_{H^3(F)}(T) \ldots L_{H^{2n-1}(F)}(T)},$

in order to get (3.1.1) it is sufficients to prove

$$T\frac{L'_X(T)}{L_X(T)} = \sum_{m\geqslant 1} \nu_m(X)T^m \tag{3.2.7}$$

But we have

$$T\frac{L'_X(T)}{L_X(T)} = \sum_{i=0}^{2n} (-1)^i T \frac{L'_{H^i(F)}(T)}{L_{H^i(F)}(T)} =$$

$$= \sum_{i=0}^{2n} (-1)^i \sum_{j\geqslant i} T_r(H^i(F^j))T^j =$$

(3.2.8)

$$= \sum_{j\geqslant 1}\left(\sum_{i=0}^{2n} (-1)^i\, T_r(H^i(F^j))\right) T^j =$$

$$= \sum_{j\geqslant 1} \mathrm{Lef}(F^j)T^j = \sum_{j\geqslant 1} \nu(F^j)T^j = \sum_{j\geqslant 1} \nu_j(X)T^j$$

since $\nu(F^j) = \nu_j(X)$.

3.3. *The concept of reduction of an algebraic variety.* In order to define this concept in its full generality and precision, the general machinery of Grothendieck schemes is necessary. We shall consider only a particular case, namely that of hypersurfaces, allowing a very elementary t reatment.

To this end, let K be a commutative field, endowed with a discrete valuation

$$v : K^* \to \mathbb{Z},$$

and denote by A the associated valuation ring, $A = \{x \in K^* | \, v(x) \geqslant \geqslant 0\} \cup \{0\}$, by $\underline{m}$ the maximal ideal of A, and by k the residual field A/m.

Let V be a projective variety over K, defined by the homogeneous polynomials

$$P_i(X_0, X_1, \ldots, X_n) \in K[X_0, X_1, \ldots, X_n].$$

We can obviously suppose that the coefficients of these polynomials belong to A and that not all of them belong to $\underline{m}$.

On the other hand, the canonical homomorphism

$$A \to A/\underline{m} = k$$

induces a homomorphism

$$A[X_0, X_1, \ldots, X_n] \to k[X_0, X_1, \ldots, X_n],$$

$$P(X_0, \ldots, X_n) \mapsto \overline{P}(X_0, X_1, \ldots, X_n).$$

The algebraic set V defined over k by the polynomials

$$\overline{P}_i(X_0, \ldots, X_n) \in k[X_0, X_1, \ldots, X_n]$$

s said to be obtained from V by reduction modulo $\underline{m}$.

3.4. *Second Weil conjecture. Let K be a discrete valuation field, which is a subfield of the complex field $\mathbb{C}$ and whose residual field is finite. Let V be a projective non-singular algebraic variety defined over k, obtained by reduction modulo $\underline{m}$ from a projective non-singular algebraic variety over K. The variety V being a fortiori defined over $\mathbb{C}$, let $V(\mathbb{C})$ be its associated complex variety, and denote by $H^i(V(\mathbb{C}), \mathbb{C})$ the usual cohomology groups with complex coefficients of $V(\mathbb{C})$. Under these conditions, the following relation holds:*

$$\deg(P_i^V(T)) = \dim_{\mathbb{C}} H^i(V(\mathbb{C}),\mathbb{C}).$$

The proof of this conjecture follows as a consequence of a general theorem of Grothendieck and Artin, establishing a relation between the étale cohomology and the usual cohomology of an algebraic variety defined over the field of complex numbers.

3.5. *Third Weil conjecture. With the notation and assumptions from* 2.2, 3.1, *the coefficients of the polynomials $P_i^X(T)$ are rational.*

This conjecture has not been proved (or disproved) hitherto. P. Deligne has recently proved a weaker form of it, namely:

The coefficients of the polynomials $P_i^X(T)$ are algebraic.

Grothendieck and Bombieri have proved that the third conjecture of Weil is a consequence of a conjecture from the theory of algebraic cycles:

Let X be an algebraic variety, denote by $Z^p(X)$ the group of its algebraic cycles of codimension p and assume we are given a Weil cohomology theory (cf. 2.1) and a homomorphism

$$\mathrm{cl}_X: Z^p(X) \otimes_{\mathbb{Z}} \mathbb{Q} \to H^{2p}(X)$$

satisfying the following conditions:

(I) If P is a point, then

$$\mathrm{cl}_P: Z^0(P) \otimes_{\mathbb{Z}} \mathbb{Q} = \mathbb{Q} \to H^0(P) = K$$

is the canonical inclusion.

(II) If U and V are two algebraic cycles of X whose algebraic intersection is defined, then

$$\mathrm{cl}_X(U \cap V) = \mathrm{cl}_X(U) \cup \mathrm{cl}_X(V).$$

(III) If U_1 is algebraically equivalent to U_2, then

$$\mathrm{cl}_X(U_1) = \mathrm{cl}_X(U_2).$$

We denote by $A^i(X)$ the image of $Z^i(X) \otimes_{\mathbb{Z}} \mathbb{Q}$ through the homomorphism cl_X. The elements of these images are called classes of algebraic cohomology and they form a ring whose multiplication is defined by means of the cup-product.

Now we are able to formulate a conjecture from the theory of algebraic cycles, which implies the validity of the third conjecture of Weil.

To this end, let $\Delta(X) \subset X \times X$ be the diagonal of $X \times X$.
Applying the Künneth formula, we get

$$\mathrm{cl}_X(\Delta(A)) = \sum_{i=0}^{2n} \pi_i, \quad n = \dim X,$$

$$\pi_i \in H^i(X) \otimes_K H^{2n-1}(X) \subset H^{2n}(X \times X).$$

The conjecture Z. The elements π_i are classes of algebraic cohomology.

We prove only that if this conjecture is true, then the coefficients of the polynomials P_i are rational numbers:

If $F: X \to X$ is the Frobenius morphism, it is sufficient to show that $\mathrm{Tr}_K(H^i(F^S))$ is a rational number. But denoting by Γ_{F^S} the graph of F^S, a classical calculation shows that

$$\mathrm{Tr}_K(H^i(F^S)) = (-1)^i \mathrm{cl}_{X\times X}(\Gamma_{F^S}) \cup \pi_i \in H^{4n}(X\times X) = K.$$

According to the conjecture Z, there exists an algebraic cycle π_i of $X\times X$ such that $\pi_i = \dfrac{1}{m}\mathrm{cl}_{X\times X}(\pi_i)$, $m \in \mathbb{Z}$, and from the property I of $\mathrm{cl}_{X\times X}$ it follows that $\mathrm{Tr}_K(H^i(F^S))$ is a rational number.

3.6. *Fourth Weild conjecture (Riemann hypothesis).* With the notation and assumptions from 2.2, 3.1, if ξ is a root of $P_i(T)$, then

$$\left|\frac{1}{\xi}\right| = q^{1/2}. \tag{3.6.1}$$

This conjecture has first been checked for various classes of algebraic varieties: Grasmann varieties, complete intersections, etc., before being recently proved in its general form by P. Deligne and J. P. Serre.

As in the case of the third conjecture, one can show that the fourth conjecture is a consequence of natural Lefschetz- and Hodge-type conjectures upon algebraic cycles.

To this end, consider a non-singular projective variety X and let $\xi \in H^2(X)$ be the class associated with a hyperplane section. We have the homomorphism

$$L_\xi = L : H^i(X) \to H^{i+2}(X), \qquad a \mapsto a \cup \xi \tag{3.6.2}$$

and, more generally, the homomorphisms

$$L^{n-i} : H^i(X) \to H^{2n-i}(X), \qquad a \mapsto a \cup \xi^{n-i} \tag{3.6.3}$$

defined for every $i \leqslant n$. Taking $i = 2j$, we get the commutative diagram

$$\begin{array}{ccc} H^{2j}(X) & \xrightarrow{L^{n-2j}} & H^{2n-2j}(X) \\ \uparrow & & \uparrow \\ A^j(X) & \longrightarrow & A^{n-j}(X) \end{array}$$

The Lefschetz conjecture can be stated as follows:

(a) For every $i \leqslant n$, the homomorphism L^{n-i} is an isomorphism;

(b) if $i = 2j$, the homomorphism

$$A^j(X) \to A^{n-j}(X)$$

induced by L^{2n-j} is an isomorphism.

Part a of the Lefschetz conjecture has been proved in the characteristic zero.

3.7. *Primitive decompositions.* Consider

$$P^i(X) = \operatorname{Ker}(L^{n-i+1})$$

$$P^i_{\text{alg}}(X) = P^{2i}(X) \cap A^i(X)$$

The elements of $P^i(X)$ (or of $P^i_{\text{alg}}(X)$) are called primitive elements of $H^i(X)$ (or of $A^i(X)$).

PROPOSITION 3.7.1. *If the Lefschetz conjecture is true, then for every i, $0 \leqslant i \leqslant 2n$, we have the "primitive decompositions"*

$$H^i(X) = \bigoplus_{r \geqslant 0} L^r P^{i-2r}(X), \tag{3.7.1.1}$$

$$A^i(X) = \bigoplus L^r P^{i-2r}(X). \tag{3.7.1.2}$$

Proof. The proof of (3.7.1.2) is completely similar to that of (3.7.1.1). In order to prove (3.7.1.1), we argue by induction on i. First suppose that $i \leqslant n$. The case $i = 0$ being trivial, consider the diagram

$$\begin{array}{ccc} H^i(X) & & \\ \uparrow L & \searrow^{L^{n-i+1}} & \\ H^{i-2}(X) & \xrightarrow[\sim]{L^{n-i+2}} & H^{2n-i+2}(X) \end{array}$$

Since the Lefschetz conjecture has been assumed to be true, L^{n-i+2} is an isomorphism, which yields the decomposition

$$H^i(X) = P^i(X) \oplus L(H^{i-2}(X))$$

from which the required decomposition follows by the induction hypothesis. The case $i > n$ is immediately reduced to the previous one.

3.8. *The Hodge conjecture*. Using the cup-product and the notation from 3.6 and 3.7, consider for every $j \leqslant n$ the symmetric bilinear form

$$\begin{aligned} &<,> : P^j_{\text{alg}}(X) \times P^j_{\text{alg}}(X) \to \mathbb{Q}, \\ &(x, y) \mapsto (-1)^j x \cup y \cup \xi^{n-2j}. \end{aligned} \tag{3.8.1}$$

The Hodge conjecture asserts that (3.8.1) is a positively definite form. This conjecture has been proved in the characteristic zero (cf. A. Weil, *Variétés Kählériennes*, Hermann, Paris, 1958).

Now we sketch the proof of the fact that the Lefschetz and Hodge conjectures imply the fourth conjecture of Weil.

First of all, using the conjectures of Lefschetz and Hodge the bilinear form

$$\begin{aligned} &A^i(X) \times A^i(X) \to Q, \\ &(a, b) \mapsto \sum_{r \geqslant 0} (-1)^{i-r} < \xi^{n-2i+2r} a_r, b_r >, \end{aligned} \tag{3.8.2}$$

where

$$a = \sum_{r \geqslant 0} L^r a_r, \; a_r \in p^{i-2r}_{\text{alg}}(X)$$

$$b = \sum_{s \geqslant 0} L^s b_s, \; b_s \in P^{i-2s}_{\text{alg}}(X)$$

is easily seen to be positively definite, hence it determines a positively definite hermitic form

$$Q : (A^i(X) \otimes_Q \mathbb{C}) \times (A^i(X) \otimes_Q \mathbb{C}) \to \mathbb{C} \tag{3.8.3}$$

LEMMA 3.8.4. *Let*

$$g : A(X) \otimes_Q \mathbf{C} \to A(X) \otimes_Q \mathbf{C}$$

be an algebra endomorphism leaving invariant the class ξ *and the fundamental class. Under this assumption, all eigenvalues of* g *are of module* 1.

Proof. The hermitic form Q depends only on the multiplicative structure of $A(X)$, on the class ξ and on the fundamental class of X. It follows that g is unitary with respect to Q, hence its eigenvalues are of module 1.

In order to prove the fourth conjecture of Weil, we have to check that all eigenvalues of the endomorphism

$$A^i(F)/(\sqrt{q})^i : A(X) \otimes_Q \mathbf{C} \to A(X) \otimes_Q \mathbf{C}$$

induced by F (the Frobenius endomorphism) are of module 1, which follows from Lemma 3.8.4.

Chapter VII

Elementary theory of non-standard real numbers

Let X be a topological space and let $\mathbb{R}(X)$ be the $\mathbb{R}$-algebra of continuous real valued functions defined on X. In the following we intend to present some properties of the topological space Spec $(\mathbb{R}(X))$.

1. For every point $x \in X$, the map

$$\mathbb{R}(X) \overset{v_n}{\to} \mathbb{R}$$

$$f \mapsto v_x(f) = f(x)$$

is a surjective algebra homomorphism. Consequently $\underline{m}_x = \mathrm{Ker}(v_x)$ is a maximal ideal of $\mathbb{R}(X)$, and we get a map

$$i_X : X \to \text{Spec } (\mathbb{R}(X))$$

$$x \mapsto i_X(x) = \underline{m}_x.$$

Proposition 1.1. *The map i_X is continuous.*

Proof. Consider an element $f \in \mathbb{R}(X)$ and the corresponding open set

$$D(f) = \{p \in \text{Spec } (\mathbb{R}(X)) | f \notin p\}.$$

It is sufficient to show that $i_X^{-1}(D(f))$ is an open subset of X.
But we have

$$i_X^{-1}(D(f)) = \{x \in X | f \notin \underline{m}_x\} = \{x \in X | f(x) \neq 0\} =$$

$$= f^{-1}(\mathbb{R} \setminus \{0\}).$$

2. Let X be a totally regular space, i.e. for every closed subset F of X, and every point $u \in X \setminus F$, there exists a continuous map

$$f: X \to R,$$

such that

$$f(x) = 0 \text{ for } x \in F,$$

$$f(u) = 1.$$

In this case, the map i_X is obviously injective; more precisely we have the following result:

PROPOSITION 2.1. *If X is a totally regular space, then the map i_X yields a homeomorphism between X and the subspace $i_X(X)$ of the topological space* Spec $(\mathbb{R}(X))$.

Proof. Taking into account Proposition 1.1, it is sufficient to show that for every open set U of X, the set $i_X(U)$ is the intersection between $i_X(X)$ and an open set of Spec $(\mathbb{R}(X))$. This will follow if we show that for every point $u \in U$ there exists an open set D_u of Spec $(\mathbb{R}(X))$, such that

$$i_X(u) \in (D_u \cap i_X(x)) \subset i_X(U).$$

To this end, let f_u be a continuous real-valued function defined on X, such that

$$f_u(u) = 1,$$

$$f_u(x) = 0 \quad \text{for} \quad x \notin U.$$

We can obviously take $D_u = D(f_u)$.

3. Denote by Spec max $(\mathbb{R}(X))$ the subspace of Spec $(\mathbb{R}(X))$ consisting of the maximal ideals of $\mathbb{R}(X)$. The map i_X obviously admits the factorization

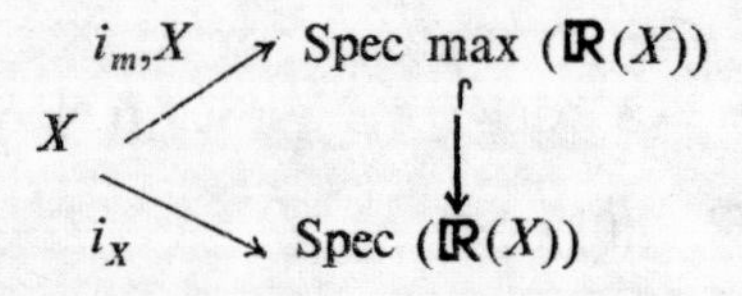

PROPOSITION 3.1. *The space X is compact if and only if*

$$i_{m,X}(X) = \text{Spec max}(\mathbb{R}(X)).$$

Proof. Let X be compact and let $\underline{m}$ be a maximal ideal of the ring $\mathbb{R}(X)$. We show there exists a point $x_0 \in X$ such that

$$\underline{m} = \underline{m}_{x_0} = \{f \in \mathbb{R}(X) \mid f(x_0) = 0\}.$$

Indeed, assume by *reductio ad absurdum* that such a point does not exist.

In this case, for every point $x \in X$, there exists a continuous map $f_x : X \to \mathbb{R}$ such that

$$f_x(x) \neq 0,$$

$$f_x \in \underline{m}.$$

Consequently, for every point $x \in X$, there exists an open set U_x such that

$$U_x \ni x,$$

$$f_x(u) \neq 0 \text{ for } u \in U_x.$$

By the compacity of X, this yields a finite covering $U_{x_1}, U_{x_2}, \ldots, U_{x_n}$ of X. The map

$$f = f_{x_1}^2 + f_{x_2}^2 + \ldots + f_{x_n}^2$$

has the following properties:

$$f \in \underline{m},$$

$$f(x) > 0 \text{ for every point } x \in X.$$

Consequently f is invertible, which is absurd since $\underline{m}$ is a proper deal of the ring $\mathbb{R}(X)$.

Now assume that $i_{m,x}$ satisfies the condition

$$i_{m,x}(X) = \text{Spec max }(\mathbb{R}(X)).$$

Then the compactness of X follows from the following result:

LEMMA 3.2. *For every ring A, the topological space Spec max (A) is compact.*

Proof. If $(D(f_i))_{i\in I}$ is a covering of Spec max (A), then $(D(f_i))_{i\in I}$ is actually a covering of Spec(A). Indeed, if $\underline{p} \in \text{Spec}(A)$, then

$$\underline{p} \subset \underline{m} \in \text{Spec max } (A),$$

$$\underline{m} \in D(f_{i_0}) f_{i_0} \notin \underline{m}$$

hence $f_{i_0} \notin \underline{p}$, i.e. $\underline{p} \in D(f_{i_0})$

Since it is well known that Spec(A) is compact, the lemma is proved.

PROPOSITION 3.3. *Let X be a totally regular space and let $\underline{p} \in$ Spec $(\mathbb{R}(X))$. The set*

$$Z(\underline{p}) = \{x \in X \mid f(x) = 0, \quad \forall f \in \underline{p}\}$$

has at most one element.

Proof. If x, y are distinct points of X, there exist $f_x, f_y \in \mathbb{R}(X)$ satisfying the conditions:

$$f_x(x) = 1, f_x(y) = 0,$$

$$f_y(y) = 1, f_y(x) = 0,$$

$$f_x \cdot f_y = 0.$$

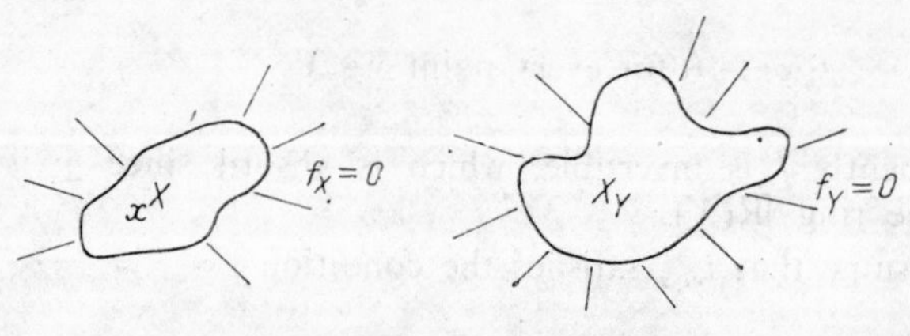

Fig. 3.3.3.1

By *reductio ad absurdum*, assume that $Z(\underline{p})$ contains two distinct points x, y. It is sufficient to check the existence of a map $f \in \underline{p}$ which does not vanish at one of these points. But, with the above notation, the relation $f_x \cdot f_y = 0$ implies the fact that either f_x or f_y belongs to $\underline{p}$.

COROLLARY 3.4. *If X is a compact space, then every prime ideal of $\mathbb{R}(X)$ is included in a unique maximal ideal of $\mathbb{R}(X)$.*

Proof. Indeed, if the prime ideal p of $\mathbb{R}(X)$ would be included in the maximal ideals $\underline{m}_x$, $\underline{m}_y$(cf. Proposition 3.1), then we would have $Z(p) \supset \{x, y\}$.

Remark 3.4.1. Corollary 3.4 holds for an arbitrary totally regular space.

3.5. *A proof of the Tichonov theorem*

LEMMA 3.5.1. *If $f: X \to Y$ is continuous and if Y is compact, then there exists a continuous map*

$$\text{Spec max } (\mathbb{R}(X)) \to \text{Spec max } (\mathbb{R}(Y))$$

such that the following diagram be commutative:

$$\begin{array}{ccc} \text{Spec max } (R(X)) & \longrightarrow & \text{Spec max } (R(Y)) \\ \uparrow{\scriptstyle i_X} & & \uparrow{\scriptstyle i_Y} \\ X & \longrightarrow & Y \end{array}$$

Proof. Using Corollary 3.4, we can define the following maps:

$$\mathbb{R}(Y) \xrightarrow{R(f)} \mathbb{R}(X)$$

$$\text{Spec } (\mathbb{R}(X)) \to \text{Spec } (\mathbb{R}(Y))$$

$$\text{Spec max } (\mathbb{R}(X)) \to \text{Spec}(\mathbb{R}(Y)) \to \text{Spec max}(\mathbb{R}(Y)).$$

The required map is the composition of the last two morphisms. In order to prove that this map is continuous, we have to check the continuity of the map

$$\text{Spec }(\mathbb{R}(Y)) \to \text{Spec max}(\mathbb{R}(Y)).$$

But we have

$$\underline{p} \in \text{Spec }(R(Y)) \Rightarrow \underline{p} \subset \underline{m} \in \text{Spec max }(R(Y)),$$

$$\underline{m} \in D(u),\ u \in R(Y) \Rightarrow \underline{p} \in D(u).$$

LEMMA 3.5.2. *If* $(X_i)_{i \in I}$ *is a family of compact spaces, there exists a continuous map*

$$\text{Spec max }(\mathbb{R}(\prod_{i \in I} X_i)) \to \prod_{i \in I} \text{Spec max }(\mathbb{R}(X_i)). \tag{3.5.2.1}$$

The proof is straightforward.

The proof of the Tichonov theorem now follows from the surjectivity of the map (3.5.2.1) and from the fact that Spec max (A) is compact for every ring A.

Remark 3.5.2.2. Let as above X be a totally regular space.

Denote by $\mathbb{R}_b(X)$ the $\mathbb{R}$-algebra of continuous and bounded $\mathbb{R}$-valued functions defined on X.

The canonical inclusion

$$i_{m,X}: X \hookrightarrow \text{Spec max }(\mathbb{R}_b(X))$$

shows that Spec max $(\mathbb{R}_b(X))$ is the Stone-Čech compactification of X.

4. DEFINITION 4.1. A topological space is said to be *normal* if for every couple F_1, F_2 of closed disjoint subsets of X there exists a continuous map

$$f: X \to \mathbb{R}$$

such that

$$f(x) = 1 \text{ for } x \in F_1$$

$$f(x) = 0 \text{ for } x \in F_2.$$

Obviously, every normal space is totally regular.

PROPOSITION 4.2. *If X is a totally regular topological space, then for every open set U of X and every $x \in U$ there exists an open set V such that*

$$x \in V \subset \overline{V} \subset U.$$

Proof. Let $F = \complement U$. There exists a continuous map

$$f \colon X \to \mathbb{R}$$

such that the following relations hold:

$$f(x) = 0$$

$$f(y) = 1 \text{ for } y \in F.$$

It is sufficient to take for V the set

$$V = \left\{ z \in X \mid f(z) < \frac{1}{2} \right\}.$$

If X is an arbitrary topological space and if $\underline{m}_x$ denotes the maximal ideal of $\mathbb{R}(X)$ associated with a point $x \in X$, we can consider two rings in fact, two $\mathbb{R}$-algebras):

$\mathbb{R}(X)_{\underline{m}_x}$, the ring of the quotients of $\mathbb{R}(X)$ with respect to the multiplicatively closed system $(\mathbb{R}X) \setminus m_x$.

$\mathscr{C}_x$, the ring of the germs of real-valued functions at x.

We obviously have a ring homomorphism (in fact, an algebra homomorphism)

$$\mathbb{R}(X)_{\underline{m}_x} \xrightarrow{\gamma_x} \mathscr{C}_x.$$

PROPOSITION 4.3. *If X is a normal space, the homomorphism γ_x is an isomorphism.*

Proof. Surjectivity of γ_x. Let $g \colon U \to \mathbb{R}$ be a continuous map defined on a neighborhood U of x. It is sufficient to show the existence of a map $h \colon X \to \mathbb{R}$ and of a neighborhood $W \ni x$ of x such that

$$W \subset U,$$

$$h|_W = g|_W.$$

To this end, let $V \ni x$ be a neighborhood of x (cf. Proposition 4.2) such that

$$x \in V \subset \overline{V} \subset U$$

and let $W \ni x$ be a neighborhood of x such that

$$x \in W \subset \overline{W} \subset V.$$

The closed subsets $\overline{W}$, $\complement V$ are disjoint. Consequently, there exists a continuous map

$$\varphi: X \to \mathbb{R}$$

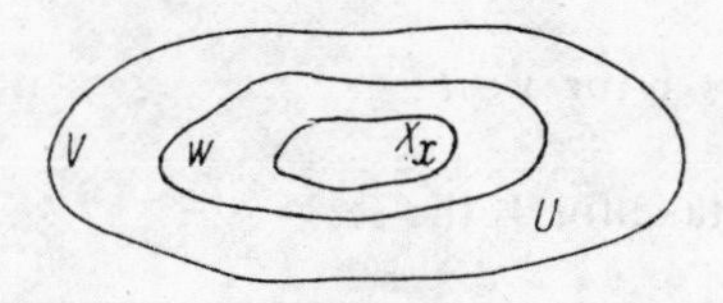

Fig. 4.3.2

such that

$$\varphi(z) = 1 \quad \text{for } z \in \overline{W},$$

$$\varphi(z) = 0 \quad \text{for } z \in \complement V.$$

Now consider the map $h: X \to \mathbb{R}$ defined as follows:

$$h(y) = \begin{cases} \varphi(y)\, g(y) & \text{for} \quad y \in V, \\ 0 & \text{for} \quad y \in \complement V. \end{cases}$$

This map is continuous and it takes on W the same values as g.

Injectivity of γ_x. Let $r, s \in \mathbb{R}(X)$ be such that $s(x) \neq 0$ and $\gamma_x\left(\dfrac{r}{s}\right) = 0.$

Since s is non-vanishing in some neighborhood U of x, it follows that r vanishes on U. Let $t: X \to \mathbb{R}$ be a continuous map such that:

$$t(x) = 1,$$

$$t(z) = 0 \quad \text{for } z \in \complement U.$$

We obviously have $t \notin \underline{m}_x$, consequently

$$\frac{r}{s} = \frac{tr}{ts}$$

But for every $z \in X$, we have

$$(tr)(z) = 0,$$

i.e. $tr = 0$. Consequently $\frac{r}{s} = 0$.

PROPOSITION 4.4. *The ring $\mathscr{C}_x$ is Henselian.*

Let L be a lattice. A non-empty subset $\mathscr{F}$ of L is said to be a filter on L if the following conditions are fulfilled:

I. $\emptyset \in \mathscr{F}$ ($\emptyset$ is the initial element of L).

II. $F_1, F_2 \in \mathscr{F} \Rightarrow F_1 \cap F_2 \in \mathscr{F}$.

III. $(F_1 < F_2)$ and $(F_1 \in \mathscr{F}) \Rightarrow (F_2 \in \mathscr{F})$.

The set of all filters on L is ordered by inclusion. A maximal object of this set is called an ultrafilter on L.

A filter $\mathscr{F}$ on L is said to be prime if the following condition is fulfilled:

If A and B are two elements of L such that $AVB \in \mathscr{F}$, then either A or B belongs to $\mathscr{F}$.

PROPOSITION 4.5. *Every ultrafilter on L is a prime filter.*

Proof. Let $A, B \in L$ be such that $A \vee B \in \mathscr{F}$. By *reductio ad absurdum,* assume that neither A nor B belongs to $\mathscr{F}$. Consider the subset $\mathscr{G}$ of L defined by

$$\mathscr{G} = \{C \in L \mid A \vee C \in \mathscr{F}\}.$$

It is readily shown that

1. $B \in \mathscr{G}$.
2. $\mathscr{G}$ is a filter on L.
3 $\mathscr{G} \supset \mathscr{F}$.

Consequently $\mathscr{G} = \mathscr{F}$ since $\mathscr{F}$ is an ultrafilter, which contradicts relation 1 in view of the assumption $B \notin \mathscr{F}$.

Remark 4.5.1. If L is the lattice $\mathscr{P}(E)$ of the subsets of a set E, then the converse of Proposition 4.5 is also true, i.e. in this case every prime filter is an ultrafilter. More generally, if L is a Boole algebra, then every prime filter on L is an ultrafilter.

Example 4.5.2. Let X be a topological space and let $Z(X)$ be the set of the subsets of X defined by

$$Z(X) = \{F \in \mathscr{P}(X) \mid F = f^{-1}(0),\ f \in \mathbb{R}(x)\}.$$

$Z(X)$ has an obvious lattice structure induced by the inclusion relation.

Denote by Filt $(Z(X))$ the set of the filters defined on $Z(X)$ and by $\mathrm{Id}(R(X))$ the set of the ideals of the algebra $\mathbb{R}(X)$.

For every filter $\mathscr{F} \in$ Filt $(Z(X))$, let $i(\mathscr{F})$ be the ideal of $\mathbb{R}(X)$ defined by

$$i(\mathscr{F}) = \{f \in \mathbb{R}(x) \mid f^{-1}(0) \in \mathscr{F}\}$$

and for every ideal $\underline{a}$ of $\mathbb{R}(X)$, consider the filter

$$\varphi(\underline{a}) = [f^{-1}(0) \mid f \in \underline{a}\}.$$

The maps

$$\varphi : \mathrm{Id}(\mathbb{R}(X)) \to \mathrm{Filt}(Z(X))$$

$$i : \mathrm{Filt}\ (z(X)) \to |d(\mathbb{R}(X))$$

satisfy the relations

(*) $i(\varphi)(\underline{a})) \supseteq \underline{a}),\ \forall \underline{a} \in \mathrm{Id}(\mathbb{R}(X)),$

(**) $\varphi(i(\mathscr{F})) = \mathscr{F},\ \forall \mathscr{F} \in \mathscr{F}\text{ilt}\ (Z(X)).$

The first relation becomes an equality if either the topology of X is discrete or the ideal $\underline{a}$ is maximal.

If $\mathscr{F}$ is a prime filter, then $i(\mathscr{F})$ is a prime ideal. If $\underline{a}$ is a prime ideal, then $\varphi(\underline{a})$ is a prime filter.

The following somewhat different case can be dealt with in a similar way.

5. Let I be an arbitrary set let $(K_i)_{i \in I}$ be a family of commutative fields and let $A = \prod_{i \in I} K_i$. We intend to describe the prime ideals of A. For every filter $\mathscr{F}$ on I and for every ideal $\underline{a} \subset A$, consider the ideal $i(\mathscr{F})$ of A, defined by

$$i(\mathscr{F}) = \{(f_i)_{i \in I} \in A | \{i/f_i = 0\} \in \mathscr{F}\}$$

and the filter $\varphi(\underline{a})$ on I, defined by

$$\varphi(\underline{a}) = \{X \subset I | \exists (f_i)_{h \in I} \in \underline{a},\ X = \{i \in I \mid f_i = 0\}\}.$$

One shows that φ establishes a bijection between ultrafilters on I and prime ideals of A and that every prime ideal of A is maximal.

If $\mathscr{U}$ is an ultrafilter on I, the field $\prod_{i \in I} K_i / i(\mathscr{U})$ is called the ultraproduct of the family $(K_i)_{i \in I}$ with respect to the ultrafilter $\mathscr{U}$ and is denoted by $\prod_{\mathscr{U}} K_i$.

Remark 5.1. The fact that every prime ideal of a ring that is a product of fields is maximal follows from the fact that such a product is a von Neumann regular ring.

DEFINITION 5.2. A commutative ring A is a *von Neumann regular ring* if for every $x \in A$ there exists $y \in A$ such that $x^2 y = x$.

PROPOSITION 5.3. *Every ring* $A = \prod_{i \in I} K_i$ *which is a product of fields is a von Neumann regular ring.*

Proof. If $x = (x_i)_{i \in I}$, then we can take $y = (y_i)_{i \in I}$, where

$$y_i = \begin{cases} x_i^{-1}, & \text{if} \quad x_i \neq 0, \\ 0 & \text{if} \quad x_i = 0. \end{cases}$$

PROPOSITION 5.4. *In a von Neumann regular ring every prime ideal is maximal.*

Proof. Let $\underline{p} \subset A$ be a prime ideal of the von Neumann regular ring A. We show that $A/\underline{p}$ is a field. Consider an element $x \in A \setminus \underline{p}$.

We have $x(xy - 1) = 0$, hence $xy = 1 \bmod \underline{p}$ i.e. the class of x is invertible in $A/\underline{p}$.

Example 1. Take $I = \mathbb{N}$ (the set of natural numbers), $K_i = \mathbb{R}$ for every $i \in \mathbb{N}$, and let $\mathscr{U}_{\mathscr{C}}$ be an ultrafilter on $\mathbb{N}$, containing the Cauchy filter

$$\mathscr{C} = \{X \subset \mathbb{N} \mid \complement X \text{ is finite}\}.$$

The field $\prod_{\mathscr{U}_{\mathscr{C}}} \mathbb{R}$ is a real-closed non-Archimedean extension of $\mathbb{R}$, called the field of non-standard real numbers, in which a calculus with infinitesimal elements in the sense of Leibniz can be introduced (cf. A. Robinson, *Non-standard Analysis)*.

PROPOSITION 5.5. *If $\mathscr{F}$ is a prime filter on $Z(X)$ then $i(\mathscr{F})$ is a prime ideal of $\mathbb{R}(X)$, and the integrity domain $\mathbb{R}(X)/i(\mathscr{F})$ is a totally ordered ring. In particular the field of the quotients $Q(\mathbb{R}(X)/i(\mathscr{F}))$ of the integrity domain $\mathbb{R}(X)/i(\mathscr{F})$ is a totally ordered field.*

Proof. Let $f, g \in \mathbb{R}(X)$ be such that $(fg)^{-1}(0) \in \mathscr{F}$. We have the relation

$$(fg)^{-1}(0) = f^{-1}(0) \cup g^{-1}(0)$$

and since $\mathscr{F}$ is prime, it follows that either f or g belongs to $i(\mathscr{F})$.

For the second part of the proposition, we first show that we can introduce an order relation on $\mathbb{R}(X)/i(\mathscr{F})$ such that the canonical homomorphism

$$\mathbb{R}(X) \to \mathbb{R}(X)/i(\mathscr{F})$$

be a monotone map with respect to the natural ordering of $\mathbb{R}(X)$. We have to show that if $h, l \in \mathbb{R}(X)$ and if $0 \leqslant h \leqslant l$, $l \in i(\mathscr{F})$, then $h \in i(\mathscr{F})$, which follows from the inclusion

$$h^{-1}(0) \supseteq l^{-1}(0).$$

In order to show that the integrity domain $\mathbb{R}(X)/i(\mathscr{F})$ is totally ordered, it is sufficient to prove the following two lemmas:

LEMMA 5.6. *Let $f \in \mathbb{R}(X)$. The relation $f \bmod i(\mathscr{F}) \geqslant 0$ holds if and only if there exists $F \in \mathscr{F}$ such that $f(x) \geqslant 0$ for every $x \in F$.*

Proof. Assume that f mod $i(\mathscr{F}) \geqslant 0$. By definition, there exists $g \in i(\mathscr{F})$ such that $f+g \geqslant 0$. It is sufficient to take $F = g^{-1}(0)$. Now assume that there exists $F = g^{-1}(0)$, $g \in i(\mathscr{F})$, such that $f(x) \geqslant 0$ for every $x \in F$. We have to show the existence of a function $h \in i(\mathscr{F})$ such that $f+h \geqslant 0$. We can obviously assume that $g \geqslant 0$. Consider the function

$$\max (f, g) : X \to \mathbb{R}.$$

We obviously have

$$\max (f, g) \geqslant 0,$$

$$\max (f, g)(x) = f(x) \text{ for } x \in F,$$

$$f\text{-}\max (f, g) \in i(\mathscr{F}).$$

LEMMA 5.7. *For every function $f \in \mathbb{R}(X)$, there exists $F \in \mathscr{F}$ such that either $f(x) \geqslant 0$ for every $x \in F$, of $f(x) \leqslant 0$ for every $x \in F$.*

Proof. If $f \in \mathbb{R}(X)$, then obviously

$$\max (f, 0) \cdot \min (f, 0) = 0.$$

Since $i(\mathscr{F})$ is a prime ideal, it follows that either $\max (f, 0) \in i(\mathscr{F})$, or $\min (f, 0) \in i(\mathscr{F})$.

In the first case, $F = (\max (f, 0))^{-1}(0)$ and $f \leqslant 0$ on F. In the second case, $F = (\min (f, 0))^{-1}(0)$ and $f \geqslant 0$ on F.

We state without proof the following propositions.

PROPOSITION 5.8. *For every prime filter $\mathscr{F}$ on the lattice $Z(X)$, the field of the quotients $\mathbb{R}_{\mathscr{F}} = Q(\mathbb{R}(X)/i(\mathscr{F})$ of the integrity domain $\mathbb{R}(X)/i(\mathscr{F})$ is a real-closed field.*

PROPOSITION 5.9. *Assume that X is discrete and that $\mathscr{F}$ is a prime filter (in fact an ultrafilter) on $Z(X)$. Then $\mathbb{R}_{\mathscr{F}} = R(X)/i(\mathscr{F})$ is a field with the following property:*

A relation of the theory of ordered commutative fields is valid in $\mathbb{R}$ if and only if it is valid in $\mathbb{R}_{\mathscr{F}}$.

In the following we give some ideas concerning the way in which an infinitesimal calculus can be introduced on the field $\mathbb{R}_{\mathscr{U}_{\mathscr{C}}}$.

First of all, if α is an element of $\mathbb{R}_{\mathscr{U}_\mathscr{C}}$, another element $|\alpha| \in \mathbb{R}_{\mathscr{U}_\mathscr{C}}$ can be defined as follows:

If $\alpha = \dot{f}$, $f\colon \mathbb{N} \to \mathbb{R}$, then $|\alpha|$ is defined by

$$|\alpha| = |\dot{f}|,$$

where $|f|\colon \mathbb{N} \to \mathbb{R}$ is the function

$$|f|(n) = |f(n)|.$$

This yields a map

$$||\colon \mathbb{R}_{\mathscr{U}_\mathscr{C}} \to \mathbb{R}^+_{\mathscr{U}_\mathscr{C}}$$

with the usual properties

$$|\alpha + \beta \leqslant |\alpha| + |\beta|, \forall \alpha, \beta \in \mathbb{R}_{\mathscr{U}_\mathscr{C}}.$$

$$|\alpha\beta| = |\alpha|\,|\beta|, \forall \alpha, \beta \in \mathbb{R}_{\mathscr{U}_\mathscr{C}}.$$

DEFINITION 5.10. An element $w \in \mathbb{R}_{\mathscr{U}_\mathscr{C}}$ is a *little infinitesimal* if for every positive real number $w \in \mathbb{R}^+$, we have

$$|w| < \varepsilon.$$

Consequently 0 is a little infinitesimal. In order to see the existence of little infinitesimals which are distinct from 0, it is sufficients to consider the element of $\mathbb{R}_{\mathscr{U}_\mathscr{C}}$ defined by the function

$$w(n) = \begin{cases} \dfrac{1}{n}, & \text{if } n \neq 0; \\ 0, & \text{if } n = 0. \end{cases}$$

An element $\alpha \in \mathbb{R}_{\mathscr{U}_\mathscr{C}}$ is called a *great infinitesimal* if for every positive real number $v \in R^+$, we have

$$|\alpha| > v.$$

The element of $\mathbb{R}_{\mathcal{U}_{\mathcal{C}}}$ defined by the function

$$\mathbb{N} \to \mathbb{R},\ n \mapsto n$$

is obviously a great infinitesimal.

In particular, $\mathbb{R}_{\mathcal{U}_{\mathcal{C}}}$ is a non-Archimedean field.

Every function

$$f\colon \mathbb{R} \to \mathbb{R}$$

extends to a function

$$\tilde{f}\colon \mathbb{R}_{\mathcal{U}_{\mathcal{C}}} \to \mathbb{R}_{\mathcal{U}_{\mathcal{C}}}$$

such that the following diagram be commutative:

$$\begin{array}{ccc} \mathbb{R} & \xrightarrow{\ f\ } & \mathbb{R} \\ {\scriptstyle i}\downarrow & & \downarrow{\scriptstyle i} \\ \mathbb{R}_{\mathcal{U}_{\mathcal{C}}} & \xrightarrow{\ \tilde{f}\ } & \mathbb{R}_{\mathcal{U}_{\mathcal{C}}} \end{array}$$

where i is the canonical inclusion of $\mathbb{R}$ in $\mathbb{R}_{\mathcal{U}_{\mathcal{C}}}$, defined by

$$i(r) = \tilde{r} \bmod i((\mathcal{U}_{\mathcal{C}}),$$
$$\tilde{r}(n) = r, \quad \forall n \in \mathbb{N}.$$

Indeed, if $\alpha = a \bmod i(\mathcal{U}_{\mathcal{C}})$, $a\colon \mathbb{N} \to \mathbb{R}$, we put

$$\tilde{f}(\alpha) = f_0 a \bmod i(\mathcal{U}_{\mathcal{C}}).$$

PROPOSITION 5.11. *A function $f\colon \mathbb{R} \to \mathbb{R}$ is continuous at x_0 if and only if for every little infinitesimal $w \in \mathbb{R}_{\mathcal{U}_{\mathcal{C}}}$, the element $\tilde{f}(x_0+w) - \tilde{f}(x_0) \in \mathbb{R}_{\mathcal{U}_{\mathcal{C}}}$ is a little infinitesimal.*

Proof. Let f be continuous at $x_0 \in \mathbb{R}$, let $w \in \mathbb{R}_{\mathcal{U}_{\mathcal{C}}}$ be a little infinitesimal defined by the function $w\colon \mathbb{N} \to \mathbb{N}$ and, by *reductio ad absurdum*, assume that $\tilde{f}(n_0 + w) - \tilde{f}(x_0)$ is not a little infinitesimal. In this case, an $\varepsilon > 0$

would exist in $\mathbb{R}$, such that the set

$$U_\varepsilon = \{n \mid |f(x_0 + w(n)) - f(x_0)| > \varepsilon\}$$

would belong to the ultrafilter $\mathcal{U}_\mathcal{C}$. On the other hand, f being continuous at x_0, there exists $\delta(\varepsilon) \in \mathbb{R}$, $\delta(\varepsilon) > 0$, such that for every $h \in \mathbb{R}$, $|h| < \delta$, we have

$$|f(x_0 + h) - f(x_0)| \leqslant \varepsilon.$$

The element $\dot{w} \in \mathbb{R}_{\mathcal{U}_\mathcal{C}}$ being a little infinitesimal, we have

$$V_{\delta(\varepsilon)} = \{n \in \mathbb{N} \mid |w(n)| < \delta(\varepsilon)\} \in \mathcal{U}_\mathcal{C}.$$

The intersection $U_\varepsilon \cap V_{\delta(\varepsilon)}$ is non-empty. If $n_0 \in U_\varepsilon \sim V_{\delta(\varepsilon)}$ we get

$$|w_{n_0}| < \delta(\varepsilon),$$

$$|f(x_0 + w(n_0) - f(x_0)| > \varepsilon$$

which is absurd.

Conversely, assume that f satisfies the condition of our proposition. In order to show that f is continuous, it will obviously suffice to prove that if $(h_n)_{n \in \mathbb{N}}$ is a convergent sequence of real numbers with $\lim_{n \to \infty} h_n = 0$, then the function

$$h : \mathbb{N} \to \mathbb{R}$$

$$d \mapsto h_n$$

defines a little infinitesimal $h \in \mathbb{R}_{\mathcal{U}_\mathcal{C}}$, and to use the fact that $f(x_0 + h) - f(x_0)$ is a little infinitesimal.

6.1. Let E be a totally ordered set. We recall that a topology τ_0 associated to the ordered relation on E can be defined as follows:

A subset V of E is open with respect to τ_0 if for every $x \in V$ there exists an open interval of E which contains x and is included in V.

6.2. Let K be an ordered field. The field K is said to satisfy the property D_1 if every sequence of intervals $(I_n)_{n \geqslant 1}$ with $I_n \supset I_{n+1}$ has a non-empty intersection.

Assume that $I_n = [a_n, b_n]$ and $(b_n - a_n)_{n \geq 1}$ tends to 0 in the topology defined by the order relation of K, where K has the property D_1. Under these conditions, $\bigcap_{n \geq 1} I_n$ is a set with a single element.

We say that the field K has the property D_2 if every upper-bounded subset of K has a least upper bound.

Condition D_2 is readily seen to imply D_1. Using the usual trick of dividing intervals, we see that the converse is also true. An ordered field satisfying the equivalent conditions D_1, D_2 is called a complete ordered field.

Example 6.2.1. *The field* $\mathbb{R}$ *of real numbers is complete.*

Example 6.2.2. *The field* $\mathbb{R}^*$ *of non-standard real numbers is not complete.*

Indeed, consider the subset $\mathbb{R} \subset \mathbb{R}^*$. This subset is obviously upper bounded in $\mathbb{R}^*$ since every great infinitesimal is greater than every standard real number. Nevertheless, $\mathbb{R}$ has no least upper bound in $\mathbb{R}^*$, for, assuming by *reductio ad absurdum* that such an element, say α, exists, then $\alpha \notin \mathbb{R}$.

On the other hand, $\frac{\alpha}{2}$ is also an upper bound for $\mathbb{R}$ since the relation $\frac{\alpha}{2} \leqslant a$ (for some $a \in \mathbb{R}$) would imply $\alpha \leqslant 2a$ which is absurd. We obviously have $\frac{\alpha}{2} < \alpha$, which contradicts the definition of α.

7. *The Čech-Stone compactification.* Let X be a topological space. An important problem of topology is that of finding a compact space $\gamma(X)$ "sufficiently close" to X and a canonical injection

$$i_\gamma : X \to \gamma(X).$$

The meaning of the expression "sufficiently close" needs some specifications.

It is natural to expect that it should imply the uniqueness of $\gamma(X)$ up to an isomorphism. Under these assumptions, we say that $\gamma(X)$ is a compactification of X.

Examples (a) *The Alexandrov compactification.* Let X be a non-compact locally compact space. There exists a compact space $\alpha(X)$ and a continuous injection

$$i_\alpha : X \to \alpha(X)$$

such that the set $\alpha(X) - i_\alpha(X)$ consists of a single point. The space $\alpha(X)$ is called the Alexandrov compactification of X and it is determined by X up to an isomorphism.

(b) The space $\mathbb{R}^2$ with the usual topology can be compactified in various ways such that it be everywhere dense in the compactification. For instance:

(1) The sphere S^2 (in fact the Alexandrov compactification

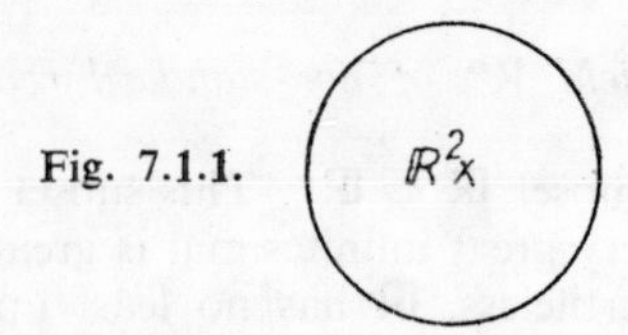

Fig. 7.1.1.

(2) The disc (Fig. 7.2.1)

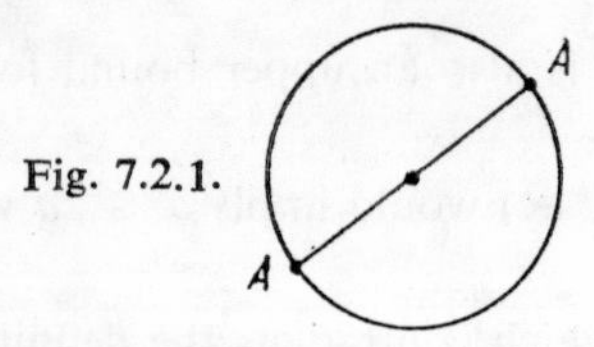

Fig. 7.2.1.

(3) The Möbius band (Fig. 7.3.1)

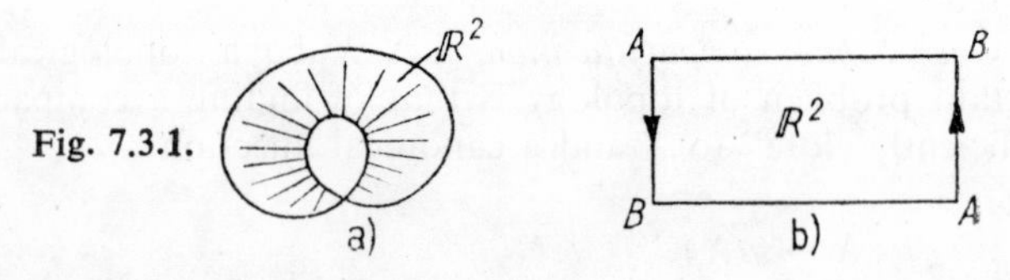

Fig. 7.3.1.

(4) The projective plane (Fig. 7.4.1)

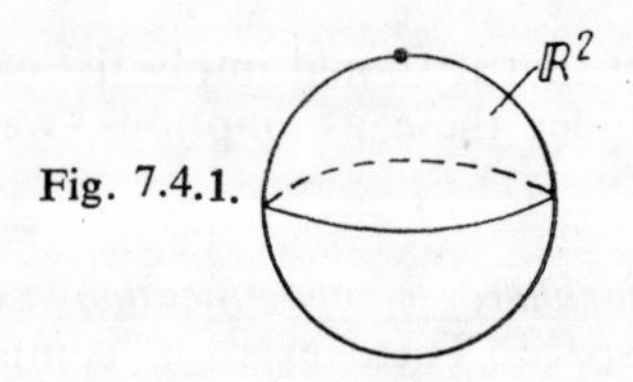

Fig. 7.4.1.

(5) The previous examples are particular cases of the following one: Consider the disc D, whose frontier is the circumference S^1. On the set D, consider any equivalence relation such that the equivalence class of a point from the interior of D consists of a single point. The quotient space has the required property.

In these examples, the expression "$\gamma(X)$ is sufficiently close to X" means that X is everywhere dense in X. In particular, we see that this condition does not ensure the uniqueness of $\gamma(X)$. On the other hand, very important invariants of X are not preserved after passing to $\gamma(X)$. For instance if X is a variety, it may happen that $\gamma(X)$ be not. Another invariant which is essentially changed after passing to $\gamma(X)$ is the ring $\mathbb{R}(X)$.

Consider the inclusion

$$i_\gamma : X \hookrightarrow \gamma(X).$$

This inclusion induces an algebra homomorphism

$$\mathbb{R}(i_\gamma) : \mathbb{R}(\gamma(X)) \to \mathbb{R}(X)$$

which can be factorized through

$$\mathbb{R}_b(i_\gamma) : \mathbb{R}(\gamma(X) \to \mathbb{R}_b(X),$$

where $\mathbb{R}_b(X)$ is the algebra of continuous and bounded functions on X.

The Čech-Stone compactification problem consists in the following: given a topological space X, find a topological space $\beta(X)$ and a continuous injection

$$i_\beta : X \to i_\beta(X)$$

yielding a homeomorphism of X onto the subspace $i_\beta(X)$ of $\beta(X)$, such that $R_b(i_\beta)$ be a surjection.

If X is identified to $i_\beta(X)$, then the Čech-Stone compactification requires that every continuous and bounded real-valued function on X could be extended to $\beta(X)$.

In fact, taking the adherence of X in $\beta(X)$ and using the fact that a closed subspace of a compact set is compact, the Čech-Stone compactification problem can be formulated by requiring that $R_b(i_\beta)$ be an isomorphism.

PROPOSITION 7.1. *Let X be a totally regular space. There exists a compact space $\beta(X)$ and a continuous injection*

$$i_\beta : X \to \beta(X)$$

such that the following conditions be fulfilled:

(I) *i_β yields a homeomorphism between X and $i_\beta(X)$.*

(II) $\mathbb{R}_b(i_\beta) : \mathbb{R}(\beta(x)) \to \mathbb{R}_b(X)$ *is an algebra isomorphism.*

The injection i_β is determined up to an isomorphism.

Proof. Existence. Take $\beta(X) = \text{Spec max } (\mathbb{R}_b(X)$ and

$$i_\beta = i_{b,X} : X \hookrightarrow \text{Spec max } (\mathbb{R}_b(X)) = \beta(X).$$

It is readily seen that $\mathbb{R}(\beta(X)) = \mathbb{R}_b(X)$, which, using Proposition 2.1, shows that conditions I, II are fulfilled.

Uniqueness. Assume we are given a diagram

$$\begin{array}{ccc} & \overset{i_\beta}{\nearrow} & \beta(X) \\ X & & \\ & \underset{i_{\beta'}}{\searrow} & \beta'(X) \end{array}$$

such that i_β, i_β satisfy conditions I, II. This yields a canonical algebra isomorphism

$$\mathbb{R}(\beta'(X)) \to \mathbb{R}(\beta(X))$$

inducing by Proposition 2.1 a homeomorphism

$$\beta(X) \to \beta'(X)$$

such that the following diagram commutes:

$$\begin{array}{ccc} & \overset{i_\beta}{\nearrow} & \beta(X) \\ X & & \downarrow \\ & \underset{i_{\beta'}}{\searrow} & \beta'(X) \end{array}$$

On local rings with the approximation property

We denote by Alg/A the category of unitary commutative algebras over the commutative ring A.

Let $(X_i)_{i\in I}$ be an inductive system of A-algebras, indexed by the ordered, right filtering set I.

We obviously have a canonical map

$$\varinjlim_{i\in I} \mathrm{Hom}_{\mathrm{Alg}/A}(B, X_i) \xrightarrow{\varphi} \mathrm{Hom}_{\mathrm{Alg}/A} B, \varinjlim_{i\in I} X_i).$$

PROPOSITION 1. *Assuming that the ring A is Noetherian, the map φ is surjective for every inductive system $(X_i)_{i\in I}$ if and only if B is an A-algebra of finite type.*

Proof. If B is an A-algebra of finite type, then φ is obviously surjective.

Now, let $(X_i)_{i\in I}$ be the inductive system of all A-subalgebras of finite type of B. Obviously, we have $B = \varinjlim_i X_i$. Since φ is surjective and $\mathrm{id}_B \in \mathrm{Hom}_{\mathrm{Alg}/A}(B, \varinjlim_{i\in I} X_i)$, it follows that there exists $u: B \to X_{i_0}$ such that the composition of the morphisms

$$B \xrightarrow{u} X_{i_0} \hookrightarrow B$$

is the identity map of B. It follows that u is an isomorphism, q.e.d.

DEFINITION 2. Let

$$F: \mathrm{Alg}/A \to \mathrm{Ens}$$

be a functor and let $(B_i)_{i \in I}$ be an inductive system in Alg/A, indexed upon an ordered right-filtering set. We say that F is locally of finite type if the canonical map

$$\varinjlim_{i \in I} F(B_i) \to F(\varinjlim_{i \in I} B_i)$$

is surjective.

PROPOSITION 2. *Let* (A, m) *be a local ring with the approximation property and let*

$$F: \text{Alg}/A \to \text{Ens}.$$

be a functor which is locally of finite type. If $\eta \in F(\hat{A})$ *and if* $C \in \mathbb{N}$, *there exists* $\xi \in F(A)$ *such that*

$$\eta \equiv \xi \bmod \underline{m}^c \hat{A}.$$

Proof. From the hypothesis it follows that there exist an A-algebra imbedding

$$B = A[X_1, \ldots, X_s]/(P_1, \ldots, P_r) \overset{\alpha}{\hookrightarrow} \hat{A}$$

and an element $y \in P(B)$, such that

$$F(\alpha)(y) = \eta.$$

In order to prove Proposition 2, it is sufficient now to remark that due to the approximation property of the local ring A, there exists a morphism β:

$$A[X_1, \ldots, X_s]/(P_1, \ldots, P_r) \overset{\beta}{\to} A$$

such that the following diagram

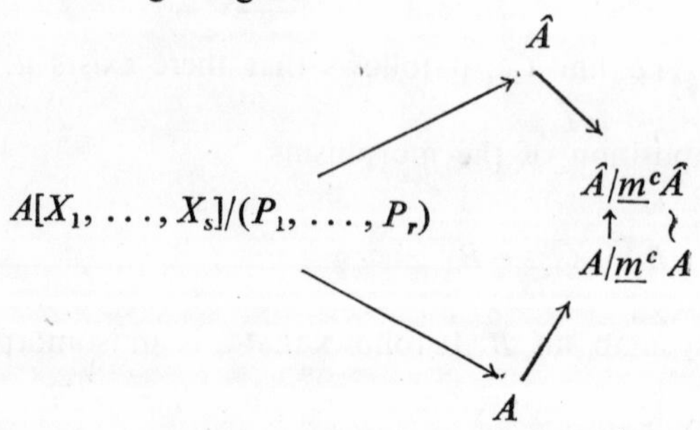

be commutative.

Consequently, we can take $\xi = F(\beta)(y)$.

Subject index

Affine scheme III, 1
algebra of propositions II, 3
algorithm II, 6
arithmetical structure IV, 1

Boole algebra II, 2
Boole ring II, 2
Brouwer algebra II, 2

Calculable function II, 7
Cantor's diagonal method II, 6
Chevalley scheme III, 3
Church's thesis II, 7
classifying object IV, section II
Clifford algebra I, section I
closed formal system II, 4
cohomology theory VI, 2
complete formal system II, 4
constructive real number II, 11

Decision problem for a plane curve II, 8
demonstrative text II, 3
descent data system II, 3
discriminant of a quadratic form I, section I
divisor V, section II

Effective equivalence relation IV, 2
elementary
  term II, 4
  function II, 5
elimination of quantifiers V, 4
elliptic function V, section I
enumeration of finite polyhedra II, 1
equivalence relation in a category IV, 2
étale topology IV, 4

Filter III, 2
formal system II, 4

Grassmann variety VI, 1
Grothendieck
  group I, section I
  topology IV, 4
group of units of a p-adic field II, 2
groupoid I, section I

Hensel-Greenberg approximation conditions I, 3
Hensel Lemma I, 3
Hilbert "symbol" I, 1
Hilbert's Nullstellensatz II, 4
homogeneous ideal III, 4
hyperbolic plane I, section I

Isotropic vector I, section I
integral scheme III, 5

Lawvere-Tierney topos IV, section II
Lefschetz conjecture VI, 3
Lefschetz's principle II, 4
Legendre "symbol" I, 1
logical formula II, 4

Machine with code II, 7
mathematical structure II, 4
model of a formal system II, 4

Non-standard
  analysis II, 4
  real number IV, section II
normal system II, 4

Peano axioms II, 1
prenex relation II, 4
primitive recursive function II, 5
product formula I, 5

Quadratic module
    module I, section I
    reciprocity law I, 1
quantified system II, 4

Radical of a quadratic form I, section I
recursive
    decidability II, 8
    function II, 5
reduction of an algebraic variety VI, 3
Riemann hypothesis VI, 3
Russel's paradox II, 4

Scheme III, 3
semantic system II, 4
semialgebraic set II, 8
spectrum of a commutative ring III, 2
Stone's representation theorem III, 2
strictly epimorphic family IV, 4

Theorem
    of Artin-Schreier II, 8
    of Gödel II, 4
    of Minkowski-Hasse I, 7
    of Tichonov VII
    of Witt I, section I

Ultrafilter II, 4
ultrapower II, 4
univalent system II, 4

Weil's diophantine conjectures vi, 3
Whitehead group I, section I
Witt
    group I, section I
    splitting of a quadratic module I, section I

PRINTED IN ROMANIA